Networked Sensing Information and Control

Venkatesh Saligrama
Editor

Networked Sensing Information and Control

Springer

Edited by:

Venkatesh Saligrama
Department of Electrical
and Computer Engineering
Boston University
8 Saint Mary's Street
Boston, MA 02215
USA

ISBN 978-0-387-68843-5 e-ISBN 978-0-387-68845-9

Library of Congress Control Number: 2007935795

Printed on acid-free paper.

9 8 7 6 5 4 3 2 1

springer.com

Preface

This book grew out of a two day workshop that was held in May 2006 and was funded by the Engineering division of the U.S. National Science Foundation (NSF) under the Control, Networks and Computational Intelligence program. The purpose of this workshop was to bring together key contributors to the field of networked sensing to discuss the state-of-the-art in research, the main mathematical issues in the design and deployment of sensor networks, and the problems solved and the problems remaining. At the end of the workshop the participants had agreed on writing assignments. The intended audience for this book are graduate students, engineers and scientists in the fields of signal processing, control, information theory and statistics. The book is typical of many recent edited collections on emerging topics in engineering. The chapters were written by some of the principal architects of recent advances in mathematical aspects of sensory networks.

Boston, MA

Venkatesh Saligrama
August, 2007

Contents

List of Contributors

Laura Balzano
University of California, Los Angeles, CA

Francesco Bullo
University of California, Santa Barbara, CA

David Castanon
Boston University, Boston, MA

Jose Cortes
University of California, Santa Cruz, CA

Alexandros Dimakis
University of California, Berkeley

Geir Dullerud
University of Illinois, Urbana-Champaign, IL

J. A. Fummeler
University of Illinois, Urbana-Champaign, IL

Anurag Ganguli
University of California, Santa Barbara, CA

Michael Gastpar
University of California, Berkeley, CA

Alfred Hero III
University of Michigan, Ann Arbor, MI

Prakash Ishwar
Boston University, Boston, MA

P. R. Kumar
University of Illinois, Urbana-Champaign, IL

Nan Ma
Boston University, Boston, MA

David Neuhoff
University of Michigan, Ann Arbor, MI

Robert Nowak
University of Wisconsin, Madison, WI

Sandeep Pradhan
University of Michigan, Ann Arbor, MI

Raviv Raich
University of Michigan, Ann Arbor, MI

Kannan Ramachandran
University of California, Berkeley, CA

Raghuram Rangarajan
University of Michigan, Ann Arbor, MI

Craig Robinson
University of Illinois, Urbana-Champaign, IL

Venkatesh Saligrama
Boston University, Boston, MA

R. Srikant
University of Illinois, Urbana-Champaign, IL

John Tsitsiklis
Massachusetts Institute of Technology, Cambridge, MA

Venu Veeravalli
University of Illinois, Urbana-Champaign, IL

Ye Wang
Boston University, Boston, MA

Wee Pee Wang
Massachusetts Institute of Technology, Cambridge, MA

Lei Ying
University of Illinois, Urbana-Champaign, IL

Manqi Zhao
Boston University, Boston, MA

Introduction

Venkatesh Saligrama

Boston University

Recent advances in sensor and computing technologies has provided the impetus for deploying distributed sensing systems. Distributed networks are envisioned in such diverse applications as building safety, environmental monitoring, power systems, manufacturing as well as military and space applications. Distributed sensing systems hold the promise of providing an inexpensive, non-intrusive means to understand phenomena that exhibit spatial and temporal variations at multiple scales.

While significant effort over the last decade in sensor development physical layer transmission and networking infrastructure has laid the initial groundwork for practical deployment, realization of such distributed sensing systems is still in its infancy. The principle challenges can be attributed to fundamental system-level difficulties. These arise due to the difficulty in monitoring a dynamic uncertain environment through an underlying power/bandwidth constrained ad-hoc networked infrastructure. The main challenge can be summarized as follows: *How to make decisions under uncertainty, which arises from spatially distributed dynamic information when sharing distributed data is limited by networking constraints.*

Over the last few years there has been significant interest in developing systematic techniques to synthesize interactive and reconfigurable distributed sensing systems that are capable of performing effective inferencing and control tasks under overall resource constraints. This can be seen from a number of different control, information theory and signal processing conferences where several tracks and sessions have been dedicated to such topics. Nevertheless, the research to date in this area is fragmented possibly due to the fact that the fundamental challenges intersect many of the traditional areas. The purpose of this book is to present broad trends in the mathematical aspects of Networked Sensing, Information and Control by some of the principal architects working in this area.

The book chapters are grouped under four major themes: (a) Blind localization, which accounts for uncertain sensor locations and attributes; (b) Distributed computation in the context of detection, function computation and data dissemination; (c) Fundamental issues in reconstruction of spatially distributed phenomena with rate constrained sensor networks; (d) Networked Estimation and Control of dynamical systems.

From a systems perspective every "piece" of "raw" information in a sensor network has typically a state-overhead information such as spatial location, time of origin, resolution, and attributes, e.g., sensing-modality, associated with it. Different classes of sensor network applications require different amounts of this "state" information from one spatio-temporal volume of the network to be available at another spatio-temporal volume in order to induce some optimal level of performance in both the network scaling and non-scaling regimes. For example, at one end of the spectrum are applications that require only simple network-wide order statistics, e.g., max, min, median, for which it is unnecessary to transport the spatial location information overheads associated with each piece of data. At the other extreme are applications such as field-mapping, e.g., for environmental monitoring, for which the spatio-temporal state information of different pieces of data is critical for good performance. Intermediate applications which also require the spatio-temporal state include singularity and boundary estimation and evolution, e.g., detecting and tracking the boundaries of convection currents and toxic plumes including classical target tracking. Classical problems of inference such as classification, estimation, and filtering need to be systematically revisited under distributed constraints.

Blind Localization

The first two chapters deal with sensing issues pertaining to unknown sensor locations/attributes. This is a fundamental issue encountered in sensor networks. The first chapter by Balzano and Nowak considers the problem of blindly calibrating sensor response using routine sensor network measurements. They show that as long as the sensors slightly oversample the signals of interest, then unknown sensor gains can be perfectly recovered. Exploiting incoherence conditions between measurement model and sensor observations they show surprisingly that neither a controlled stimulus nor a dense deployment is required. The second chapter by Rangarajan, Raich, and Hero considers the problem of target localization when sensor locations are unknown. They use the distributed weighted multidimensional scaling (dwMDS) algorithm to obtain estimates of the sensor positions. By exploiting the fact that target motion is manifested only at a sparse set of sensors they are able to effectively localize target position.

Distributed Computation over Unreliable Communication Networks

The next set of chapters address distributed computation techniques on unreliable communication networks. These consider sensing scenarios that require only simple network-wide order statistics and global knowledge of sensor attributes or locations are not a major concern. These set of chapters are focused on distributed computation and fundamental limits to communication message complexity. The first chapter in this part by Tang and Tsitsiklis study the problem of decentralized detection in a sensor network consisting of nodes sending information to a fusion center which is located at the root of a tree. Such a tree configuration, although effective from an energy perspective could possibly have worse performance, compared to other distributed architectures such as a parallel configuration. Surprisingly, they show that the optimal error exponent is often the same as that corresponding to a parallel configuration. The following chapter by Ying, Srikant and Dullerud, focus on symmetric function computation over an unreliable communication network. By using multi-reception diversity to combat channel noise and data aggregation they reduce the number of transmissions. They propose a distributed algorithm whose energy consumption is only a factor of $\log \log n$ more than the lower bound. The last chapter in this part by Dimakis and Ramachandran focuses on distributed storage. This is an important issue in long term deployments of sensor networks. These networks must log important information and store it until the information can be recovered. Since storage nodes are expected to fail, redundancy is necessary to guarantee the required reliability and the authors describe distributed, scalable and energy-efficient algorithms to generate and dynamically maintain encoded information representations in networks.

Rate Constrained Field Reconstruction

The third part of the book deals with fundamental issues in reconstruction of spatially distributed phenomena with rate constrained sensor networks. The first of these chapters by Wang, Ma, Zhao, Ishwar and Saligrama study distributed field reconstruction within a target distortion D using a dense network of noisy one–bit randomized scalar quantizers in the presence of additive observation noise of unknown distribution. They construct an order optimal scheme and show that when the noise, sensor placement pattern, and the sensor schedule satisfy certain minimal technical requirements, it is possible to drive the MSE to zero with increasing sensor density at points of field continuity while ensuring that the per–sensor bitrate and sensing–related network overhead rate simultaneously go to zero. The following chapter by Neuhoff and Pradhan study a similar question in a Bayesian setting. They take an information theoretic perspective and characterize fundamental tradeoffs between sensor density, mean-squared distortion and encoding rate. They pose and answer the following questions: What happens to the encoding rate, in bits per unit area, produced by the encoders as the sensors become more numerous and dense? Does the increasing density of sensors cause the encoding rate to increase without limit? Or does the increasing correlation between neighboring sensor measurements sufficiently mitigate the increasing density to permit encoding rate to remain bounded as density increases? The last chapter in this section by Gastpar takes yet another perspective on the problem. In contrast to the first two chapters' focus on source coding the last chapter focuses on implication of wireless multi-access networks. Shannon's theory established that bits are a universal currency of information in point-to-point communication. Nevertheless, whether bits form a universal currency is far from clear in networked settings. The chapter presents a rich collection of examples illustrating the different classes of problems: (a) networked communication problems for which bits are a universal currency of information, (b) *counter-examples* where bits are far from being universal, and (c) *approximate* separation theorems, where bits are somehow close to universal.

Dynamics and Control over Communication Networks

The final part deals with networked real-time control and considers dynamic sensed environments. The first chapter by Robinson and Kumar considers the design of transport layer, one of five layers in the internet protocol suite, for networked control problems. Traditional data network transport protocols fail to deal with the time value of, or the relationships between, the data being transmitted. They introduce a new method, that they call Linear Temporal Coding (LTC), for potential use in wireless networked control systems. In LTC, instead of retransmitting dropped packets, linear combinations of packet contents are formed based on the previous messages' delivery status. They illustrate advantages of this protocol over conventional protocol, which favors transmission of the most recent packet. The second chapter by Saligrama and Castañón expands on this theme and considers distributed tracking with a sensor network subjected to random intermittent communication connectivity. The problem they consider is to design efficient encoding and fusion rules to optimally track moving objects with intermittent communications. The emphasis in this chapter is on designing lossless protocols, i.e., protocols which recover centralized performance. They derive scalable communication protocols that transmit a summary of all past measurements that achieve lossless performance. The third chapter in this part by Fummeler and Veeravalli focuses on sensor management in tracking applications. Specifically, they consider the problem of tracking with low-power sensors and develop efficient dynamic management strategies for conserving battery power. In particular, they develop dynamic sleep-wake schedules for sensors that optimizes their lifetimes while ensuring small tracking errors. Unfortunately, open loop policies with fixed or random duty cycles are not efficient. Their key idea is that the location of the object (if known) at the time when the sensor is put to sleep would be useful in determining the sleep duration of the sensor; the closer the object, the shorter the sleep duration should be. The last chapter by Ganguli, Cortés and Bullo investigates multi-agent robotic systems. Such systems have recently gained importance as basic components of complex networks intended to perform a wide variety of tasks such as search and rescue, exploration, environmental monitoring, location-aware computing, and the maintaining of structures. In this chapter they develop algorithms for *visually-guided agents,* i.e., mobile robotic agents with line-of-sight sensing and communication capabilities, to solve a distributed version of the *Art Gallery Problem.* This problem involves placing guards inside an art gallery such that every point is visible by at least one guard. By exploiting a combination of "geometric structure + information management + navigation algorithms" they are able to control individual agents to explore and traverse nonconvex polygons only based on local sensing and communication to achieve the deployment objectives.

The motivating examples in this book are drawn from specific domains such as national defense, environmental monitoring and security. However, as pointed out in several of these chapters, the theory and methods are general and applicable to a far wider range of networked sensing applications. We hope that this book will generate additional interest in the general area of sensor networks.

Part I

Blind Localization

1

Blind Calibration of Networks of Sensors: Theory and Algorithms

Laura Balzano and Robert Nowak

University of California, Los Angeles and University of Wisconsin, Madison

1.1 Introduction

With the wide variety of sensor network applications being envisioned and implemented, it is clear that in certain situations the applications need more accurate measurements than uncalibrated, low-cost sensors provide. Arguably, calibration errors are one of the major obstacles to the practical use of sensor networks [3], because they allow a user to infer a difference between the readings of two spatially separated sensors when in fact that difference may be due in part to miscalibration. Consequently, automatic methods for jointly calibrating sensor networks in the field, without dependence on controlled stimuli or high-fidelity groundtruth data, is of significant interest. We call this problem *blind calibration*.

One approach to blind sensor network calibration is to begin by assuming that the deployment is very dense, so that neighboring nodes should (in principle) have nearly identical readings [4]. Unfortunately, many existing and envisioned sensor network deployments may not meet the density requirements of such procedures.

The difference in our approach is that it leverages correlation in the collection of sensors without requiring a dense deployment, making it much more suitable for practical applications. We assume this correlation is defined and that the readings from n sensors lie in a subspace of n-dimensional Euclidean space.

We assume a linear model for the sensor calibration functions. This means that the sensor readings are calibrated up to an unknown gain and offset (bias) for each sensor, possibly after applying a suitable and fixed transformation to the raw sensor readings, e.g., taking the logarithm or applying the original factory calibration transformation.

This work makes three main contributions. First, we propose a novel automatic sensor calibration procedure that requires solving a linear system of constraints involving routine sensor measurements. By "routine" we mean that actual signal measured by the sensor network is uncontrolled and unknown.

This is why we refer to the problem as blind calibration. The constraint equations are based on mild assumptions that guarantee that the sensor measurements are at least slightly correlated over space, i.e., the network oversamples the underlying signals of interest. Second, we prove the rather surprising fact that these assumptions, which are commonly met in practice, suffice to perfectly recover unknown sensor gains. That is, *it is possible to blindly calibrate the gains using only the routine readings made by the sensors.* Third, we prove that the sensor offsets (biases) can also be partially recovered from routine readings; they can be completely recovered with some additional overhead.

To give a preview of our approach, suppose we are measuring a temperature field with an array of n sensors. Temperature fields tend to vary smoothly, and so they may be considered to be bandlimited. The Nyquist theorem dictates a minimum spacing between sensors in order to adequately sample a bandlimited signal. If sensors are spaced more closely than the minimum requirement, then we are "oversampling" the signal. In this case, the underlying bandlimited signal will lie in a lower dimensional (low frequency) subspace of the n dimensional measurement space. This condition provides a useful constraint for blind calibration. Correctly calibrated signals must lie in the lower dimensional subspace, and this leads to a system of linear equations which can be used to solve for the gain and offset calibration parameters.

1.2 Problem Formulation

Consider a network of n sensors. At a given time instant, each sensor makes a measurement, and we denote the vector of n measurements by $\boldsymbol{x} = [x(1), \ldots, x(n)]'$, where $'$ denotes the vector transpose operator (so that $\boldsymbol{x}$ is an $n \times 1$ column vector). We will refer to $\boldsymbol{x}$ as a "snapshot." When necessary, we will distinguish between snapshots taken at different times using a subscript (e.g., $\boldsymbol{x}_s$ and $\boldsymbol{x}_t$ are snapshots at times s and t).

Each sensor has an unknown gain and offset associated with its response, so that instead of measuring $\boldsymbol{x}$ the sensors report

$$y(j) = \frac{x(j) - \beta(j)}{\alpha(j)}, \; j = 1, \ldots, n$$

where $\boldsymbol{\alpha} = [\alpha(1), \ldots, \alpha(n)]'$ are the sensors' gain calibration factors and $\boldsymbol{\beta} = [\beta(1), \ldots, \beta(n)]'$ are the sensors' calibration offsets. It is assumed that $\alpha(j) \neq 0, j = 1, \ldots, n$. With this notation, the sensor measurement $y(j)$ can be calibrated by the linear transformation $x(j) = \alpha(j)y(j) + \beta(j)$. We can summarize this for all n sensors using the vector notation

$$\boldsymbol{x} = \boldsymbol{Y}\boldsymbol{\alpha} + \boldsymbol{\beta} \, , \tag{1.1}$$

where $\boldsymbol{Y} = \text{diag}(\boldsymbol{y})$ and the diag operator is defined as

$$\text{diag}(\boldsymbol{y}) = \begin{bmatrix} y(1) & & \\ & \ddots & \\ & & y(n) \end{bmatrix}.$$

The blind calibration problem entails the recovery of $\boldsymbol{\alpha}$ and $\boldsymbol{\beta}$ from *routine* uncalibrated sensor readings such as $\boldsymbol{y}$.

In general, without further assumptions, blind calibration appears to be an impossible task. However, it turns out that under mild assumptions that may often hold in practice, quite a bit can be learned from raw (uncalibrated) sensor readings like $\boldsymbol{y}$. Assume that the sensor network is slightly "oversampling" the phenomenon being sensed. Mathematically, this means that the calibrated snapshot $\boldsymbol{x}$ lies in a lower dimensional subspace of n-dimensional Euclidean space. Let $\mathcal{S}$ denote this "signal subspace" and assume that it is r-dimensional, for some integer $0 < r < n$. For example, if the signal being measured is bandlimited and the sensors are spaced closer than required by the Shannon-Nyquist sampling rate, then $\boldsymbol{x}$ will lie in a lower dimensional subspace spanned by frequency basis vectors. If we oversample (relative to Shannon-Nyquist) by a factor of 2, then $r = n/2$. Basis vectors that correspond to smoothness assumptions, such as low-order polynomials, are another potentially relevant example. In general, the signal subspace may be spanned by an arbitrary set of r basis vectors. The calibration coefficients $\boldsymbol{\alpha}$ and $\boldsymbol{\beta}$ and the signal subspace $\mathcal{S}$ may change over time, but here we assume they do not change over the course of blind calibration. As we will see, this is a reasonable assumption, since the network may be calibrated from very few snapshots.

Let $\boldsymbol{P}$ denote the orthogonal projection matrix onto the orthogonal complement to the signal subspace $\mathcal{S}$. Then every $\boldsymbol{x} \in \mathcal{S}$ must satisfy the constraint

$$\boldsymbol{P}\boldsymbol{x} = \boldsymbol{P}(\boldsymbol{Y}\,\boldsymbol{\alpha} + \boldsymbol{\beta}) = 0 \tag{1.2}$$

This is the key idea behind our *blind calibration* method. Because the projection matrix $\boldsymbol{P}$ has rank $n-r$, the constraint above gives us $n-r$ linearly independent equations in $2n$ unknown values ($\boldsymbol{\alpha}$ and $\boldsymbol{\beta}$). If we take snapshots from the sensor network at k distinct times, $\boldsymbol{y}_1, \ldots, \boldsymbol{y}_k$, then we will have $k(n-r)$ equations in $2n$ unknowns. For $k \geq 2n/(n-r)$ we will have more equations than unknowns, which is a hopeful sign. In fact, as we will show, in many cases it is possible to blindly recover $\boldsymbol{\alpha}$ and $\boldsymbol{\beta}$ from a sufficient number of uncalibrated sensor snapshots.

1.3 Initial Observations

1.3.1 Blind Calibration in No Noise

Given k snapshots at different time instants $\boldsymbol{y}_1, \ldots, \boldsymbol{y}_k$, the subspace constraint (1.2) results in the following system of $k(n-r)$ equations:

$$\boldsymbol{P}(\boldsymbol{Y}_i \boldsymbol{\alpha} + \boldsymbol{\beta}) = 0 \ , \ i = 1, \ldots, k \tag{1.3}$$

The true gains and offsets must satisfy this equation, but in general the equation may be satisfied by other vectors as well. Establishing conditions that guarantee that the true gains and/or offsets are the only solutions is the main theoretical contribution of this work.

It is easy to verify that the solutions for $\boldsymbol{\beta}$ satisfy

$$\boldsymbol{P}\boldsymbol{\beta} = -\boldsymbol{P}\bar{\boldsymbol{Y}}\boldsymbol{\alpha} \tag{1.4}$$

where $\bar{\boldsymbol{Y}} = \frac{1}{k}\sum_{i=1}^{k} \boldsymbol{Y}_i$, the time-average of the snapshots. One immediate observation is that the constraints only determine the components of $\boldsymbol{\beta}$ (in terms of the data and $\boldsymbol{\alpha}$) in the signal "nullspace" (the orthogonal complement to $\mathcal{S}$). The component of the offset $\boldsymbol{\beta}$ that lies in the signal subspace is unidentifiable. This is intuitively very easy to understand. Our only assumption is that the signals measured by the network lie in a lower dimensional subspace. The component of the offset in the signal subspace is indistinguishable from the *mean* or average signal. Recovery of this component of the offset requires extra assumptions, such as assuming that the signals have zero mean, or additional calibration resources, such as the non-blind calibration of some of the sensor offsets. We discuss this further in Section 1.5.

Given this characterization of the $\boldsymbol{\beta}$ solutions, we can re-write the constraints (1.3) in terms of $\boldsymbol{\alpha}$ alone:

$$\boldsymbol{P}(\boldsymbol{Y}_i - \bar{\boldsymbol{Y}})\boldsymbol{\alpha} = 0 \ , \ i = 1, \ldots, k \tag{1.5}$$

If $\widehat{\boldsymbol{\alpha}}$ is a solution to this system of equations, then every vector $\boldsymbol{\beta}$ satisfying $\boldsymbol{P}\boldsymbol{\beta} = -\boldsymbol{P}\bar{\boldsymbol{Y}}\widehat{\boldsymbol{\alpha}}$ is a solution for $\boldsymbol{\beta}$ in the original system of equations (1.3). In other words, for a given $\widehat{\boldsymbol{\alpha}}$, the value of the component of the offset in the nullspace is $\boldsymbol{P}\bar{\boldsymbol{Y}}\widehat{\boldsymbol{\alpha}}$.

Another simple but very important observation is that there is one degree of ambiguity in $\boldsymbol{\alpha}$ that can never be resolved blindly using routine sensor measurements alone. The gain vector $\boldsymbol{\alpha}$ can be multiplied by a scalar c, and it cannot be distinguished whether this scalar multiple is part of the gains or part of the true signal. We call this scalar multiple the global gain factor. A constraint is needed to avoid this ambiguity, and without loss of generality we will assume that $\alpha(1) = 1$. This constraint can be interpreted physically to mean that we will calibrate all other sensors to the gain characteristics of sensor 1. The choice of sensor 1 is arbitrary and is taken here simply for convenience.

In Sections 1.4 and 1.5, we show that with noiseless measurements and perfect knowledge of the subspace $\boldsymbol{P}$, there is exactly one solution to 1.5 and it is possible to calibrate the gain calibration factors. We also show that we are able to calibrate the offset calibration factors with further information.

1.3.2 Blind Calibration in Noise

If noise, mismodeling effects, or other errors are present in the uncalibrated sensor snapshots, then a solution to (1.5) may not exist. There are many methods for finding the best possible solution, and in this work we employ singular value decomposition and standard least squares techniques.

First, note that the constraints can be expressed as

$$\mathbf{C}\,\boldsymbol{\alpha} = 0 \tag{1.6}$$

where the matrix $\mathbf{C}$ is given by

$$\mathbf{C} = \begin{bmatrix} \boldsymbol{P}(\boldsymbol{Y}_1 - \bar{\boldsymbol{Y}}) \\ \vdots \\ \boldsymbol{P}(\boldsymbol{Y}_k - \bar{\boldsymbol{Y}}) \end{bmatrix} \tag{1.7}$$

In the ideal case, there is always at least one solution[1] to the constraint $\mathbf{C}\boldsymbol{\alpha} = 0$, since the true gains must satisfy this equation. As we will show in Section 1.4, under certain conditions there will always be exactly one solution for $\boldsymbol{\alpha}$.

On the other hand, if the sensor measurements contain noise or if the assumed calibration model or signal subspace is inaccurate, then a solution may not exist. That is, the matrix $\mathbf{C}$ may have full column rank and thus will not have a right nullspace. A reasonable *robust* solution in such cases is to find the right singular vector of $\mathbf{C}$ associated with the smallest singular value. This vector is the solution to the following optimization.

$$\begin{aligned} \widehat{\boldsymbol{\alpha}} = \;& \arg\min_{\boldsymbol{\alpha}} \|\mathbf{C}\boldsymbol{\alpha}\|_2^2 \\ & \text{subject to } \|\boldsymbol{\alpha}\|_2^2 = 1 \end{aligned} \tag{1.8}$$

In other words, we find the vector of gains such that $\mathbf{C}\widehat{\boldsymbol{\alpha}}$ is as close to zero as possible. This vector can be efficiently computed in numerical computing environments, such as Matlab, using the economy size singular value decomposition (svd)[2]. Note that in the ideal case (no noise or error) the svd solution satisfies (1.5). Thus, this is a general-purpose solution method.

Blind calibration of the gains can also be implemented by solving a system of equations in a least squared sense as follows. Recall that we have one constraint on our gain vector, $\boldsymbol{\alpha}(1) = 1$. This can be interpreted as knowing the gain coefficient for the first sensor. We can use this knowledge as an additional constraint on the solution.

$$\begin{aligned} \widehat{\boldsymbol{\alpha}} = \;& \arg\min_{\boldsymbol{\alpha}} \|\mathbf{C}\boldsymbol{\alpha}\|_2^2 \\ & \text{subject to } \boldsymbol{\alpha}(1) = 1 \end{aligned} \tag{1.9}$$

---

[1] This does not include the trivial solution $\boldsymbol{\alpha} = 0$. With no noise and perfect knowledge of $\boldsymbol{P}$, there is always one vector in the nullspace of $\mathbf{C}$.

[2] The Matlab command is svd($\mathbf{C}$, 0).

An equivalent optimization without constraints can be derived as follows. If we let $\mathbf{c_1}, \ldots, \mathbf{c_n}$ be the columns of $\mathbf{C}$, let $\tilde{\boldsymbol{\alpha}}$ be the gain vector with $\alpha(1)$ removed, and let $\tilde{\mathbf{C}}$ be the matrix $\mathbf{C}$ with the first column removed, we can rewrite the system of equations as $\tilde{\mathbf{C}}\,\tilde{\boldsymbol{\alpha}} = -\mathbf{c_1}$. The robust solution is the value of $\tilde{\boldsymbol{\alpha}}$ that minimizes the LS criterion $\|\tilde{\mathbf{C}}\,\tilde{\boldsymbol{\alpha}} + \mathbf{c_1}\|_2^2$.

More generally, we may know several of the gain coefficients for what we call *partially blind calibration*. Let $\mathbf{h}$ be the sum of the $\boldsymbol{\alpha}(i)\mathbf{c_i}$ corresponding to the known gains, let $\tilde{\boldsymbol{\alpha}}$ be the gain vector with the known gains $\boldsymbol{\alpha}(i)$ removed, and let $\tilde{\mathbf{C}}$ be the matrix $\mathbf{C}$ with those columns $\mathbf{c_i}$ removed. Now we have $\tilde{\mathbf{C}}\,\tilde{\boldsymbol{\alpha}} = -\mathbf{h}$ and the robust solution is the minimizer of

$$\|\tilde{\mathbf{C}}\,\tilde{\boldsymbol{\alpha}} + \mathbf{h}\|_2^2 \tag{1.10}$$

We can solve this optimization in a numerically robust manner by avoiding the squaring of the matrix $\tilde{\mathbf{C}}$ that is implicit in the conventional LS solution, $\tilde{\boldsymbol{\alpha}} = (\tilde{\mathbf{C}}'\tilde{\mathbf{C}})^{-1}\tilde{\mathbf{C}}'(-\mathbf{h})$. This "squaring" effectively worsens the condition number of the problem and can be avoided by using QR decomposition techniques[3].

1.3.3 Estimation of the Subspace

The key component of the problem formulation proposed in Section 1.2 is the true signal subspace $\boldsymbol{P}$. In some cases, the phenomenon behavior has enough structure that the subspace can be derived. In other cases, the deviation of the true signals from an assumed subspace is small enough that, with a robust implementation of blind calibration, the results will still be good. Or, if a subspace does not change over time, it can be estimated early in the deployment with calibrated sensors and then used later to identify how the sensor calibration has drifted.

¿From experience, we have seen that there are further cases where blind calibration is sensitive to errors in the subspace. It is beyond the scope of this work to provide extensive tools for identifying the true signal subspace, however here we briefly describe estimation of the subspace rank r from the measurements $y(i)$, $i = 1, \ldots, k$.

Given a particular measurement vector $\boldsymbol{y}$, we can subtract the mean of the k measurement vectors $\bar{\boldsymbol{y}}$ and get

$$\boldsymbol{y} - \bar{\boldsymbol{y}} = \frac{\boldsymbol{x} - \bar{\boldsymbol{x}}}{\boldsymbol{\alpha}} \tag{1.11}$$

where division of the vector is element-wise. Without loss of generality, we can assume that $\bar{\boldsymbol{x}} = 0$. Letting $\boldsymbol{A} = \mathrm{diag}(\boldsymbol{\alpha})$ we can rewrite this as $\boldsymbol{y} = \boldsymbol{A}^{-1}\boldsymbol{x}$.

Each signal $\boldsymbol{x}$ lies in an r-dimensional subspace $\mathcal{S}$, where $r < n$ but is unknown. Let $\boldsymbol{\phi}_1, \ldots, \boldsymbol{\phi}_r$ denote a basis for $\mathcal{S}$. Then $\boldsymbol{x} = \sum_{i=1}^{r} \theta_i \boldsymbol{\phi}_i$, for certain coefficients $\theta_1, \ldots, \theta_r$. Therefore, $\boldsymbol{y} = \sum_{i=1}^{r} \theta_i \boldsymbol{A}^{-1}\boldsymbol{\phi}_i$.

---

[3] We used $\boldsymbol{\alpha} = \tilde{\mathbf{C}}\backslash{-\mathbf{h}}$ in Matlab.

We can find the unknown r by building the covariance matrix of the measurements.

$$\mathbf{Cov}_{\boldsymbol{y}} = \frac{\mathbf{1}}{\mathbf{k}} \sum_{\mathbf{j}=\mathbf{1}}^{\mathbf{k}} (\boldsymbol{y}_{\mathbf{j}} - \bar{\boldsymbol{y}})(\boldsymbol{y}_{\mathbf{j}} - \bar{\boldsymbol{y}})^{\mathbf{T}} \tag{1.12}$$

As long as $k > r$ and the vectors $\boldsymbol{x}_i$, $i = 1, \ldots, k$, are linearly independent, the rank of this covariance matrix will be r.

1.4 Gain Calibration

This section theoretically characterizes the existence of unique solutions to the gain calibration problem. As pointed out in Section 1.3.1, the gain calibration problem can be solved independently of the offset calibration task, as shown in (1.5), which corresponds to simply removing the mean snapshot from each individual snapshot. Therefore, it suffices to consider the case in which the snapshots are zero-mean and to assume that $\bar{\boldsymbol{Y}} = 0$, in which case the gain calibration equations may be written as

$$\boldsymbol{P}\boldsymbol{Y}_i\boldsymbol{\alpha} \; = \; 0 \; , \; i = 1, \ldots, k \tag{1.13}$$

The results we present also hold for the general case in which $\bar{\boldsymbol{Y}} \neq 0$. We first consider general conditions guaranteeing the uniqueness of the solution to (1.13) and then look more closely at the special case of bandlimited subspaces.

1.4.1 General Conditions

The following conditions are sufficient to guarantee that a unique solution to (1.13) exists.

A1. **Oversampling:** Each signal $\boldsymbol{x}$ lies in a known r-dimensional subspace $\mathcal{S}$, $r < n$. Let $\boldsymbol{\phi}_1, \ldots, \boldsymbol{\phi}_r$ denote a basis for $\mathcal{S}$. Then $\boldsymbol{x} = \sum_{i=1}^{r} \theta_i \boldsymbol{\phi}_i$, for certain coefficients $\theta_1, \ldots, \theta_r$.

A2. **Randomness:** Each signal is randomly drawn from $\mathcal{S}$ and has mean zero. This means that the signal coefficients are zero-mean random variables. The joint distribution of these random variables is absolutely continuous with respect to Lebesgue measure (i.e., a joint r-dimensional density function exists). For any collection of signals $\boldsymbol{x}_1, \ldots, \boldsymbol{x}_k$, $k > 1$, the joint distribution of the corresponding kr coefficients is also absolutely continuous with respect to Lebesgue measure (i.e., a joint kr-dimensional density function exists).

A3. **Incoherence:** Define the $nr \times n$ matrix

$$\mathbf{M}_{\Phi} = \begin{bmatrix} \boldsymbol{P}\,\mathrm{diag}(\boldsymbol{\phi}_1) \\ \vdots \\ \boldsymbol{P}\,\mathrm{diag}(\boldsymbol{\phi}_r) \end{bmatrix} \tag{1.14}$$

and assume that $\text{rank}(\mathbf{M}_{\Phi}) = n - 1$. Note that $\mathbf{M}_{\Phi}$ is a function of the basis of the signal subspace. The matrix $\boldsymbol{P}$, the orthogonal projection matrix onto the orthogonal complement to the signal subspace $\mathcal{S}$, can be written as $\boldsymbol{P} = \mathbf{I} - \boldsymbol{\Phi}\boldsymbol{\Phi}'$, where $\mathbf{I}$ is the $n \times n$ identity matrix and $\boldsymbol{\Phi} = [\phi_1, \dots, \phi_r]$.

Assumption A1 guarantees that the calibrated or true sensor measurements are correlated to some degree. This assumption is crucial since it implies that measurements must satisfy the constraints in (1.3) and that, in principle, we can solve for the gain vector $\boldsymbol{\alpha}$. Assumption A2 guarantees the signals are not too temporally correlated (e.g., different signal realizations are non-identical with probability 1). Also, the zero-mean assumption can be removed, as long as one subtracts the average from each sensor reading. Assumption A3 essentially guarantees that the basis vectors are sufficiently incoherent with the canonical sensor basis, i.e., the basis that forms the columns of the identity matrix. It is easy to verify that if the signal subspace basis is *coherent* with the canonical basis, then $\text{rank}(\mathbf{M}_{\Phi}) < n - 1$. Also, note that $\mathbf{M}_{\Phi}\mathbf{1} = 0$, where $\mathbf{1} = [1, \dots, 1]'$, which implies that $\text{rank}(\mathbf{M}_{\Phi})$ is at most $n - 1$. In general, assumption A3 only depends on the assumed signal subspace and can be easily checked for a given basis. In our experience, the condition is satisfied by most signal subspaces of practical interest, such as lowpass, bandpass or smoothness subspaces.

Theorem 1. *Under assumptions A1, A2 and A3, the gains $\boldsymbol{\alpha}$ can be perfectly recovered from any $k \geq r$ signal measurements by solving the linear system of equations (1.5).*

The theorem is proved in the Appendix. The theorem demonstrates that the gains are *identifiable* from routine sensor measurements; that is, in the absence of noise or other errors, the gains are perfectly recovered. In fact, the proof shows that under A1 and A2, the condition A3 is both necessary and sufficient. When noise and errors are present, the estimated gains may not be exactly equal to the true gains. However, as the noise/errors in the measurements tend to zero, the estimated gains tend to the true gains.

1.4.2 Bandlimited Subspaces

In the special case in which the signal subspace corresponds to a frequency domain subspace, a slightly more precise characterization is possible which shows that even fewer snapshots suffice for blind calibration. As stated above, assumption A3 is often met in practice and can be easily checked given a signal basis $\boldsymbol{\Phi}$. One case where A3 is automatically met is when the signal subspace is spanned by a subset of the Discrete Fourier Transform (DFT) vectors:

$$\phi_m = [1, e^{\frac{-i2\pi m}{n}}, \dots, e^{\frac{-i2(n-1)\pi m}{n}}]'/\sqrt{n}, \quad m = 0, \dots, n-1$$

In this case we show that only $\lceil \frac{n-1}{n-r} \rceil + 1$ snapshots are required. This can be significantly less than r, meaning that the time over which we must assume that the subspace and calibration coefficients are unchanging is greatly reduced. The following assumptions and theorem summarize this result.

B1. **Oversampling:** Assume that each signal $\boldsymbol{x}$ lies in a bandlimited r-dimensional subspace $\mathcal{S}$, $r < n$, spanned by $\boldsymbol{\phi}_{m_1}, \ldots, \boldsymbol{\phi}_{m_r}$, where $m_1, \ldots, m_r$ are distinct integers from the set $\{0, \ldots, n-1\}$. Furthermore, assume that these integers are aperiodic in the following sense. Let $\mathbf{s}$ denote the vector with one at locations $m_1, \ldots, m_r$ and zero otherwise. This vector indicates the *support* set of the signal subspace in the DFT domain. The integers $m_1, \ldots, m_r$ are called aperiodic if every circular (mod n) shift of $\mathbf{s}$ is distinct. It is easy to check this condition and, in fact, most bandlimited subspaces have aperiodic support sets.

B2. **Randomness:** Note that each signal $\boldsymbol{x} \in \mathcal{S}$ can be written as $\boldsymbol{x} = \sum_{j=1}^{r} \theta_j \boldsymbol{\phi}_{m_j}$, for certain coefficients $\theta_1, \ldots, \theta_r$. Each signal is randomly drawn from $\mathcal{S}$ and has mean zero. This means that the signal coefficients are zero-mean random variables. The joint distribution of these random variables is absolutely continuous with respect to Lebesgue measure (i.e., a joint r-dimensional density function exists). Also assume that multiple signal observations are statistically uncorrelated.

Assumption B1 guarantees that the calibrated or true sensor measurements are spatially correlated to some degree. As before, this assumption is crucial since it implies that measurements must satisfy the constraints in (1.3) and that, in principle, we can solve for the gain vector $\boldsymbol{\alpha}$. The rationale behind the assumption that the frequency support set is aperiodic is less obvious, but its necessity is due to the 2π-periodicity of the DFT (see [1] for further details). Assumption B2 guarantees the signals are not temporally correlated (analogous to A2, above). The following theorem characterizes the identifiability of the sensor gains in this situation.

Theorem 2. *It is necessary to make at least $k \geq \lceil \frac{n-1}{n-r} \rceil$ signal measurements in order to determine the gains, where $\lceil z \rceil$ denotes the smallest integer greater than or equal to z. Moreover, under assumptions B1 and B2, the gains $\boldsymbol{\alpha}$ can be perfectly recovered from any $k = \lceil \frac{n-1}{n-r} \rceil + 1$ signal measurements by solving the linear system of equations (1.5).*

The theorem is proved in the Appendix. The proof takes advantage of the special structure of the DFT basis. Alternatively, one could apply Theorem 1 in this case to obtain a slightly weaker result; namely that under B1 and B2 $k \geq r$ observations suffice to perfectly recover the gains.

1.5 Offset Calibration

The component of the offset in the signal subspace is generally unidentifiable, but in special cases it can be determined. For example, if it is known that the phenomenon of interest fluctuates symmetrically about zero (or some other known value), then the average of many measurements will tend to zero (or the known mean value). In this situation, the average

$$\frac{1}{k}\sum_{i=1}^{k}\boldsymbol{y}_i = \left(\frac{1}{k}\sum_{i=1}^{k}\boldsymbol{x}_i - \boldsymbol{\beta}\right)/\boldsymbol{\alpha} \approx -\boldsymbol{\beta}/\boldsymbol{\alpha}$$

where the the division operation is taken element-by-element. This follows since $\frac{1}{k}\sum_{i=1}^{k}\boldsymbol{x}_i \approx 0$ for large enough k. Thus we can identify the offset simply by calculating the average of our measurements. More precisely, we can identify $\widetilde{\boldsymbol{\beta}} = \boldsymbol{\beta}/\boldsymbol{\alpha}$, which suffices since we can equivalently express the basic relationship (1.1) between calibrated and uncalibrated snapshots as $\boldsymbol{x} = (\boldsymbol{Y} + \widetilde{\boldsymbol{\beta}})\,\boldsymbol{\alpha}$.

Another situation in which we can determine (or partially determine) the component of the offset in the signal subspace is when we have knowledge of the correct offsets for a subset of the sensors. We call this *partially blind* offset calibration. Suppose that we are able to directly measure the offsets at $m < n$ sensors, indexed by m distinct integers $1 \leq \ell_1, \ldots, \ell_m \leq n$. Let $\boldsymbol{\beta}_m$ denote an $m \times 1$ vector these offsets. Let $\boldsymbol{T}$ be an $m \times n$ "selection" matrix that when applied to an arbitrary $n \times 1$ vector produces an $m \times 1$ vector of the elements at locations $\ell_1, \ldots, \ell_m$ from the original vector. With this notation, we can write $\boldsymbol{\beta}_m = \boldsymbol{T}\boldsymbol{\beta}$. Also note that

$$\boldsymbol{\beta}_m = \boldsymbol{T}\boldsymbol{\beta} = \boldsymbol{T}(\boldsymbol{P}\boldsymbol{\beta} + (\boldsymbol{I}-\boldsymbol{P})\boldsymbol{\beta}) = \boldsymbol{T}\boldsymbol{P}\boldsymbol{\beta} + \boldsymbol{T}(\boldsymbol{I}-\boldsymbol{P})\boldsymbol{\beta},$$

where $(\boldsymbol{I}-\boldsymbol{P})\boldsymbol{\beta}$ is the offset component in the signal subspace and $\boldsymbol{P}\boldsymbol{\beta}$ is the offset component in the orthogonal complement to the signal subspace.

As pointed out in Section 1.3.1, we can determine the component of the offset in the nullspace using $\boldsymbol{P}\boldsymbol{\beta} = -\boldsymbol{P}\bar{\boldsymbol{Y}}\widehat{\boldsymbol{\alpha}}$. Let us assume that this component is known (from the estimated calibration gains), and define $\boldsymbol{\beta}_\Delta = \boldsymbol{T}(\boldsymbol{I} - \boldsymbol{P})\boldsymbol{\beta}$, the signal subspace component of the offset at sensors $\ell_1, \ldots, \ell_m$. This component satisfies the relation

$$\boldsymbol{\beta}_\Delta = \boldsymbol{\beta}_m - \boldsymbol{T}\boldsymbol{P}\boldsymbol{\beta} = \boldsymbol{\beta}_m + \boldsymbol{T}\boldsymbol{P}\bar{\boldsymbol{Y}}\widehat{\boldsymbol{\alpha}}. \tag{1.15}$$

The projection matrix corresponding to the signal subspace, $(\boldsymbol{I}-\boldsymbol{P})$, can be written in terms of a set of orthonormal column vectors, $\boldsymbol{\phi}_1, \ldots, \boldsymbol{\phi}_r$, that span the signal subspace. Let $\boldsymbol{\Phi} = [\boldsymbol{\phi}_1 \cdots \boldsymbol{\phi}_r]$ denote an $n \times r$ matrix whose columns are the basis vectors. Then $(\boldsymbol{I}-\boldsymbol{P}) = \boldsymbol{\Phi}\boldsymbol{\Phi}'$ and so we can also write $\boldsymbol{\beta}_\Delta = \boldsymbol{T}\boldsymbol{\Phi}\boldsymbol{\Phi}'\boldsymbol{\beta}$. Note that the offset component in the signal subspace is completely determined by the r parameters $\boldsymbol{\theta} = \boldsymbol{\Phi}'\boldsymbol{\beta}$. Defining $\boldsymbol{\Phi}_T = \boldsymbol{T}\boldsymbol{\Phi}$, we can write $\boldsymbol{\beta}_\Delta = \boldsymbol{\Phi}_T\boldsymbol{\theta}$. If $\boldsymbol{\Phi}_T$ is invertible, then using (1.15) the parameters $\boldsymbol{\theta}$ can be uniquely determined by $\boldsymbol{\theta} = \boldsymbol{\Phi}_T^{-1}(\boldsymbol{\beta}_m + \boldsymbol{T}\boldsymbol{P}\bar{\boldsymbol{Y}}\widehat{\boldsymbol{\alpha}})$. Thus, if $\boldsymbol{\Phi}_T$ is invertible, then the complete offset vector $\boldsymbol{\beta}$ can be determined from the subset of offsets $\boldsymbol{\beta}_m$ and the estimated gains $\widehat{\boldsymbol{\alpha}}$. If $\boldsymbol{\Phi}_T$ has rank $q < r$, then we can determine the signal subspace offset component up to a remaining unidentifiable component in a smaller $r-q$ dimensional subspace of the signal subspace.

The rank of $\boldsymbol{\Phi}_T$ cannot be greater than m, the number of known sensor offsets, which shows that to completely determine the offset component in the signal subspace we require at least $m = r$ known offsets. In general, knowing the offsets for an arbitrary subset of m sensors may not be sufficient (i.e., $\boldsymbol{\Phi}_T$ may not be invertible), but there are important special cases when it is. First note the $\boldsymbol{\Phi}$, by construction, has full rank r. Also note that the selection matrix $\boldsymbol{T}$ selects the m rows corresponding to the known calibration offsets and eliminates the remaining $n-m$ rows. So, we require that the elimination of any subset of $n-m$ rows of $\boldsymbol{\Phi}$ does not lead to a linearly dependent set of $(m \times 1)$ columns. This requirement is known as an *incoherence condition* , and it is satisfied as long as the signal basis vectors all have small inner products with the natural or canonical sensor basis ($n \times 1$ vectors that are all zero except for a single non-zero entry). For example, frequency vectors (e.g., Discrete Fourier Transform vectors) are known to satisfy this type of incoherence condition [5]. This implies that for subspaces of bandlimited signals, $\boldsymbol{\Phi}_T$ is invertible provided $m \geq r$.

1.6 Evaluation

In order to evaluate whether this theory of blind calibration is possible in practice, we explore its performance in simulation under both measurement noise and the mis-characterization of the projection matrix $\boldsymbol{P}$. Additionally, we show the performance of the algorithm on two temperature sensor datasets, one dataset from a controlled experiment where the sensors are measuring all the same phenomenon and thus lie in a 1-dimensional subspace, and the other from a deployment in a valley at a nature preserve called the James Reserve[4], where the true dimension of the spatial signal is unknown.

1.6.1 Simulations

To test the blind calibration methods on simulated data, we simulated both a field and snapshots of that field. We generated gain and offset coefficients, measurement noise, and most importantly, a projection matrix $\boldsymbol{P}$.

We simulated a smooth field by generating an 256×256 array of pseudo-random Gaussian noises (i.e., a white noise field) and then convolving it with the smooth impulse response function $h(i,j) = e^{(-s((i-l/2)^2+(j-l/2)^2)}$, $s > 0$. Figure 1.1 shows an example field with the smoothing parameter $s = 1$, which could represent a smoothly varying temperature field, for example. We simulated sensor measurements by sampling the field on a uniform 8×8 grid of $n = 64$ sensors. For gains, we drew uniformly from $\boldsymbol{\alpha} \in [0.5, 1.5]$ and for

[4] `http://www.jamesreserve.edu`

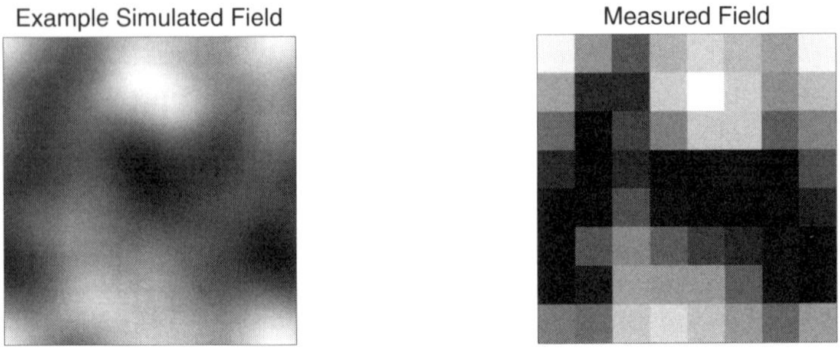

Fig. 1.1. Two example simulated square fields. On the left, a 256×256 field generated with a basic smoothing kernel, which represents a true continuous field. On the right, an 8×8 grid of measurements of the same field. The fields can be quite dynamic and still meet the assumptions for blind calibration. The fields are shown in pseudo-grayscale, with black denoting the minimum valued regions and white denoting the maximum valued regions.

offsets from $\boldsymbol{\beta} \in [-.5, .5]$. After applying $\boldsymbol{\alpha}$ and $\boldsymbol{\beta}$ to the measurements, we then added Gaussian noise, with mean zero and variance σ.

Separately, we created $\boldsymbol{P}$ to be a low-pass DFT matrix. We kept 3 frequencies in $2d$, which means with symmetries we have an $r = 49$-dimensional subspace[5]. With this setup, we can adjust the parameters of the smoothing kernel, while keeping $\boldsymbol{P}$ constant, to test robustness of blind calibration to an assumed subspace model that may over- or under-estimate the dimension of the subspace of the true field. The smoothing kernel and projection $\boldsymbol{P}$ both characterize lowpass effects, but the smoothing operator is only approximately described by the projection operator, even in the best case. We can also create our field by projecting the random field onto the r-dimensional subspace using $\boldsymbol{P}$; this represents the case where the true subspace is known exactly.

Estimates of the gains and offsets were calculated using the methods discussed above and described in more detail below. The graphs show error in gain, or equivalently in offset, as a fraction of the vector magnitude as follows:

$$err_{\alpha} = \frac{\|\boldsymbol{\alpha} - \hat{\boldsymbol{\alpha}}\|_2}{\|\boldsymbol{\alpha}\|_2} \tag{1.16}$$

Error in the uncalibrated signal was calculated in the same way, assuming the estimate for gain, $\hat{\boldsymbol{\alpha}}$, is the all-ones vector and the estimate for offset, $\hat{\boldsymbol{\beta}}$, is the all-zeros vector. Error in the predicted signal was calculated by first applying the gain and offset factors to each measured signal $\boldsymbol{y}_i$ ($\hat{\boldsymbol{x}}_i = \boldsymbol{y}_i \circ \hat{\boldsymbol{\alpha}} + \hat{\boldsymbol{\beta}}$ where $\circ$ is the hadamard product) and then calculating error as:

[5] If the 2-dimensional signal has p frequencies, then the subspace is of rank $r = (2p+1)^2$.

$$err_{\boldsymbol{x}_i} = \frac{\|\boldsymbol{x}_i - \hat{\boldsymbol{x}}_i\|_2}{\|\boldsymbol{x}_i\|_2}, \qquad i = 1, \ldots, k \tag{1.17}$$

Finally, the error was then averaged over the k measurements.

Error Results using SVD

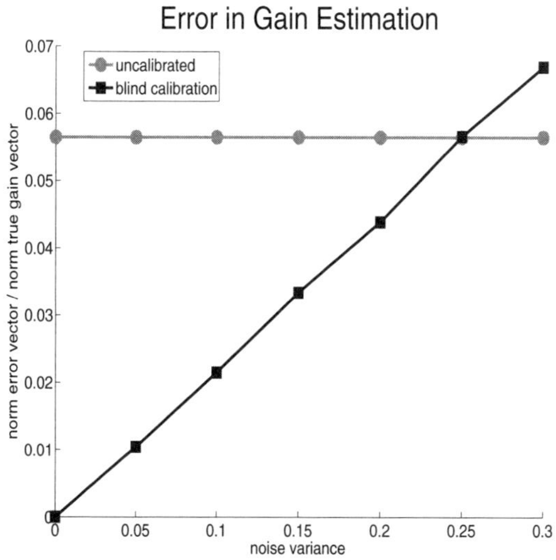

Fig. 1.2. Gain error performance with exact knowledge of $\boldsymbol{P}$ and increasing measurement noise. Error averaged over 100 simulation runs.

We simulated blind calibration with the described simulation set-up. We first generated mean-zero fields using our smoothing kernel and took snapshot measurements of each field. We used $k = 5r$ snapshots (slightly more than the theoretical minimum of $k = r$) in order to improve our zero-mean signal assumption and to provide added robustness to noise and modeling errors. Then we constructed the matrix $\mathbf{C}$ from equation (1.7) and took the minimum right singular vector as the estimate of the gains $\boldsymbol{\alpha}$ as described in Section 1.3.2. We then estimated $\boldsymbol{\beta} = -\bar{\boldsymbol{Y}}\boldsymbol{\alpha}$.

Results from totally blind calibration in simulation, under the burden of increasing noise variance using exact knowledge of the subspace defined by $\boldsymbol{P}$, are shown in Figures 1.2, 1.3, and 1.4. That is, the fields in these simulations were created by projecting random signals into the space defined by projection matrix $\boldsymbol{P}$. The maximum value in the signals was 1, and therefore the noise variance can be taken as a percentage; i.e., variance of 10^{-2} represents 1%

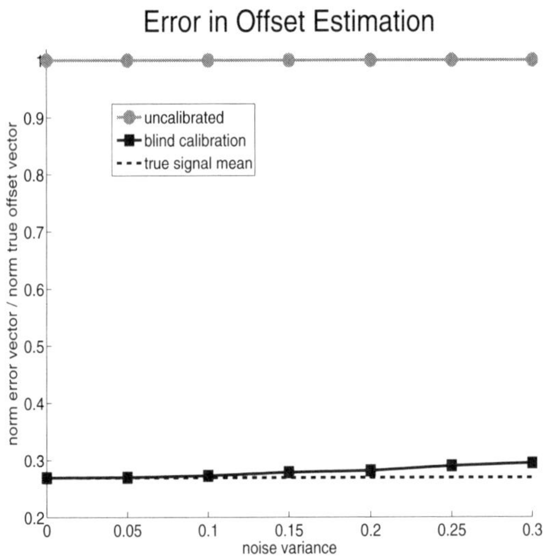

Fig. 1.3. Offset error performance with exact knowledge of $\boldsymbol{P}$ and increasing measurement noise. Error averaged over 100 simulation runs.

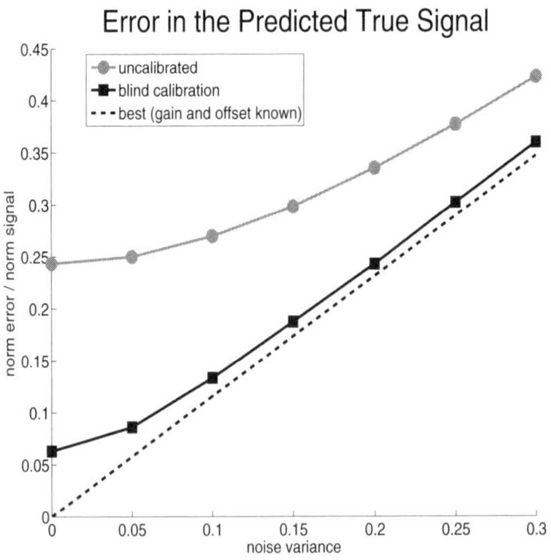

Fig. 1.4. Error in the estimate of the signal x with exact knowledge of $\boldsymbol{P}$ and increasing measurement noise. Error averaged over 100 simulation runs.

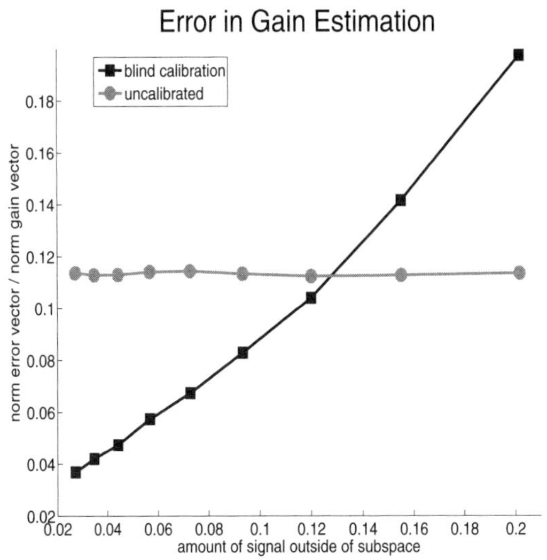

Fig. 1.5. Gain error performance with uncertainty in the subspace $\boldsymbol{P}$. Error averaged over 100 simulation runs.

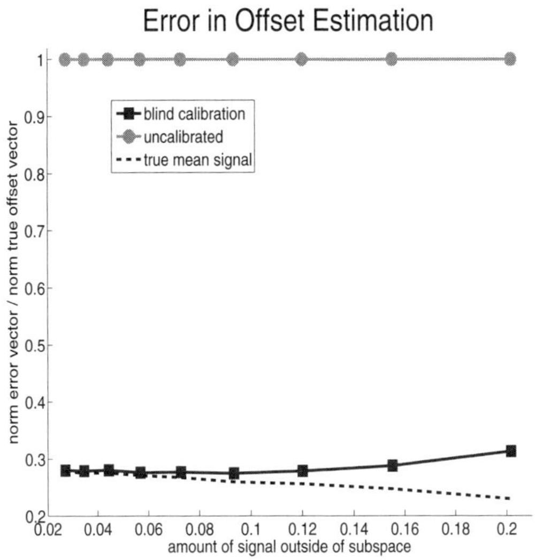

Fig. 1.6. Offset error performance with uncertainty in the subspace $\boldsymbol{P}$. Error averaged over 100 simulation runs.

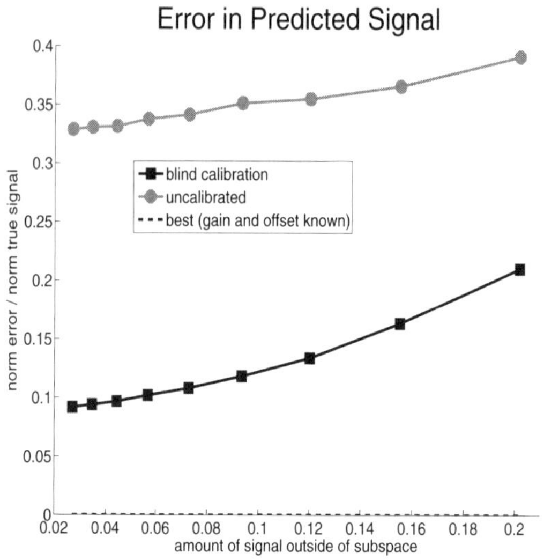

Fig. 1.7. Error in the estimate of the signal x with with uncertainty in the subspace $\boldsymbol{P}$. Error averaged over 100 simulation runs.

noise in the signal. Figure 1.2 shows the error in gain estimation as compared to the error in the uncalibrated signal. Figure 1.3 shows the error in offset estimation, along with the uncalibrated error and the energy of the mean true signal. Figure 1.4 shows error in the resulting estimate of the signals $\boldsymbol{x}$, after applying the estimates of $\boldsymbol{\alpha}$ and $\boldsymbol{\beta}$ to $\boldsymbol{y}$ in order to recover $\hat{\boldsymbol{x}}$.

Blind calibration did very well in this scenario. Not until the noise in the signal was 25% did the uncalibrated gains outperform the blindly calibrated gains. As for offset, the blindly calibrated estimate was always much better than the uncalibrated, and in fact stayed very close to the theoretical minimum of the mean true signal. The combination is illustrated nicely in Figure 1.4, where we can see that the error in the blindly calibrated signal remains very close to the case where we know the corect gain and offset. The figures show mean error over 100 simulation runs.

Knowing the true subspace exactly is possible in practice only when performing blind calibration in a very well-known environment, such as an indoor factory. Even in this case, there will be some component of the true signals which is outside of the subspace defined by the chosen $\boldsymbol{P}$. Figure 1.7 shows how gain and offset error are affected by out-of-subspace components in the true signals. We used a basic smoothing kernel to control smoothness of the true field and kept $\boldsymbol{P}$ constant as described above with $r = 49$. The smoothing kernel and the projection operator are both low-pass operators, but even in the best case, some of the smoothed field will be outside of the space defined by the projection matrix $\boldsymbol{P}$. We defined the error in $\boldsymbol{P}$ as $\|\boldsymbol{x} - \boldsymbol{P}\boldsymbol{x}\|_2 / \|\boldsymbol{x}\|_2$. The x-axis value in the figure is the average error in $\boldsymbol{P}$ over 100 random fields smoothed with a given smoothness parameter. The figure shows mean and median error in gain and offset estimates over these 100 simulation runs. Again the results are compelling. The gain estimation error was around 10^{-2} even when 10% of the signal was outside of the subspace. The offset estimation as well was still very accurate, below 7×10^{-3} even when 20% of the signal was outside of the subspace.

Comparison of Techniques

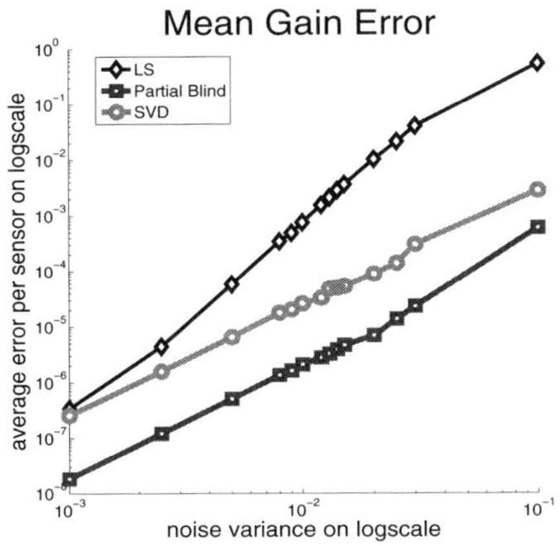

Fig. 1.8. Gain error performance for SVD, blind LS, and partially blind LS. Results show mean error over 50 simulation runs.

Here we compare the SVD technique to the LS technique and the totally blind calibration to partially blind calibration, where we know some of the calibration coefficients ahead of time.

The methods each work as follows. *SVD* performs gain estimation using the minimum right singular vector of the svd, i.e. by finding the solution to Equation (1.8), and normalizes assuming $\boldsymbol{\alpha}(1) = 1$. Offsets are then estimated using $\boldsymbol{\beta} = -\bar{\boldsymbol{Y}}\boldsymbol{\alpha}$. *Totally blind LS* performs gain estimation by solving Equation (1.9) in a least-squares sense and assuming knowledge only of $\boldsymbol{\alpha}(1) = 1$. Offsets are estimated as in SVD. *Partially blind LS* performs gain estimation by solving equation (1.10) in the least-squared sense, but now assuming we know some number of true gains and offsets. Offsets are then estimated as described in Section 1.5 for non-zero mean signals, i.e. using $\boldsymbol{\beta}_{\Delta} = \boldsymbol{T\Phi\Phi}'\boldsymbol{\beta}$ to solve for $\theta = \boldsymbol{\Phi}'\boldsymbol{\beta}$ and thus $\boldsymbol{\beta}$.

For partially blind LS we use enough of the true offsets such that we can solve for the complete component of $\boldsymbol{\beta}$ in the signal subspace. The fields we simulated are nearly bandlimited subspaces, and so the theory would imply that r true offsets are enough to estimate $\boldsymbol{\beta}$. In order to be robust to noise,

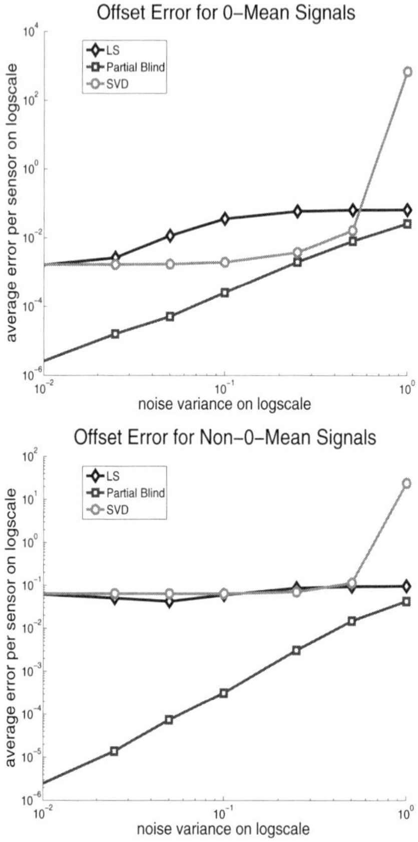

Fig. 1.9. Offset error performance for SVD, blind LS, and partially blind LS. The top graph shows offset error for zero-mean signals, and the bottom graph is for non-zero-mean signals. Results show mean error over 50 simulation runs.

we used knowledge of the offsets of $r + 5$ sensors, again slightly more than the bare minimum suggested by the theory.

A comparison of the techniques is quite interesting. First, as we expect, the partially blind estimation does better than the other two methods in all cases; this follows from the fact that it is using more information. In Figure 1.8 you can see in the gain estimation, the SVD method out-performs totally blind LS, but partially blind LS has the lowest error of all the methods.

In the case of offset error, the SVD and totally blind LS techniques out-perform one another depending on the noise variance and whether or not the signals are zero-mean.

Figure 1.9 shows offset error for all three techniques. The partially blind LS method is unaffected by non-zero mean signals, which follows because method for estimating the offsets does not change with a zero-mean assumption. The other methods, on the other hand, capture the mean signal as part of their offset estimates, and as we can see, estimation error using the non-zero-mean signals is higher than using zero-mean signals.

The most intriguing part of these results is that totally blind LS performs slightly better than SVD for the offset estimate in non-zero-mean signals, despite the fact that it is using a gain estimate with more error from the first step in order to estimate the offsets. This implies that if calibration offset is the most important for calibration of your system, and you have non-zero-mean signals, you might prefer the totally blind LS method over the SVD.

1.6.2 Evaluation on Sensor Datasets

We evaluate blind calibration on two sensor network datasets, which we call the *calibration dataset* and the *cold air drainage transect dataset*.

Calibration Dataset

The *calibration dataset* was collected in September 2005 [2] along with data from a reference-caliber instrument in order to characterize the calibration of the thermistors used for environmental temperature measurement at the James Reserve. From the experiment, the conclusion was drawn that after the factory-supplied calibration was applied to the raw sensor measurements, the sensors differed from the reference thermocouple linearly, i.e. by only a gain and offset. Thus these sensors are suitable for evaluating the work we have done thus far on blind calibration. The data is available in the NESL CVS repository[6].

The setup of this experiment consisted of nine[7] temperature sensors. These sensors were placed in a styrofoam box along with a thermocouple attached to a datalogger, providing ground truth temperature readings. Therefore, all sensors were sensing the same phenomenon, and so the subspace spanned by the nine measurements is rank one. Thus, for $\boldsymbol{P}$ we used a lowpass dct matrix which kept only the dc frequency space. To illustrate, we used the following commands in Matlab:

```
r = 1; n = 9;
I = eye(n);
U = dct(I);
U(r+1:n,:) = 0;
P = idct(U);
```

We calibrated these data using snapshots from the dataset and the SVD method. Figure 1.10 shows the calibration coefficient estimates and reconstructed signals for the sensors in the experiment. The gains and offsets were

[6] This data is available at `http://www.ee.ucla.edu/~sunbeam/bc/`

[7] The experiment had ten sensors, one of which was faulty. In this analysis we used data from the nine functional sensors.

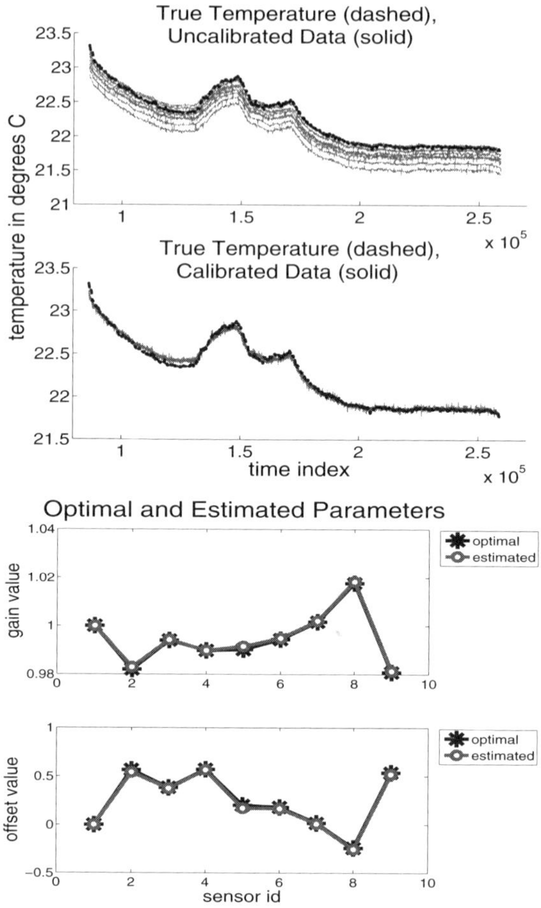

Fig. 1.10. Results of blind calibration on the *calibration dataset.*

recovered with very little error. The uppermost plot shows the data before and after calibration, along with the ground truth measurement in blue. The lower plot shows the true and estimated gains and offsets. This clearly demonstrates the utility of blind calibration.

Cold Air Drainage Dataset

The *cold air drainage transect dataset* consists of data from an ongoing deployment at the James Reserve. The deployment measures air temperature and humidity in a valley in order to characterize the predawn cold air drainage. The sensors used are the same as the sensors in the *calibration dataset*, and thus again the factory calibration brings them within an offset and gain of one another. The data we used for evaluation is from November 2, 2006, and

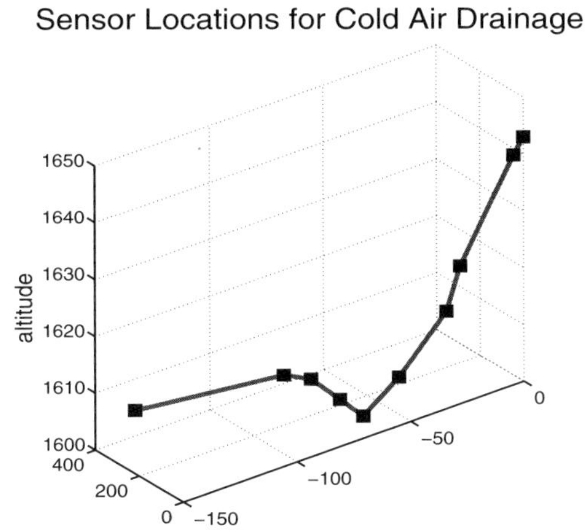

Fig. 1.11. The mica2 motes in the cold air drainage transect run down the side of a hill and across a valley. The mote locations pictured are those that we used in the evaluation of blind calibration.

it is available in the sensor data repository called SensorBase[8]. On this same day, we visited the James Reserve with a reference-caliber sensor and took measurements over the course of the day in order to get the true calibration parameters for comparison.

The deployment consists of 26 mica2 motes which run from one side of a valley to the other (Figure 1.11) across a streambed and in various regions of tree and mountain shade. Each mote has one temperature and one humidity sensor. For our purposes, we collected calibration coefficients from 10 of the temperature sensors.

The signal subspace in this application does not correspond to a simple lowpass or smooth subspace, since sensors at similar elevations may have similar readings, but can be quite distant from each other. In principle, the signal subspace could be constructed based on the geographic positions and elevations of the sensor deployment. However, since we have the calibrated sensor data in this experiment, we can use these data directly to infer an approximate signal subspace. We constructed the projection $\boldsymbol{P}$ using the subspace associated with the four largest singular values of the calibrated signal data matrix.

We performed totally blind calibration using SVD. We constructed $\mathbf{C}$ using 64 snapshots taken over the course of the morning along with $\boldsymbol{P}$ as described. Figure 1.12 shows the results. The gain error was very small, only .0053 average per sensor, whereas if we were to assume the gain was 1 and not calibrate the sensors at all, the error would be .0180 average per sensor. On the other

[8] `http://sensorbase.org`

hand, the offset error was only slightly better with blind calibration than it would have been without: we saw .3953 average error per sensor as compared to 0.4610 error if the offsets were assumed to be zero. We believe that the offset estimation did not perform well due primarily to the fact that the mean signal is not zero in this case (e.g., the average sensor readings depend on elevation). Better offset estimates could be obtained using knowledge of one or more of the true sensor offset values.

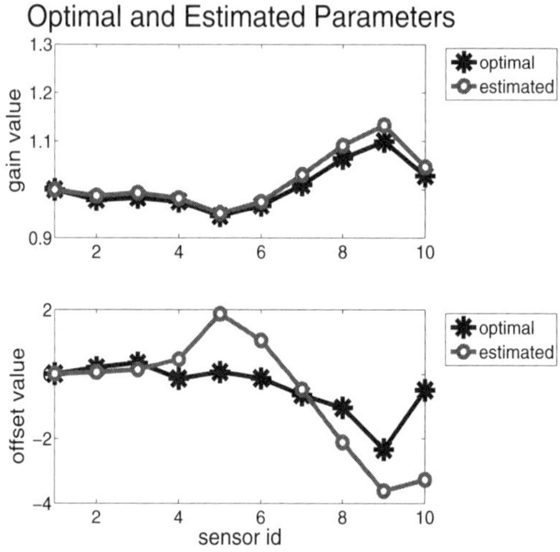

Fig. 1.12. True and estimated gains and offsets for the *cold air drainage transect dataset.*

1.7 Related Work

The most straightforward approach to calibration is to apply a *known* stimulus $\boldsymbol{x}$ to the sensor network and measure the response $\boldsymbol{y}$. Then using the groundtruth input $\boldsymbol{x}$ we can adjust the calibration parameters so that (1.1) is achieved. We call this *non-blind* calibration, since the true signal $\boldsymbol{x}$ is known. This problem is called inverse linear regression; mathematical details can be found at [9]. Non-blind calibration is used routinely in sensor networks [11, 14], but may be difficult or impossible in many applications.

As for blind calibration in sensor networks, the problem of relating measurements such as received signal strength or time delay to distance for localization purposes has been studied extensively [10, 13]. This problem is quite different from the blind calibration problem considered in this chapter, which assumes that the measurements arise from external signals (e.g., temperature) and not from range measurements between sensors. In [15], the problem of calibrating sensor range measurements by enforcing geometric constraints in a system-wide optimization is considered. Calibration using geometric and physical constraints on the behavior of a point light source is considered in [6]. The constraint that proximal sensors in dense deployments make very similar measurements is leveraged in [4]. In this work, our constraint is simply that the phenomenon of interest lies in a subspace. This is a much more general constraint and hopefully therefore it can be widely applicable.

Blind equalization and blind deconvolution [12] are related problems in signal processing. In these problems, the observation model is of the form $\boldsymbol{y} = \mathbf{h} * \boldsymbol{x}$, where $*$ is the convolution operator, and both $\mathbf{h}$ and $\boldsymbol{x}$ must be recovered from $\boldsymbol{y}$. Due to the difference between the calibration and convolution models, there are significant differences between blind deconvolution and blind calibration. However, in certain circumstances, blind calibration is mathematically equivalent to multi-channel blind deconvolution [7, 8]. Blind calibration involves observing multiple unknown signals through one unknown calibration function. Multi-channel blind deconvolution involves observing one unknown signal through multiple unknown channels. In the deconconvolution set-up, a common assumption is that the multiple channels have finite impulse response functions. This places the channels in a lower dimensional subspace (of the space of all impulse responses), and singular value decomposition techniques, similar to those proposed for blind calibration in this chapter, have been devised for solving the blind deconvolution problem in this setting [7, 8]. The blind calibration problem is mathematically equivalent to the multi-channel blind deconvolution problem, in the following special case. First observe that the multiplicative calibration gain can be expressed as a convolutional operation in the frequency domain. Now suppose that the signal subspace is a lowpass frequency subspace spanned by low-frequency DFT vectors. In this case, blind gain calibration is exactly equivalent to multi-FIR-channel blind deconvolution (the two problems are related by the DFT). In general, however, the signal subspace may not be a low-frequency subspace, and the two problems are quite different. In this sense, blind calibration is much more general than multi-channel blind deconvolution, but the relation between the two problems suggests that more sophisticated solution methods, such as IQML [8], might be applicable to blind calibration.

1.8 Extensions and Future Work

There are many issues in blind calibration that could be explored further. The two main areas ripe for study are the choice of the subspace $\boldsymbol{P}$ and the implementation of blind calibration. There are many possible choices for a suitable subspace, including frequency subspaces and smoothness subspaces. How to choose the subspace when faced with a sensor deployment where the true signals are unknown is an extremely important question for blind calibration. Methodologies for creating a $\boldsymbol{P}$ would be extremely useful to the more general application of blind calibration, especially ones which could incorporate trusted measurements or the users' knowledge of the physical space where the sensors are deployed. At the same time, implementations of blind calibration that are robust to model error in the subspace would allow users to be more liberal in the choice of $\boldsymbol{P}$.

The theoretical analysis shown here is done under noiseless conditions and with a perfect model. Future work includes both noisy analysis to find analytical bounds that can be compared to simulation results and sensitivity analysis for our system of linear equations. Our experience is that solutions are robust to noise and mismodeling in some cases, and sensitive in others; we do not have a good understanding of the robustness of the methodology at this time.

Extending the formulation to handle non-linear calibration functions would be useful in cases where a raw non-linear sensor response must be calibrated. We believe that many of the techniques developed in this work can be extended to more general polynomial-form calibration functions. Other interesting topics include distributed blind calibration and blind calibration in the presence of faulty sensors.

1.9 Conclusions

The problem of sensor calibration is central to the practical use of sensor networks. The blind calibration formulation and methods developed in this work use only routine sensor measurements, and thus give an extremely promising formulation for the mass calibration of sensors. We have shown that calibration gains are identifiable. We have proved how many measurements are necessary and sufficient to estimate the gain factors, and we have shown necessary and sufficient conditions to estimate the offsets. We have demonstrated a working implementation on simulated and real data, which uncovered interesting relationships between implementation and blind calibration performance. Overall, we have demonstrated that blind calibration has great potential to be possible in practice, and we feel that the proposed formulation merits further investigation.

References

[1] L. Balzano and R. Nowak. Blind calibration for signals with bandlimited subspaces. Technical report, Information Sciences Laboratory at the University of Wisconsin-Madison, February 2007.

[2] L. Balzano, N. Ramanathan, E. Graham, M. Hansen, and M. B. Srivastava. An investigation of sensor integrity. Technical Report UCLA-NESL-200510-01, Networked and Embedded Systems Laboratory, 2005.

[3] P. Buonadonna, D. Gay, J. Hellerstein, W. Hong, and S. Madden. Task: Sensor network in a box. Technical Report IRB-TR-04-021, Intel Research Berkeley, January 2005.

[4] V. Bychkovskiy, S. Megerian, D. Estrin, and M. Potkonjak. A collaborative approach to in-place sensor calibration. *Lecture Notes in Computer Science*, 2634:301–316, 2003.

[5] E. Candes and J. Romberg. Quantitative robust uncertainty principles and optimally sparse decompositions. *Foundations of Computational Mathematics*, 2006.

[6] J. Feng, S. Megerian, and M. Potkonjak. Model-based calibration for sensor networks. *Sensors*, pages 737 – 742, October 2003.

[7] M. Gurelli and C. Nikias. Evam: An eigenvector-based algorithm for multichannel blind deconvolution of input colored signals. *IEEE Transactions on Signal Processing*, 43:134–149, January 1995.

[8] G. Harikumar and Y. Bresler. Perfect blind restoration of images blurred by multiple filters: Theory and efficient algorithms. *IEEE Transactions on Image Processing*, 8(2):202 – 219, February 1999.

[9] B. Hoadley. A bayesian look at inverse linear regression. *Journal of the American Statistical Association*, 65(329):356 – 369, March 1970.

[10] A. Ihler, J. Fisher, R. Moses, and A. Willsky. Nonparametric belief propagation for self-calibration in sensor networks. In *Proceedings of the Third International Symposium on Information Processing in Sensor Networks*, 2004.

[11] N. Ramanathan, L. Balzano, M. Burt, D. Estrin, T. Harmon, C. Harvey, J. Jay, E. Kohler, S. Rothenberg, and M.Srivastava. Rapid deployment with confidence: Calibration and fault detection in environmental sensor networks. Technical Report CENS TR 62, Center for Embedded Networked Sensing, 2006.

[12] O. Shalvi and E. Weinstein. New criteria for blind deconvolution of nonmimimum phase systems (channels). *IEEE Trans. on Information Theory*, IT-36(2):312 – 321, March 1990.

[13] C. Taylor, A. Rahimi, J. Bachrach, H. Shrobe, and A. Grue. Simultaneous localization, calibration, and tracking in an ad hoc sensor network. In *IPSN '06: Proceedings of the Fifth International Conference on Information Processing in Sensor Networks*, pages 27–33, 2006.

[14] G. Tolle, J. Polastre, R. Szewczyk, D. Culler, N. Turner, K. Tu, S. Burgess, T. Dawson, P. Buonadonna, D. Gay, and W. Hong. A macroscope in the redwoods. In *Proceedings of Sensys*, 2005.

[15] K. Whitehouse and D. Culler. Calibration as parameter estimation in sensor networks. In *Proceedings of the 1st ACM International Workshop on Wireless Sensor Networks and Applications*, pages 59–67, 2002.

[16] W. M. Wonham. *Linear Multivariable Control.* Springer-Verlag, New York, 1979.

Appendix

Theorem 1:

Proof. First note that the case where the signal subspace is one-dimensional ($r = 1$) is trivial. In this case there is one degree of freedom in the signal, and hence one measurement coupled with the constraint that $\alpha(1) = 1$ suffices to calibrate the system. For the rest of the proof we assume that $1 < r < n$ and thus $2 \leq k < n$.

Given k signal observations $\boldsymbol{y}_1, \ldots, \boldsymbol{y}_k$, and letting $\widehat{\boldsymbol{\alpha}}$ represent our estimated gain vector, we need to show that the system of equations

$$\begin{bmatrix} \boldsymbol{P}\boldsymbol{Y}_1 \\ \vdots \\ \boldsymbol{P}\boldsymbol{Y}_k \end{bmatrix} \widehat{\boldsymbol{\alpha}} = 0 \tag{1.18}$$

has rank $n-1$, and hence may be solved for the $n-1$ degrees of freedom in $\widehat{\boldsymbol{\alpha}}$. Note each subsystem of equations, $\boldsymbol{P}\boldsymbol{Y}_j$, has rank less than or equal to $n-r$ (since $\boldsymbol{P}$ is rank $n-r$). Therefore, if $k < \frac{n-1}{n-r}$, then the system of equations certainly has rank less than $n-1$. This implies that it is necessary that $k \geq \frac{n-1}{n-r}$. Next note that $\boldsymbol{Y}_j = \boldsymbol{X}_j\boldsymbol{A}$, where $\boldsymbol{X}_j = \mathrm{diag}(\boldsymbol{x}_j)$ and $\boldsymbol{A} = \mathrm{diag}([1, 1/\alpha(2), \ldots, 1/\alpha(n)]')$. Then write

$$\begin{bmatrix} \boldsymbol{P}\boldsymbol{X}_1 \\ \vdots \\ \boldsymbol{P}\boldsymbol{X}_k \end{bmatrix} \boldsymbol{d} = 0 \tag{1.19}$$

where $\boldsymbol{d} = \boldsymbol{A}\widehat{\boldsymbol{\alpha}}$. The key observation is that satisfaction of these equations requires that $\boldsymbol{X}_j\boldsymbol{d} \in \mathcal{S}$, for $j = 1, \ldots, k$. Any $\boldsymbol{d}$ that satisfies this relationship will imply a particular solution for $\widehat{\boldsymbol{\alpha}}$, and thus $\boldsymbol{d}$ must not be any vector other than the all-ones vector for blind calibration to be possible.

Recall that by definition $\boldsymbol{X}_j = \mathrm{diag}(\boldsymbol{x}_j)$. Also note that $\mathrm{diag}(\boldsymbol{x}_j)\boldsymbol{d} = \mathrm{diag}(\boldsymbol{d})\boldsymbol{x}_j$. So we can equivalently state the requirement as

$$\mathrm{diag}(\boldsymbol{d})\boldsymbol{x}_j \in \mathcal{S}, \; j = 1, \ldots, k \; . \tag{1.20}$$

The proof proceeds in two steps. First, A2 implies that $k \geq r$ signal observations will span the signal subspace with probability 1. This allows us to re-cast the question in terms of a basis for the signal subspace, rather than particular realizations of signals. Second, it is shown that A3 (in terms of the basis) suffices to guarantee that the system of equations has rank $n-1$.

Step 1: We will show that all solutions to (1.20) are contained in the set

$$\mathcal{D} \;=\; \{\boldsymbol{d} : \mathrm{diag}(\boldsymbol{d})\boldsymbol{\phi}_i \in \mathcal{S}, \; i = 1, \ldots, r\}.$$

We proceed by contradiction. Suppose that there exists a vector $\widetilde{\boldsymbol{d}}$ that satisfies (1.20) but does not belong to $\mathcal{D}$. Since $\widetilde{\boldsymbol{d}}$ satisfies (1.20), we know that there exists an $\boldsymbol{x} \in \mathcal{S}$ such that $\operatorname{diag}(\widetilde{\boldsymbol{d}})\boldsymbol{x} \in \mathcal{S}$. We can write $\boldsymbol{x}$ in terms of the basis, as $\boldsymbol{x} = \sum_{i=1}^{r} \theta_i \boldsymbol{\phi}_i$, and $\operatorname{diag}(\widetilde{\boldsymbol{d}})\boldsymbol{x} = \sum_{i=1}^{r} \theta_i \operatorname{diag}(\widetilde{\boldsymbol{d}})\boldsymbol{\phi}_i$. Since by assumption $\widetilde{\boldsymbol{d}}$ does not satisfy $\operatorname{diag}(\widetilde{\boldsymbol{d}})\boldsymbol{\phi}_i \in \mathcal{S}$, $i = 1, \ldots, r$, it follows that the coefficients $\theta_1, \ldots, \theta_r$ must weight the components outside of the signal subspace so that they cancel out. In other words, the set of signals $\boldsymbol{x} \in \mathcal{S}$ that satisfy $\operatorname{diag}(\widetilde{\boldsymbol{d}})\boldsymbol{x} \in \mathcal{S}$ is a proper subspace (of dimension less than r) of the signal subspace $\mathcal{S}$. However, if we make $k \geq r$ signal observations, then with probability 1 they collectively span the entire signal subspace (since they are jointly continuously distributed). In other words, the probability that all k measurements lie in a lower dimensional subspace of $\mathcal{S}$ is zero. Thus, $\widetilde{\boldsymbol{d}}$ cannot be a solution to (1.20).

Step 2: Now we characterize the set $\mathcal{D}$. First, observe that the vectors $\boldsymbol{d} \propto \mathbf{1}$, the constant vector, are contained in $\mathcal{D}$, and those correspond to the global gain factor ambiguity discussed earlier. Second, note that every $\boldsymbol{d} \in \mathcal{D}$ must satisfy $\boldsymbol{P} \operatorname{diag}(\boldsymbol{d})\boldsymbol{\phi}_i = \boldsymbol{P} \operatorname{diag}(\boldsymbol{\phi}_i)\boldsymbol{d} = 0$, $i = 1, \ldots, r$, where $\boldsymbol{P}$ denote the projection matrix onto the orthogonal complement to the signal subspace $\mathcal{S}$. Using the definition of $\mathbf{M}_{\Phi}$ given in (1.14), we have the following equivalent condition: every $\boldsymbol{d} \in \mathcal{D}$ must satisfy $\mathbf{M}_{\Phi}\boldsymbol{d} = 0$. We know that the vectors $\boldsymbol{d} \propto \mathbf{1}$ satisfy this condition. The condition $\operatorname{rank}(\mathbf{M}_{\Phi}) = n - 1$ guarantees that these are the only solutions. This completes the proof.

Theorem 2:

Proof. First note again that the theorem is trivial if the signal subspace is one-dimensional ($r = 1$), since in this case there is one degree of freedom in the signal, and hence one measurement (coupled with the constraint that $\alpha(1) = 1$) suffices to calibrate the system. For the rest of the proof we assume that $1 < r < n$ and thus $2 \leq k < n$.

As in the proof of Theorem 1, solutions must satisfy (1.19), or equivalently the equations $\boldsymbol{x}_j \bullet \boldsymbol{d} \in \mathcal{S}$, for $j = 1, \ldots, k$. Let $\boldsymbol{x}$ denote an arbitrary signal vector, and let $\boldsymbol{z} = \boldsymbol{x} \bullet \boldsymbol{d}$. We can express $\boldsymbol{z}$ in terms of the representation $\boldsymbol{x}$ in the basis of $\mathcal{S}$ as

$$\boldsymbol{z} = \sum_{j=1}^{r} \theta_j (\boldsymbol{\phi}_{m_j} \bullet \boldsymbol{d}).$$

Recall that multiplication in the time domain is equivalent to (circular) convolution in the DFT domain. Let $\boldsymbol{Z}$, $\boldsymbol{X}$, and $\boldsymbol{D}$ be $n \times 1$ vectors denoting the DFTs of $\boldsymbol{z}$, $\boldsymbol{x}$ and $\boldsymbol{d}$, respectively (e.g., $\boldsymbol{Z}(\ell) = \frac{1}{\sqrt{n}} \sum_{q=1}^{n} \boldsymbol{z}(q) e^{-j\frac{2\pi}{n}q\ell}$). Note that $\boldsymbol{X}(\ell) = \sum_{j=1}^{r} \theta_j \delta(\ell - m_j)$, $\ell = 0, \ldots, n-1$, where $\delta(k) = 1$ if $k = 0$ and 0 otherwise. Then $\boldsymbol{Z}$ is the circular convolution of $\boldsymbol{D}$ and $\boldsymbol{X}$; i.e., the ℓth element of $\boldsymbol{Z}$ is given by

$$\boldsymbol{Z}(\ell) = \sum_{q=1}^{n} \boldsymbol{D}(q)\boldsymbol{X}([\ell - q]_n)$$

where $[\ell]_n$ is equal to $\ell \bmod n$. Hence, $\boldsymbol{z} \in \mathcal{S}$ if and only if the support of $\boldsymbol{Z}$ is on the set of frequencies $m_1, \ldots, m_r$.

For each $\ell = 1, \ldots, n$, let $\boldsymbol{X}_\ell$ denote the $n \times 1$ vector with entries $\boldsymbol{X}_\ell(q) = \boldsymbol{X}([\ell - q]_n)$, $q = 0, \ldots, 1$ (i.e., $\boldsymbol{X}_\ell$ is obtained by reversing $\boldsymbol{X}$ and circularly shifting the result by ℓ). Then we can write $\boldsymbol{Z}(\ell) = \boldsymbol{X}'_\ell \boldsymbol{D}$. Thus, we can express the constraint on the support of $\boldsymbol{Z}$ as follows:

$$\boldsymbol{X}'_\ell \boldsymbol{D} = 0, \quad \text{for all } \ell \neq m_1, \ldots, m_r \tag{1.21}$$

Notice that this places $n-r$ constraints on the vector $\boldsymbol{D}$. Also observe that the $n-r$ row vectors $\boldsymbol{X}'_\ell$, $\ell \neq m_1, \ldots, m_r$, correspond to an $(n-r) \times n$ submatrix of the circulant matrix

$$\boldsymbol{\Xi} = [\boldsymbol{X}'_0; \boldsymbol{X}'_1; \cdots; \boldsymbol{X}'_{n-1}] \tag{1.22}$$

(circulant because each row $\boldsymbol{X}_\ell$ is a circularly shifted version of the others). Furthermore, because signal coefficients $\theta_1, \ldots, \theta_r$ are randomly distributed according to B2, $\boldsymbol{\Xi}$ has full rank. This follows by recalling that circulant matrices are diagonalized by the DFT, and the eigenvalues of a circulant matrix are equal to the DFT of the first row. The first row of $\boldsymbol{\Xi}$ is $\boldsymbol{X}_0$ (indexed-reversed $\boldsymbol{X}$). It is a simple exercise to see that the DFT of $\boldsymbol{X}_0$ reproduces the original signal $\boldsymbol{x}$. Since the non-zero DFT coefficients of $\boldsymbol{x}$ are randomly distributed according to a continuous density, the elements of the $\boldsymbol{x}$ are non-zero with probability 1. This implies that the eigenvalues of $\boldsymbol{\Xi}$ are non-zero with probability 1, and thus $\boldsymbol{\Xi}$ is full-rank. Consequently, the $n-r$ constraint equations in (1.21) are linearly independent. Also note that $\boldsymbol{D} \propto [1, 0, \ldots, 0]'$ (DFT of the constant vector) satisfies (1.21), so in addition to the one degree of freedom due to the intrinsic ambiguity of the global gain factor, there are $r-1$ other degrees of freedom remaining in the solutions to (1.21).

Now suppose that we make k signal observations $\boldsymbol{x}_1, \ldots, \boldsymbol{x}_k$, randomly drawn according to B2. Each signal produces a system of constraints of the form in (1.21). Let $\boldsymbol{X}_{j,0}$, denote the indexed-reversed DFT of $\boldsymbol{x}_j$, $j = 1, \ldots, k$. These vectors generate the first row of k matrices denoted $\boldsymbol{\Xi}_j$, $j = 1, \ldots, k$ (each defined analagously to $\boldsymbol{\Xi}$ above). Note that each vector $\boldsymbol{X}_{j,0}$ displays the same sparsity pattern (since all signals are assumed to lie in an r-dimensional DFT subspace and each vector has at most r non-zero entries). Since B2 assumes that the coefficients of each signal are uncorrelated, it follows that any subset of no more than r of the vectors $\{\boldsymbol{X}_{j,0}\}_{j=1}^k$ is a linearly independent set. Now consider the collective constraints generated by the k signal measurements:

$$\boldsymbol{X}'_{j,\ell} \boldsymbol{D} = 0, \quad \text{for all } \ell \neq m_1, \ldots, m_r \text{ and } j = 1 \ldots, k \tag{1.23}$$

These constraints can be expressed in matrix notation by letting $\widetilde{\boldsymbol{\Xi}}_j$ be the $(n-r) \times n$ submatrix obtained by retaining the $n-r$ rows of $\boldsymbol{\Xi}_j$ satisfying $\ell \neq m_1, \ldots, m_r$. Then let $\widetilde{\boldsymbol{\Xi}} = [\widetilde{\boldsymbol{\Xi}}_1; \cdots; \widetilde{\boldsymbol{\Xi}}_k]$. Then (1.23) can be written as

$$\widetilde{\boldsymbol{\Xi}} \boldsymbol{D} = \boldsymbol{0} \tag{1.24}$$

We know that $\boldsymbol{D} \propto [1, 0, \ldots, 0]'$ satisfies (1.23), so the number of linearly independent equations above can be at most $n-1$. In follows that the first column of $\widetilde{\boldsymbol{\Xi}}$ is zero, and thus we may eliminate the first element of the vectors $\boldsymbol{D}$ and the first column of $\widetilde{\boldsymbol{\Xi}}$. Define $\bar{\boldsymbol{D}}$ by removing the first element of $\boldsymbol{D}$, $\bar{\boldsymbol{\Xi}}_j$ by removing the first column from $\widetilde{\boldsymbol{\Xi}}_j$, and $\bar{\boldsymbol{\Xi}} = [\bar{\boldsymbol{\Xi}}_1; \cdots ; \bar{\boldsymbol{\Xi}}_k]$. The constraints can be written as

$$\bar{\boldsymbol{\Xi}}\, \bar{\boldsymbol{D}} = \mathbf{0} \tag{1.25}$$

and we wish to show that $\bar{\boldsymbol{D}} = \mathbf{0}$ is the only solution (i.e., $\bar{\boldsymbol{\Xi}}$ is full rank). The matrix dimensions imply that $\text{rank}(\bar{\boldsymbol{\Xi}}) \leq \min\{k(n-r), n-1\}$, so choosing $k \geq (n-1)/(n-r)$ is a necessary condition. The necessity of the condition that the integers (frequencies) $m_1, \ldots, m_r$ are aperiodic (see B1) can also be seen at this point. Suppose for the sake of contradiction that the frequencies were not aperiodic. Then, because the support set of one row can align with another (at a different circular shift), one of the columns of $\bar{\boldsymbol{\Xi}}$ is the zero vector, and thus $\text{rank}(\bar{\boldsymbol{\Xi}})$ would be less than $n-1$.

Now we show that $k = \lceil (n-1)/(n-r) \rceil + 1$ signal measurements suffice to recover the gains. To prove that $\bar{\boldsymbol{\Xi}}$ is full rank in this case, it suffices to show that the nullspaces of $\bar{\boldsymbol{\Xi}}_j$ are disjoint. Without loss of generality, we consider the case of a null vector of $\bar{\boldsymbol{\Xi}}_1$. Let us denote this vector by $\boldsymbol{v}$. We will show that this vector is not in the nullspace of the other submatrices $\bar{\boldsymbol{\Xi}}_j$, $j = 2, \ldots, k$. Define $\bar{\boldsymbol{\Xi}}_{/1} = [\bar{\boldsymbol{\Xi}}_2; \cdots ; \bar{\boldsymbol{\Xi}}_k]$. The non-zero entries of the matrix $\bar{\boldsymbol{\Xi}}_{/1}$ are the (random) DFT coefficients from the $k-1$ signal observations $\boldsymbol{x}_2, \ldots, \boldsymbol{x}_k$, and these are independent of $\boldsymbol{v}$, which depends only on $\boldsymbol{x}_1$. By assumption B2, these coefficients are continuous random variables. Consider the random variable $p = \boldsymbol{v}' \widetilde{\boldsymbol{\Xi}}'_{/1} \widetilde{\boldsymbol{\Xi}}_{/1} \boldsymbol{v}$. Treating $\boldsymbol{v}$ as a fixed vector, the variable p is a quadratic polynomial function of the random DFT coefficients of $\boldsymbol{x}_2, \ldots, \boldsymbol{x}_k$. There are two distinct possibilities. Either p is the zero function, or p is a non-zero polynomial function. Suppose p is the zero function, then the conditional expectation of p given $\boldsymbol{v}$ satisfies $E[p|\boldsymbol{v}] = 0$. However, note that $E[p|\boldsymbol{v}] = \boldsymbol{v}' E[\bar{\boldsymbol{\Xi}}'_{/1} \bar{\boldsymbol{\Xi}}_{/1}] \boldsymbol{v}$, and it is easy to verify that the matrix $E[\bar{\boldsymbol{\Xi}}'_{/1} \bar{\boldsymbol{\Xi}}_{/1}]$ is full rank as follows. The matrix has a block structure, with each block of having the form $E[\bar{\boldsymbol{\Xi}}'_i \bar{\boldsymbol{\Xi}}_j]$. The blocks corresponding to the same signal observation (i.e., $i = j$) are full rank because of the circulant matrix property discussed above. The blocks corresponding to two different signal observations ($i \neq j$) are exactly zero since the signals are assumed to be uncorrelated with each other and zero mean. Together these to observations show that $E[\bar{\boldsymbol{\Xi}}'_{/1} \bar{\boldsymbol{\Xi}}_{/1}]$ is full rank. Therefore, $E[p|\boldsymbol{v}] > 0$ for every non-zero vector $\boldsymbol{v}$, and it follows that p cannot be equal to the zero function. However, if p is a non-zero polynomial function, then the probability that $p = 0$ is 0, implying that $\bar{\boldsymbol{\Xi}}_{/1} \boldsymbol{v} \neq 0$. This last argument follows from the well-known fact that the probability measure of the set of zeros of a polynomial function of continuous random variables is exactly zero [16]. Thus, we have shown that with probability 1, $\bar{\boldsymbol{\Xi}} \boldsymbol{v} \neq 0$ for every $\boldsymbol{v} \neq 0$, concluding the proof.

2

Sparse multidimensional scaling for blind tracking in sensor networks

R. Rangarajan[1], R. Raich[2], and A. O. Hero III[3]

[1] University of Michigan rangaraj@eecs.umich.edu
[2] University of Michigan ravivr@eecs.umich.edu
[3] University of Michigan hero@eecs.umich.edu

2.1 Introduction

In this chapter, we consider the problem of tracking a moving target using sensor network measurements. We assume no prior knowledge of the sensor locations and so we refer to this tracking as 'blind' . We use the distributed weighted multidimensional scaling (dwMDS) algorithm to obtain estimates of the sensor positions. Since dwMDS can only find sensor position estimates up to rotation and translation, there is a need for alignment of sensor positions from one time frame to another. We introduce a sparsity constraint to dwMDS to align current time sensor positions estimates with those of the previous time frame. In the presence of a target, location estimates of sensors in the vicinity of the target will vary from their initial values. We use this phenomenon to perform link level tracking relative to the initially estimated sensor locations.

Wireless sensor networks have been deployed for a number of monitoring and control applications such as target tracking [28], environmental monitoring [29], manufacturing logistics [26], geographic routing, and precision agriculture [44]. For many target tracking applications such as anomaly detection [21, 45], species distribution and taxonomy [19], and surveillance [4], the main purpose of the sensor network is to locate and track changes in remote environments. For example, species distribution and classification are currently documented using sightings, captures, and trap locations, which involve considerable manpower, time, and effort. Deploying mobile sensors with cameras can improve remote counts of the species as they move around in the environment. For surveillance applications, the sensors must be able to locate where the intruders or the vehicles are moving in the network. Another example is the problem of locating equipment in a warehouse. The sensors that tag the equipment must register their physical locations and activate an alarm if they are about to exit the building. As another example, in secure protocol and network routing it is critical to track anomalies such as worm activity, flash crowds, outages, and denial of service attacks in the network.

Automatic self-configuration and self-monitoring of sensor networks is the key enabling technology for these tracking applications. To respond to changes in the sensor network, it is critical to know where the changes are occurring. Data measurements from the sensors must be registered to their physical locations in the network in order to make optimal decisions. For dense sensor networks, the large size makes it impractical for humans to manually enter the physical location of the sensors and it is too expensive to attach the GPS to every device in the network. The sensors must have the capabilities to automatically estimate their relative positions and detect changes in the network at low cost, e.g., with minimum battery power.

Self-localization algorithms can be broadly classified into two categories, centralized strategies and decentralized strategies. In a centralized approach, all the data collected by the sensors must be communicated to the fusion center which then makes a decision based on this information. Algorithms that use multidimensional scaling (MDS) [40], maximum likelihood estimation [30], and convex optimization [14] have been proposed for centralized estimation and have shown to perform well. However, this may be impractical when the sensors operate with limited power and bandwidth. For networks with thousands of sensors, transmission of sensor data to a fusion center overwhelm the low-bandwidth capacity of sensor networks. Furthermore, remote sensors are frequently battery operated and battery replacement may be infeasible or expensive.

The need to conserve power and bandwidth has set the stage for more efficient decentralized strategies for localization. Among the popular approaches are adaptive trilateration [32, 39] and successive refinement [9, 23] algorithms. In trilateration, each sensor gathers information about its location with respect to anchor nodes, also referred to as seeds [31], through a shortest path. Using the range estimates from the seeds, a sensor uses trilateration to estimate its location in the network. In successive refinement algorithms, each sensor localizes its position in its own coordinate system based on the information communicated from only its neighbors. Sensors refine their location estimates iteratively using updates from neighboring sensors and finally merge their local coordinates systems, effectively finding the solution to the localization problem. Recently, there has been research emphasis on localization based on a moving target, called a mobile in [6, 34, 42]. The mobile moves randomly in the network while transmitting signals thereby allowing the sensors to calibrate their range to the mobile. This provides a large number of measurements with greater diversity which helps overcome environmental obstacles and enables improved estimation of the sensor node locations.

Most localization algorithms assume the presence of anchor nodes, i.e., certain sensors which have knowledge of their positions in the network. In the absence of anchor nodes, the sensor location estimates are only accurate up to a rotation and translation. The intuition behind this result is the following: consider an ad-hoc network of N sensors. The objective is to find the N sensor locations given the $N(N-1)/2$ inter-sensor distance measurements.

The distance information depends only on the differences in the sensor locations so that the positions of the N sensors in the network can be rotated and translated without changing these distances. In this chapter, we present a sparsity constrained dwMDS algorithm, which can localize the relative positions of the sensor nodes even in the absence of anchor nodes. The dwMDS algorithm proposed in [9] is a successive refinement method, where a global cost function is divided into multiple local cost functions at each sensor location and the computational load involved in finding the sensor location estimates is divided among the sensors in a distributed fashion. The allocation of non-negative continuous weights to the measured data overcomes the problem of combining local maps to one global map, a problem that is common to other decentralized methods [23]. We call our new algorithm sparsity penalized dwMDS. More importantly, we explain how the anchorless sparse dwMDS algorithm can efficiently track changes in the network.

Sensor localization is frequently viewed as an essential prelude to the monitoring and tracking of active phenomena. Target tracking and detection has been one such motivating application of sensor networks [24, 43, 1]. Most target tracking applications assume known sensor locations or estimate the location of sensor nodes separately before employing the tracking algorithm. The standard model used for describing the state dynamics of a moving target is the linear Gaussian model [37]. When the measurement model is also Gaussian, the optimal tracker is given by the Kalman filter. For nonlinear state space and measurement models, other techniques such as Extended Kalman Filter (EKF) [24], unscented Kalman Filter (UKF) [43], and Gaussian sum approximation [1] have been proposed. Particle filtering algorithms were then formulated for target tracking, where the probability density of the state is approximated by a point mass function on a set of discrete points [13]. The discrete points are chosen through importance sampling. The advantage of particle filtering is its applicability to a large range of densities, noise processes, and measurement models. More recently, researchers have looked at the simpler problem of tracking in a binary sensing modality [2, 25]. The sensor outputs a high, when the target is within a sensing range and outputs a low, when the target falls outside its range. Based on the fusion of the sensor outputs, an approximate link level trajectory can be realized to track the target. Such a binary sensing modality has limited accuracy but requires minimal power consumption and has the advantage of analytical tractability [41]. This procedure can also be interpreted as a target detection problem implemented for multiple time steps.

Distributed target detection methods have been proposed in the literature [33] in the context of designing an optimal decision statistics at the sensor fusion center. The detection problem has also been addressed for under communication constraints, where the sensor transmitting the information needs to send an optimal summary of the gathered information to the fusion center [7]. In the context of anomaly detection in internet data, approximate density of incoming traffic is constructed for each location. Distance between densities is then used as a similarity measure in the MDS algorithm to form a map of the internet network. By performing MDS over time, it is shown that anomalies such as network scans, worm attacks, and denial of service attacks can be identified and classified [36, 16]. For wormhole detection in ad hoc sensor networks, most research efforts require mobile nodes equipped with special hardware or GPS devices [22, 5] to localize the wormhole.

In contrast to the methods proposed in the literature, we present the sparsity penalized dwMDS algorithm which localizes the sensor nodes in the absence of anchors and tracks multiple targets amongst the sensor links. The principle behind our proposed algorithm is the following: in the 'acquisition phase' or initialization, an initial estimate of sensor locations is acquired. Once the sensors have been initially localized, it is only the network topology that is critical to the problem of tracking. Hence, during the tracking phase, we introduce a sparsity constraint to the dwMDS problem formulation, which attempts to fix the alignment of the sensor network with respect to the alignment of the localized network at the previous time instance. By doing so, we keep monitoring the network with respect to a fixed geometry obtained by the localization algorithm at the first time instance ($t = 1$). The sparsity constraint only reassigns a small fraction of the sensor locations, while maintaining the locations of remaining sensors close to their previous estimates. When the sensor network is then used for tracking, only the sensors affected by the presence of a target are perturbed, while the rest of the location estimates remain unchanged. Based on the differences in the sensor location estimates between two time-frames, we propose a novel perturbation based link level tracking algorithm, which accurately localizes a target to within a small set of sensor links. Figure 2.1 shows the localization process in the absence of targets. The actual sensor locations are marked as circles and the anchor nodes are highlighted using squares. The sensors communicate among themselves and the anchor nodes to obtain location estimates indicated as crossed circles. Figure 2.2 shows the localization process in the presence of a target. The measurements of the sensor nodes closest to the target are affected and the sensors appear further apart than they are in reality. This change in the sensor location estimates can be used to perform link level tracking.

Link level tracking has many attractive features, the most important of which is that it does not require a physical model for the target, which is fundamental to most tracking algorithms in the literature [3]. Moreover, the goal of certain sensor networks is to obtain an estimate of the location of the targets, or detect changes in the network. For example, in military applications, the sensors can locate a target relative to the network and the network can activate the appropriate sensors to identify the target. For animal tracking in biological research, it is sufficient to have a low resolution tracking algorithm to monitor animal behavior and interactions with their own clan and with other species.

We introduce the sparsity constrained dwMDS algorithm for simultaneous sensor localization and link level tracking in this chapter. We give a flavor of

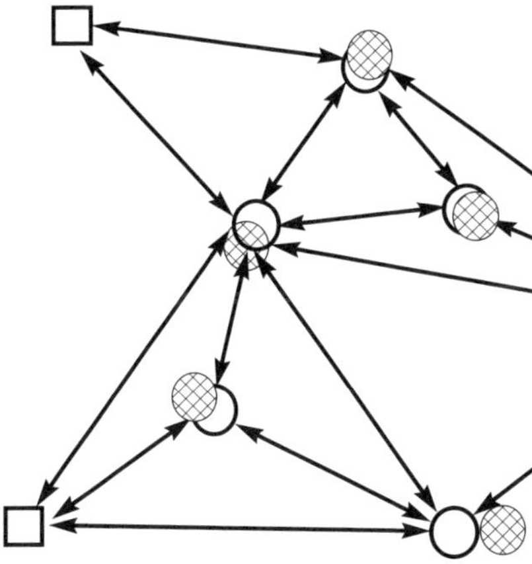

Fig. 2.1. Localization in the absence of target. Anchor nodes (square), true sensor locations (circle), estimated sensor coordinates (crossed circle).

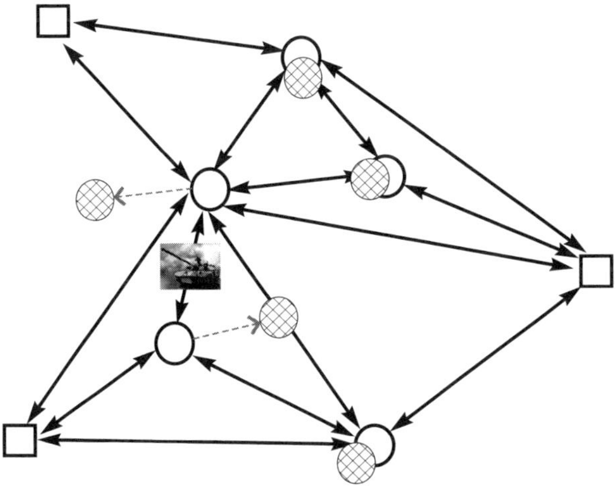

Fig. 2.2. Link level tracking based on localization in the presence of target.

how the algorithm can be extended to estimate actual target coordinates using standard tracking algorithms. Furthermore, the algorithm we present here can be used to design optimal sensor scheduling strategies for tracking to limit power consumption in sensor networks. We incorporate the sparsity constraint such that the localization algorithm is still distributed in its implementation to minimize communication and computational costs.

This chapter is organized as follows: Section 2.2 formally introduces the problem of sensor localization. Section 2.3 introduces the classical MDS algorithm and its variations. We then present our sparsity penalized dwMDS algorithm in Section 2.4. In Section 2.5, we explain how this algorithm can be applied for link level tracking. Finally, Section 2.6 concludes this chapter by discussing the extensions of this formulation for model-based multiple target tracking and sensor management strategies.

2.2 Problem formulation

We begin by introducing the nomenclature used in this chapter. We denote vectors in $\mathbb{R}^M$ by boldface lowercase letters and matrices in $\mathbb{R}^{M \times N}$ by boldface uppercase letters. The identity matrix is denoted by $\mathbf{I}$. We use $(\cdot)^T$ to denote the transpose operator. We denote the l_2-norm of a vector by $\|\cdot\|$, i.e., $\|\mathbf{x}\| = \sqrt{\mathbf{x}^T\mathbf{x}}$. A Gaussian random vector with mean $\boldsymbol{\mu}$ and covariance matrix $\mathbf{C}$ is denoted as $\mathcal{N}(\boldsymbol{\mu}, \mathbf{C})$.

The purpose of the sparsity constrained MDS algorithm is to simultaneously localize and track targets. We first formally state the sensor localization problem. Consider a network of $N = n+m$ nodes in d dimensional space. The localization algorithms can be applied to arbitrary d $(d < N)$ dimensional spaces. Since applications for localization typically occur in physical space, we will restrict our attention to $d = 2, 3$ dimensions. Let $\{\mathbf{x}_i\}_{i=1}^N, \mathbf{x}_i \in \mathbb{R}^d$ be the true location of the n sensors. The m sensor nodes $\{\mathbf{x}_i\}_{i=n+1}^{n+m}$ are anchor nodes, i.e., whose locations are known. We introduce the anchor nodes to keep the formulation as general as possible. Later, we set $m = 0$ for anchor free localization. Denote $\mathbf{X} = [\mathbf{x}_1, \mathbf{x}_2, \ldots, \mathbf{x}_N]$ as the $d \times N$ matrix of actual sensor locations. Let $\mathbf{D} = (d_{i,j})_{i,j=1}^N$ be the matrix of the true inter-sensor distances, where $d_{i,j}$ denotes the distance between sensor i and sensor j. It is common that some wireless sensor networks may have imperfect a priori knowledge about the locations of certain sensor nodes. This information is encoded through parameters r_i and $\bar{\mathbf{x}}_i$, where $\bar{\mathbf{x}}_i$ is the sensor location and r_i is the corresponding confidence weight. If $\bar{\mathbf{x}}_i$ is unavailable, then we set $r_i = 0$. The problem setting is explained through an illustration of a sensor network in Fig. 2.3. We assign weights $w_{i,j}$ for measurements between sensors i and j to indicate the accuracy of the distance estimate. In this sensor network, each sensor communicates to its three nearest neighbors and hence, the weights corresponding to links between non neighboring sensors are zero.

Sensor localization is the process of estimating the location of the n sensor nodes $\{\mathbf{x}_i\}_{i=1}^n$ given $\{\mathbf{x}_i\}_{i=n+1}^{n+m}$, $\{r_i\}$, $\{\bar{\mathbf{x}}_i\}$ and pairwise range measurements $\{\delta_{i,j}^t\}$ taken over time $t = 1, 2, \ldots, K$. The indices (i, j) run over a subset of $\{1, 2, \ldots, N\} \times \{1, 2, \ldots, N\}$. The range measurements can be obtained by sensing modalities such as time-of-arrival (TOA), received signal strength (RSS), or proximity.

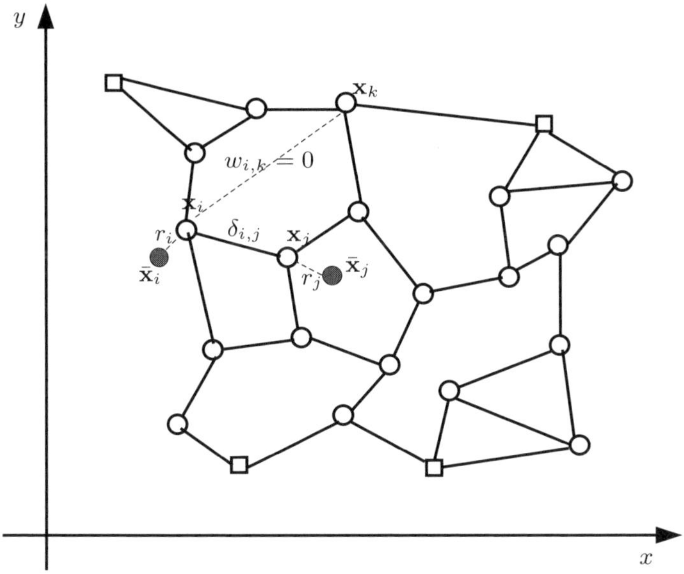

Fig. 2.3. Sensor localization setup: Anchor nodes (square), sensor nodes (circle), a priori sensor locations (blocked circle). The communicating sensors are connected using solid lines. The non neighboring sensor links have zero weight.

2.3 Classical MDS and variations

Multidimensional scaling (MDS) is a methodology for recovering underlying low dimensional structure in high dimensional data. The measured data can come from confusion matrices, group data, or any other (dis)similarity measures. MDS has found numerous applications in cognitive science, marketing, ecology, information science, and manifold learning [11, 12]. In the context of sensor localization, the goal in MDS is to discover the sensor locations (lower dimensional embedding) from the inter-sensor distances obtained by a given sensing method (high dimensional data).

Classical MDS [18] provides a closed-form solution to the sensor locations when the inter-sensor measurements are the inter-sensor Euclidean distances, i.e., in the absence of noise or nonlinear effects. We assume all pairwise range measurements are available, and so we can compute the complete matrix of distances:

$$d_{i,j} = \|\mathbf{x}_i - \mathbf{x}_j\| = \sqrt{(\mathbf{x}_i - \mathbf{x}_j)^T(\mathbf{x}_i - \mathbf{x}_j)}. \tag{2.1}$$

Denote by $\mathbf{D}^{(2)}$ the matrix of squared distances, i.e., $\mathbf{D}^{(2)} = (d_{i,j}^2)_{i,j=1}^N$. Then $\mathbf{D}^{(2)}$ can be rewritten as

$$\mathbf{D}^{(2)} = \boldsymbol{\psi}\mathbf{1}^T - 2\mathbf{X}^T\mathbf{X} + \mathbf{1}\boldsymbol{\psi}^T, \tag{2.2}$$

where $\mathbf{1}$ is an N-element vector of ones and $\boldsymbol{\psi} = [\mathbf{x}_1^T\mathbf{x}_1, \mathbf{x}_2^T\mathbf{x}_2, \ldots, \mathbf{x}_N^T\mathbf{x}_N]^T$. Let $\mathbf{H} = \mathbf{I} - (1/N)\mathbf{1}\mathbf{1}^T$. Multiplying on the left of $\mathbf{D}^{(2)}$ by $-1/2\mathbf{H}$ and the right by $\mathbf{H}$, we obtain

$$\mathbf{A} = -\frac{1}{2}\mathbf{H}\mathbf{D}^{(2)}\mathbf{H} = \mathbf{H}\mathbf{X}^T\mathbf{X}\mathbf{H}. \tag{2.3}$$

Given $\mathbf{A}$, one can discover the matrix $\mathbf{X}$ to a rotation and translation by solving the following variational problem

$$\min_{\mathbf{Y}} \|\mathbf{A} - \mathbf{Y}^T\mathbf{Y}\|_F^2, \tag{2.4}$$

where $\|\cdot\|_F$ indicates the Frobenius norm and the search space is over all full rank $d \times N$ matrices. The solution to $\mathbf{X}$ is then given by

$$\mathbf{X} = \mathrm{diag}(\lambda_1^{1/2}, \ldots, \lambda_d^{1/2})\mathbf{V}_1^T, \tag{2.5}$$

where the singular value decomposition (SVD) of $\mathbf{A}$ is given by

$$\mathbf{A} = [\mathbf{V}_1 \ \mathbf{V}_2] \ \mathrm{diag}(\lambda_1, \ldots, \lambda_d, \lambda_{d+1}, \ldots, \lambda_N) \ [\mathbf{V}_1 \ \mathbf{V}_2]^T. \tag{2.6}$$

The matrix $\mathbf{V}_1$ consists of the eigenvectors of the first d eigenvalues $\lambda_1, \ldots, \lambda_d$, while the rest of the $N - d$ eigenvectors are represented as $\mathbf{V}_2$. The term $\mathrm{diag}(\lambda_1, \ldots, \lambda_N)$ refers to a $N \times N$ diagonal matrix with λ_i as its ith diagonal element. Though the solution to the classical MDS is obtained in closed-form, the algorithm has the following deficiencies:

1. MDS requires knowledge of all inter-sensor distances. Obtaining all pairwise range measurements is prohibitive due to the size of the sensor network and the limited power of the sensors. In our problem formulation, this implies that $w_{i,j} \neq 0, \forall i, j$, which makes MDS fall under the category of a centralized approach, i.e., all the information needs to be transmitted to the fusion center which then performs the MDS algorithm. Due to power and bandwidth limitations in the sensor network, this process is infeasible.
2. The inter-sensor range measurements $\delta_{i,j}$ are corrupted by environment and receiver noise which further degrades the quality of the measurements, i.e., $\delta_{i,j}$ is only an estimate of the inter-sensor distance $d_{i,j}$.
3. MDS uses the squared distance matrix which tends to amplify the measurement noise, resulting in poor performance.

As mentioned in Section 2.1, there has been significant effort directed towards designing decentralized strategies for sensor localization. However, consistent reconstruction of the sensor locations is attainable only in the presence of anchor nodes. If the current localization algorithms are implemented for anchor free localization, the geometry of the sensor network assumes different alignments as localization is performed over various time instants. This makes it

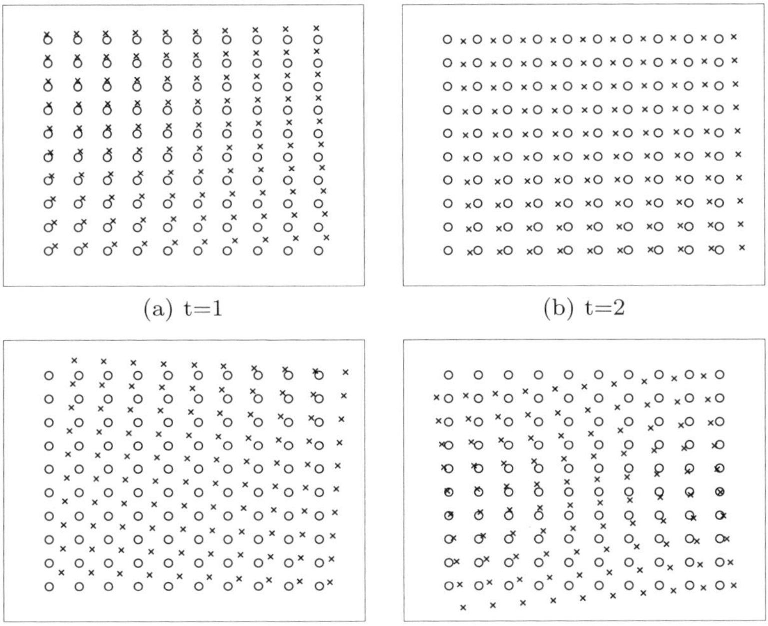

(a) t=1 (b) t=2

(c) t=3 (d) t=4

Fig. 2.4. Anchor free sensor localization by dwMDS. True sensor locations (circle), estimated sensor locations (cross).

impossible to locate where the changes are occurring in the network. To illustrate this phenomenon, we implement the dwMDS algorithm for sensor localization in the absence of anchor nodes and in the absence of target. We provide snapshots of the sensor location estimates (cross) along with their actual locations (circle) in Fig. 2.4 as a function of time. Observe that the geometry of the network is maintained, while the true locations are subject to rotation and translation. Now consider a target moving through this network. In this scenario, the localization process is affected by two factors: the lack of anchor nodes and some inaccurate inter-sensor measurements in the vicinity of the target. With anchor free localization, the process of tracking a target becomes extremely difficult. To overcome this problem, we propose our sparsity constrained dwMDS algorithm that aligns the current sensor location estimates to those of previous time frames.

2.4 Sparsity penalized MDS

Consider using the MDS algorithm independently to obtain the sensor location estimates at time t and at time $t-1$. Alignment between these two sets of points can be performed in various ways. For example, in Procrustes analysis [17] alignment is performed by finding the optimal affine transformation of one set of nodes that yields the set closest to the second set of points in the least squares sense. However, this procedure cannot guarantee that many sensor locations estimates will remain unchanged from their previously estimated values. The errors in the sensor location estimates between two time steps may accumulate over time resulting in alignment errors. In contrast, we introduce a sparseness penalty on the distances between the sensor location estimates at time t ($\mathbf{x}_i$) and at time $t-1$ ($\mathbf{x}_i^{(t-1)}$) directly to the sensor localization algorithm. Construct a vector of Euclidean distances between the location estimates at time t and at time $t-1$

$$\mathbf{g}^{(t)} = \left[\|\mathbf{x}_1 - \mathbf{x}_1^{(t-1)}\|, \ldots, \|\mathbf{x}_n - \mathbf{x}_n^{(t-1)}\|\right]^T. \tag{2.7}$$

Define the l_0-measure of a vector $\mathbf{v} = [v_1, v_2, \ldots, v_n]$ as the number of nonzero elements given by

$$\|\mathbf{v}\|_0 \triangleq \sum_{i=1}^{n} I(v_i \neq 0), \tag{2.8}$$

where $I(\cdot)$ is the indicator function. Using an l_0-constraint on the distance vector $\mathbf{g}^{(t)}$ of the form $\|\mathbf{g}^{(t)}\|_0 \leq q$, we guarantee that no more than q of the location estimates will vary from their previous time frame values. Minimizing a cost function under the l_0-constraint requires a combinatorial search which is computationally infeasible. Define the l_p-measure of a vector $\mathbf{v}$ as

$$\|\mathbf{v}\|_p \triangleq \Big(\sum_{i=1}^{n} |v_i|^p\Big)^{1/p}. \tag{2.9}$$

For a quadratic cost function, an l_p-constraint ($0 < p \leq 1$) induces a sparse solution. Among all l_p sparsifying constraints, only $p = 1$ offers a convex relaxation to the l_0-constraint [15]. To promote sparsity, we next advocate the use of the l_p-constraint as a penalty term via the Lagrange multiplier in the dwMDS algorithm to solve for the sensor location estimates. Hence the term *sparsity penalized MDS*.

The cost function of the dwMDS algorithm [9] is motivated by the variational formulation of the classical MDS, which attempts to find sensor location estimates that minimize the inter-sensor distance errors. Keeping in mind that it is the geometry of the sensor network which is crucial for tracking, we present a novel extension of the dwMDS algorithm through the addition of the sparseness inducing l_p-constraint. At any time t, we seek to minimize the overall cost function $C^{(t)}$ given by

$$C^{(t)} = \sum_{1\leq i\leq n} \sum_{i\leq j\leq n+m} \sum_{1\leq l\leq M} w_{i,j}^{(t),l} \left(\delta_{i,j}^{(t),l} - d_{i,j}(\mathbf{X})\right)^2 + \sum_{i=1}^{n} r_i \|\bar{\mathbf{x}}_i - \mathbf{x}_i\|^2 + \lambda \|\mathbf{g}^{(t)}\|_p^p. \tag{2.10}$$

The Euclidean distance $d_{i,j}(\mathbf{X})$ is defined in (2.1). For each time t, there are M range measurements $\delta_{i,j}^{(t),l}$ for each sensor link i, j. As in [9], the weights $w_{i,j}^{(t),l}$ can be chosen to quantify the accuracy of the predicted distances. When no measurement is made between sensor i and sensor j, $w_{i,j}^{(t),l} = 0$. Furthermore, the weights are symmetric, i.e., $w_{i,j}^{(t),l} = w_{j,i}^{(t),l}$, and $w_{i,i}^{(t),l} = 0$. If available, the a priori information of sensor locations is encoded through the penalty terms $\{r_i \|\bar{\mathbf{x}}_i - \mathbf{x}_i\|^2\}$. Finally, we introduce an l_p-constraint $(0 \leq p \leq 1)$ on the distances between the sensor locations at time t and the estimated sensor locations at time $t-1$. The Lagrange multiplier of the sparseness penalty is denoted as λ. We can tune the value of λ to yield the desired sparsity level in $\mathbf{g}^{(t)}$. Later, when we apply the algorithm for tracking, the sparseness will be advantageous as only those sensors which are highly affected by the target will vary from their initial positions, thereby allowing for a detection of the target through the process of relative sensor localization. To solve this optimization problem, we propose to use the successive refinement technique, where each sensor node i updates its location estimate by minimizing the global cost function $C^{(t)}$, after observing range measurements at node i and receiving position estimates from its neighboring nodes.

2.4.1 Minimizing cost function by optimization transfer

Unlike classical MDS for which we could obtain a closed-form expression for the estimates, there is no closed-form solution to minimizing $C^{(t)}$. Therefore, we solve the local nonlinear least squares problem iteratively using a quadratic majorization function similar to SMACOF (Scaling by MAjorizing a COmplicated Function [20]). This procedure can be viewed as a special case of optimization transfer algorithms through surrogate objective functions [27], e.g., the popular EM algorithm.

A majorizing function $T(\mathbf{x}, \mathbf{y})$ of $C(\mathbf{x})$ is a function $T : \mathbb{R}^d \times \mathbb{R}^d \to \mathbb{R}$, which satisfies the following properties: $T(\mathbf{x}, \mathbf{y}) \geq C(\mathbf{x})$, $\forall \mathbf{y}$ and $T(\mathbf{x}, \mathbf{x}) = C(\mathbf{x})$. In other words, the majorizing function upper bounds the original cost function. Using this property, we can formulate an iterative minimization procedure as follows: denote the initial condition as $\mathbf{x}_0$. Starting from $n = 1$, obtain $\mathbf{x}_n$ by solving

$$\mathbf{x}_n = \arg\min_{\mathbf{x}} T(\mathbf{x}, \mathbf{x}_{n-1}),$$

until a convergence criterion for $C(\mathbf{x})$ is met. We can easily observe that this iterative scheme always produces a non increasing sequence of cost functions, i.e.,

$$C(\mathbf{x}_{n+1}) \leq T(\mathbf{x}_{n+1}, \mathbf{x}_n) \leq T(\mathbf{x}_n, \mathbf{x}_n) = C(\mathbf{x}_n).$$

The first and last relations follows from the properties of majorizing functions while the middle inequality follows from the fact that $\mathbf{x}_{n+1}$ minimizes $T(\mathbf{x}, \mathbf{x}_n)$. Now the trick is to choose a majorizing function that can be minimized analytically, e.g., a linear or quadratic function. We propose a quadratic majorizing function $T^{(t)}(\mathbf{X}, \mathbf{Y})$ for the global cost $C^{(t)}(\mathbf{X})$. Minimizing $C^{(t)}(\mathbf{X})$ through the majorization algorithm is the simple task of minimizing the quadratic function $T^{(t)}(\mathbf{X}, \mathbf{Y})$, i.e.,

$$\frac{\partial T^{(t)}(\mathbf{X}, \mathbf{Y})}{\partial \mathbf{x}_i} = 0, \quad i = 1, 2, \ldots, n. \tag{2.11}$$

If we denote the estimates of the sensor nodes at iteration k as $\mathbf{X}^k$, the recursion for the update of location estimates for node i from (2.11) is given by

$$\mathbf{x}_i^k = \frac{1}{a_i}\left(c_i + \mathbf{X}^{k-1}\mathbf{b}_i^{k-1}\right), \tag{2.12}$$

where $\mathbf{b}_i^{k-1}$, a_i, and c_i are defined in (2.32)-(2.35) respectively. The details of the derivation of the sparsity penalized MDS algorithm can be found in the appendix. For each sensor i, the jth element of the vector $\mathbf{b}_i^{k-1}$ depends on the weight $w_{i,j}$. Since the weights of the nodes not in the neighborhood of the sensor are zero, the corresponding elements in the vector $\mathbf{b}_i^{k-1}$ are also zero; therefore the update rule for node i in (2.12) will depend only on the location of its nearest neighbors and not on the entire matrix $\mathbf{X}^{k-1}$. This facilitates the distributed implementation of the algorithm. The proposed algorithm is summarized in Fig. 2.5. We illustrate the majorization procedure in Fig. 2.6. The original cost function (solid) and the corresponding surrogate (dotted) is presented for every iteration, along with the track of the estimates at iteration k (circle). Our proposed algorithm introduces a sparseness penalty

Inputs: $\{\bar{w}_{i,j}^{(t)}\}$, $\{\bar{\delta}_{i,j}^{(t)}\}$, $\{r_i\}$, $\{\bar{\mathbf{x}}_i\}$, $\{\mathbf{x}_i^{(t-1)}\}$, ϵ, $\mathbf{X}^0$ (initial condition for iterations).
Set $k = 0$, compute cost function $C^{(t),0}$ and a_i from equations (2.10) and (2.34) respectively
repeat
- k=k+1
- for $i = 1$ to n
  - compute $\mathbf{b}_i^{k-1}$ from equation (2.32)
  - $\mathbf{x}_i^k = \frac{1}{a_i}\left(c_i + \mathbf{X}^{k-1}\mathbf{b}_i^{k-1}\right)$
  - compute $C_i^{(t),k}$
  - update $C^{(t),k}$ to $C^{(t),k} - C_i^{(t),k-1} + C_i^{(t),k}$
  - communicate $\mathbf{x}_i^k$ to neighbors of sensor i (nodes for which $w_{i,j} > 0$)
  - communicate $C^{(t),k}$ to next node ($(i+1) \mod n$)
- end for

until $C^{(t),k} - C^{(t),k-1} < \epsilon$

Fig. 2.5. Description of the sparsity constrained MDS algorithm.

on the distance between estimate at time $t-1$ and the current estimate. If the sparsity regularization parameter λ is not chosen properly, many sensor

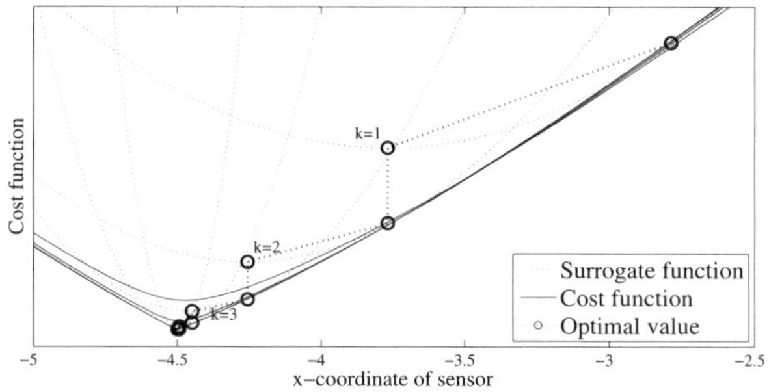

Fig. 2.6. Majorization procedure: cost function (solid curve), surrogate function (dotted curve), optimal location estimate at each iteration (circle). Only a single coordinate is updated in this picture.

positions estimates might slowly vary with time, thereby creating cumulative error in the sensor localization. An interesting way to counteract this problem would be to penalize the distance between the current estimate and the initial estimate at $t = 1$. Using such a constraint would mean that the sensors are always compared to the fixed initial frame and errors do not accumulate over time. The implementation of this algorithm would be straightforward as it would simply involve changing the index $t-1$ to 1 in the original algorithm presented in Fig. 2.5. However, using the estimate from time $t-1$ has the property that it is easily adapted to the case of mobile sensors.

2.4.2 Implementation

Weights: When RSS measurements are used to compute distance estimates, the weights are set using the locally weighted regression methods (LOESS) scheme [8] similar to one used in the dwMDS algorithm [9]. The weight assignment is given by

$$w_{i,j} = \begin{cases} \exp\left(-\delta_{i,j}^2/h_{i,j}^2\right), & \text{if i and j are neighbors} \\ 0, & \text{otherwise,} \end{cases}$$

where $h_{i,j}$ is the maximum distance measured by either sensor i or j. A naive equal weight assignment to all measurements is also shown to work well with our algorithm.

Initialization: For the successive refinement procedure, the sensor locations estimates must be initialized for every time frame. Though several initialization algorithms have been proposed in the literature, we use a naive random initialization. We would like to point out that the initialization is not a critical component to our algorithm, as we are solely interested in the alignment of sensors in the network and not on the exact locations. Irrespective of the initial estimates, the sparseness penalty will ensure that the estimated sensor locations are relatively close to those of previous time frames. Our algorithm is found to be robust with respect to the initial estimates.

Neighborhood selection: Traditionally, the neighbors are chosen based on the distance measure obtained from the RSS measurements, i.e., select all sensors within a distance R as your neighbors. When the RSS measurements are noisy, there is a significant bias in the neighborhood selection rule. This method has a tendency to select sensors which are, on average, less than the actual distances $\|\mathbf{x}_i - \mathbf{x}_j\|$. A simple two-stage adaptive neighborhood selection rule is proposed in [9] to overcome the effect of this bias. We use this selection rule in our algorithm.

Range measurement models: The inter-sensor measurements can be obtained by RSS, TOA, or proximity. Any one of these approaches can be used in our algorithm. Our sparsity constrained MDS algorithm is fairly robust to either of these measurement models. For the simulations in this chapter, we use the RSS to obtain a range measurement between two sensors. It can be shown through the central limit theorem (CLT) that the RSS is log-normal in its distribution [10], i.e., if $P_{i,j}$ is the measured power by sensor i transmitted by sensor j in milliWatts, then $10\log_{10}(P_{i,j})$ is Gaussian. Thus $P_{i,j}$ in dBm is typically modeled as

$$\begin{aligned} P_{i,j} &= \mathcal{N}(\bar{P}_{i,j}, \sigma_0^2) \\ \bar{P}_{i,j} &= P_0 - 10n_p \log\left(\frac{d_{i,j}}{d_0}\right), \end{aligned} \tag{2.13}$$

where $\bar{P}_{i,j}$ is the mean received power at distance $d_{i,j}$, σ_0 is the standard deviation of the received power in dBm, and P_0 is received power in dBm at a reference distance d_0. n_p is referred to as the path-loss exponent that depends on the multipath in the environment. Given the received power, we use maximum likelihood estimation to compute the range, i.e., distance between the sensor nodes i and j. The maximum likelihood estimator of $d_{i,j}$ is given by

$$\delta_{i,j} = 10^{((P_0 - P_{i,j})/10n_p)}. \tag{2.14}$$

Simulation of tracker without a target

The simulation parameters are chosen as follows: we deploy a 10×10 uniform grid of sensors in a network. We consider anchor free localization, i.e., $m = 0$ and we assume we make a single inter-sensor measurement ($M = 1$). We set the sparseness parameter λ to produce a change in the location estimates for only a small portion of the sensors. The value of λ will depend on the size of the network and the noise in the measurements. If the RSS measurements are very noisy, then range estimates become inaccurate which tend to vary the sensor

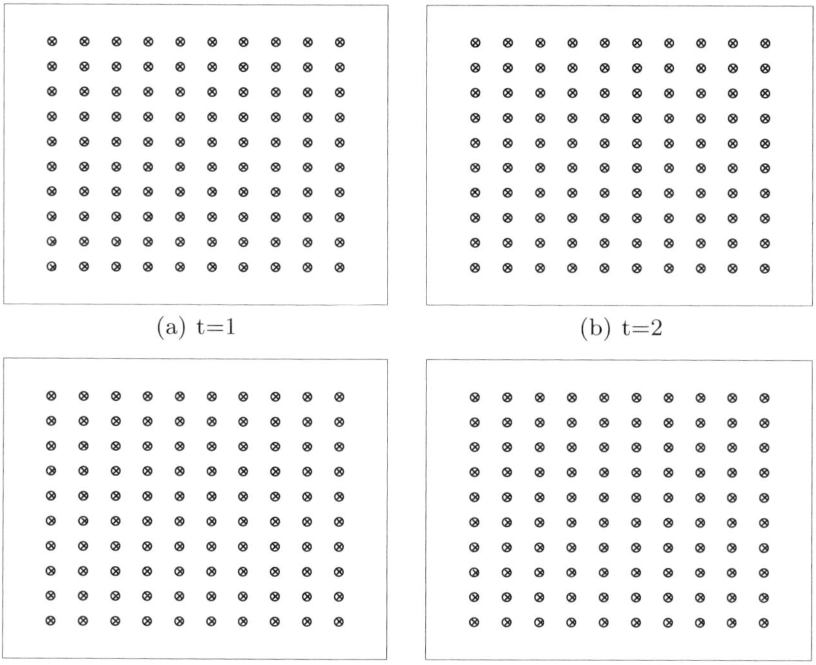

(a) t=1 (b) t=2

(c) t=3 (d) t=4

Fig. 2.7. Anchor free sensor localization by sparsity penalized MDS. True sensor locations (circle), sensor position estimates (cross).

location estimates. Hence λ is selected to ensure that sensor location estimates remain aligned with the previous time frame estimates. In this simulation, we set $\lambda = 0.1$ and the noise variance $\sigma_0 = 0.15$. Each sensor communicates with its 15 nearest neighbors. The weights of the RSS measurements were chosen based on the LOESS scheme described earlier. The weights of links for non communicating sensors were set to zero. We demonstrate the performance of the sparsity constrained MDS algorithm on this sensor network as a function of time in Fig. 2.7. The true locations are denoted as circles and the estimated locations as crosses.

2.5 Tracking using sparse MDS

Here we present an algorithm for performing link level tracking using the sparsity constrained MDS algorithm. By link level tracking, we refer to localization of targets to within a set of inter-sensor links. Link level tracking is attractive in the sense that there is no need to assume a physical model for a target. However, it is important to know the effect of the target on the inter-sensor measurements. Researchers have proposed various models for the signal strength measurements ranging from the traditional linear Gaussian model to the binary sensing model. These are approximate statistical models and the distribution of the measurements in the presence of a target remains an open question.

To model the statistics under the setting of vehicle tracking, we conducted experiments using RF sensors hardware in the presence of a target. We constructed a fine grid of locations, where the target was placed and RSS measurements were recorded between two static sensors for positions on the grid. Upon gathering the data, we fit the following statistical model in the presence of target. The RSS measurements at sensor link i, j are distributed as

$$\begin{aligned} P_{i,j}^k | \hat{P}_{i,j} &\sim \mathcal{N}(\hat{P}_{i,j}, \sigma_0^2), \text{ i.i.d}, \quad k = 1, 2, \ldots, M \\ \hat{P}_{i,j} &\sim \mathcal{N}(\bar{P}_{i,j}, \sigma_1^2), \end{aligned} \tag{2.15}$$

where $P_{i,j}^k$ is the kth inter-sensor measurement when the target is in the neighborhood of the sensors. The M sensor link measurements are correlated through the random variable $\hat{P}_{i,j}$. The values obtained from our actual experiments were $\sigma_0 \approx 0.1463$dBm and $\sigma_1 \approx 1.5$dBm. The noise variance in the measurements σ_1 was roughly an order of 10 times larger than σ_0. In other words, RSS measurements tend to have a larger variance due to scattering and attenuation of the signals in the presence of a target. A confidence measure for such a log-normal distribution of the RSS data is obtained using the Kolmogorov-Smirnov (KS) test in [35] and the model is shown to work well for sensor localization. We assume this statistical model for the RSS measurements, when the target is within a specified distance R of the sensor link i, j. The distance R depends on the reflectivity of the object. If the object is highly reflective, then the variation in the RSS measurements is detected by more links.

Given the measurement model, we formulate the optimal decision statistic to detect a presence of a target in a particular sensor link using the likelihood ratio test (LRT). For a fixed false alarm level α, the LRT for each link i, j is given by

$$\left| \frac{1}{M} \sum_{l=1}^{M} P_{i,j}^{(t),l} - P_{i,j}^{'} \right| \underset{\mathrm{H}_0}{\overset{\mathrm{H}_1}{\gtrless}} \gamma, \tag{2.16}$$

where $\gamma = (\sigma_0/\sqrt{M})Q^{-1}(\alpha/2)$ and $P_{i,j}^{'}$ is the mean received power in the sensor link estimated using an initial set of range measurements. $\{P_{i,j}^{(t),l}\}_{l=1}^{M}$ is the set of inter-sensor measurements made by link i, j at time t. We assume that the sensor network is in its steady state operation mode. We do not consider the transient effects in the measured data when it is obtained in the absence of any target. This most powerful test of level α yields the probability of correct detection

$$\beta = 2Q\left(Q^{-1}(\alpha/2)\sqrt{\frac{\sigma_0^2}{\sigma_0^2 + M\sigma_1^2}}\right). \tag{2.17}$$

A derivation of the decision rule and its performance is given in the appendix. The performance of the optimal detector is clearly dependent on the number of samples available for the inter-sensor measurements. As the number of measurements M becomes very large, β in (2.17) tends to 1. However, if only few samples are available, β may not approach 1 and misdetect type errors may become non negligible. In such a case, instead of using LRT, we can use a test on the variation of the sensor location estimates at time t from their estimates at a previous time. In other words, we can perform a simple hypothesis test for each link of the form,

$$\|d_{i,j}^t - d_{i,j}^T\| \underset{\mathrm{H}_0}{\overset{\mathrm{H}_1}{\gtrless}} \gamma_{i,j}, \tag{2.18}$$

where $T = 1$ or $T = t - 1$ depending on whether the sensors are static or mobile.

Simulation of tracker in the presence of target

We present our results by simulating moving targets in a uniform 10×10 grid of sensors. We set $m = 0$, i.e., no anchor nodes. We assume no a priori knowledge of the sensor coordinates, i.e., $r_i = 0$. Each sensor communicates only to its 15 nearest neighbors and the weights for those links were chosen by the LOESS strategy. The rest of the weights were set to zero. We obtain $M = 50$ data measurements for each communicating sensor link in the network. We set the sparseness parameter λ to produce a change in the location estimates for only a small portion of the sensors. We allow any number of targets to appear in a sensor network with probability 0.4. Though our algorithm is robust to randomly moving targets in the network, we consider a state-space model for the purposes of this simulation to produce a visually pleasing target trajectory. We apply the sparsity constrained MDS algorithm as multiple targets move through the sensor network.

The results are shown in Fig. 2.8. The true sensor locations are shown as circles and the estimated sensor locations are indicated using crosses. The sensors corresponding to those sensor links that declared a target present using the distance based target localization algorithm (DBT) in (2.18) are shown in filled circles. The target trajectories are shown as inverted triangles. We observe that as the targets move, the sparsity constrained MDS algorithm reconstructs sensor location estimates with the majority of them unchanged from the previous time step. Thus, in conjunction with sparse dwMDS, the DBT is able to localize the targets to within a small set of sensor links.

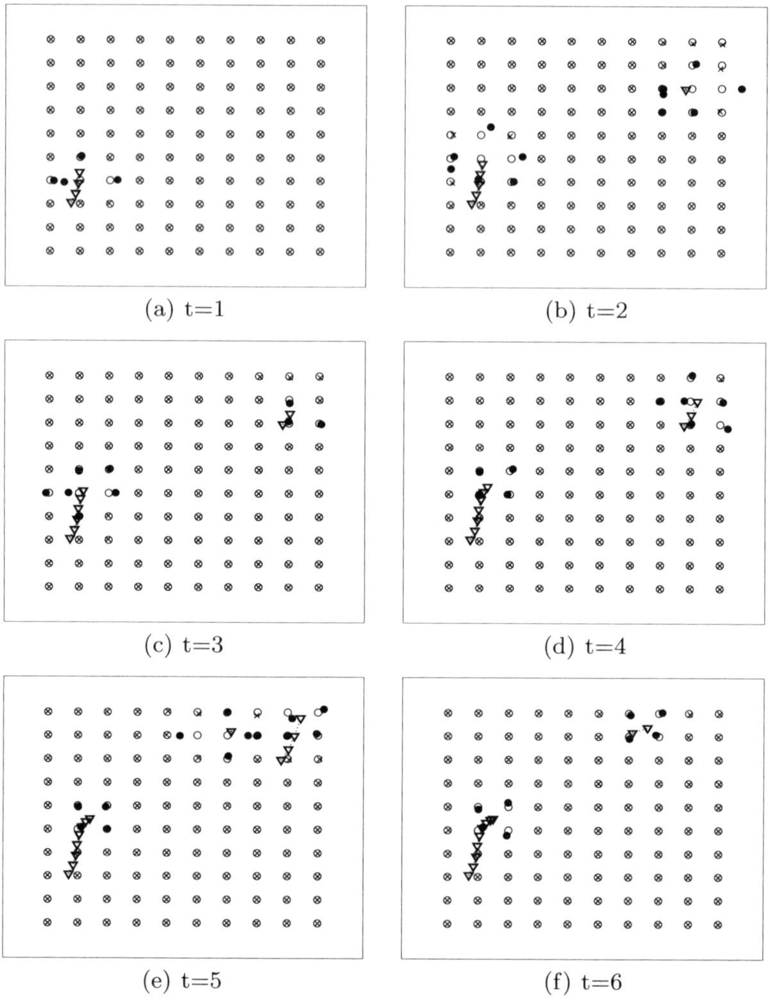

Fig. 2.8. Anchor free sensor localization by sparsity constrained MDS in the presence of targets. True sensor locations (circle), estimated sensor locations (cross), sensors localizing the target (blocked circle), target trajectory (inverted triangle).

2.5.1 Numerical Study

We analyze the performance of the localization algorithms using ROC curves. We consider the following setup: we deploy a 10×10 uniform grid of sensors in a network (see Fig. 2.11). We consider anchor free localization, i.e., $m = 0$ and make a single inter-sensor measurement ($M = 1$) at each time frame. We assume no a priori knowledge of the sensor coordinates, i.e., $r_i = 0$. Each sensor communicates only to its 8 nearest neighbors and the weights for those links were chosen by the LOESS strategy [9]. The rest of the weights were set to zero. Furthermore, we set noise variances σ_0 and σ_1 defined in (2.13) and (2.15), respectively as $\sigma_0 = 1$ and $\sigma_1 = 5\sigma_0 = 5$. Sensor links within a radius $R = 1.5$ indicate the presence of a target, i.e., follow the H_1 hypothesis. We set the reference distance d_0 defined below (2.13) to be $d_0 = 1$ and the path loss exponent $\eta = 2$. We set the sparseness parameters $\lambda = 2.5$ and $p = 1$ to produce a change in the location estimates for only a small portion ($< 10\%$) of the sensors.

We begin by considering the case of random appearance of targets in the sensor network, i.e., targets appear at different locations every time instant. For the DBT, we set $\tau = 0$ in (2.18), i.e., we compare our distance estimates to a fixed initial frame. For every time instant, the DBT and the LRT are performed on each active sensor link and the process is repeated for 5000 target locations. The resulting ROC curve is presented in Fig. 2.9. The ROC for the LRT using simulations is indicated using circles and the corresponding theoretical curve obtained from (2.17) is shown as a solid line. We observe that the simulation and the theoretical curves match for the LRT. The ROC for the DBT is shown using a dashed line. The DBT algorithm yields higher probability of correct detection than the LRT for most false alarm levels. For example, at false alarm level $\alpha = 0.3$, β for the DBT is approximately 0.89 which is 5% more than that of the LRT, which yields $\beta \approx 0.84$.

The intuition for this result is as follows: in the presence of a target, the RSS measurements of the sensor links are spatially-correlated. The presence of a target in a given link implies that with high probability the target is present in neighboring sensor links. However, the RSS model in (2.15) specifies only the distribution of the measurements independently on each link. The LRT makes complete use of the RSS measurements but is limited in its performance as the optimal decision statistic for each sensor link i, j is independent of other sensor link measurements. On the other hand, the DBT finds the active sensor links only based on the estimated distances through sparsity penalized MDS. However, since the inter-sensor distances are computed at each sensor using information from its nearest neighbors, this method makes an implicit use of the spatial correlation of the measurements in its decision statistic, which results in an improvement in performance.

Next, we consider the case of a moving target, where we assume a standard state-space target motion model (for the purpose of a visually pleasing trajectory). We repeated the same algorithms for 5000 such trajectories. The LRT based algorithm yields the same performance curve as the test is independent of whether the target is moving or not. The resulting ROC curve for the DBT is presented as a dotted line in Fig. 2.9. Since we continue to base our decision rule on the fixed initial frame ($\tau = 0$), we observe that the performance of the DBT is also similar to the case of random target appearances.

In the case of a moving target, the RSS measurements are also temporally-correlated. Given a set of sensors indicating a presence of a target at a particular time, there is a high probability that the target is in the vicinity of

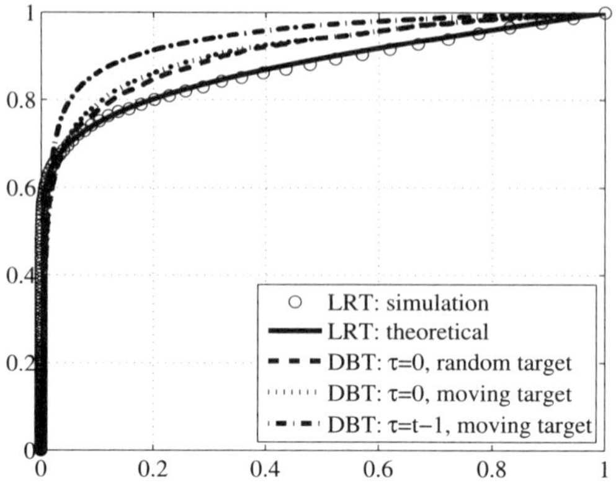

Fig. 2.9. ROC curve for the LRT and the DBT link level tracking algorithm. LRT (solid line), DBT for a random target with $\tau = 0$ (dashed), DBT for a moving target with $\tau = 0$ (dotted), and DBT for a moving target with $\tau = t - 1$ (dashed dotted).

these sensors at the next time frame. To examine the effect of the temporal correlation, we can compare the current estimated distances to the estimated distances from the previous time-frame rather than the initial frame, i.e., set $\tau = t-1$ instead of $\tau = 0$. The temporal correlation of the RSS measurements is captured in the DBT through the sparsity constraint used for aligning the sensors locations estimates. In other words, with high probability the sensor location estimates that are perturbed in the previous time-frame will also be perturbed in the current time-frame, thereby increasing the probability of detection.

The results for $\tau = t - 1$ are presented in Fig. 2.9 using a dashed dotted line. We observe that the performance gains for DBT with $\tau = t - 1$ are higher as compared to DBT with $\tau = 0$ as such a decision rule incorporates both spatial and temporal correlations of the target dynamics. For example, for $\alpha = 0.1$, β for the LRT is 0.75. The result of spatial smoothing alone yields $\beta \approx 0.79$. By performing both spatial and temporal smoothing, we can obtain $\beta \approx 0.86$ through our algorithm, which corresponds to a 15% increase in performance.

We make the following observations for the two proposed tests:

- The DBT for link level tracking outperforms the LRT as it can account for the spatial and the temporal correlations in the target motion.
- The LRT outperforms DBT for low false alarm levels ($\alpha < 0.01$) for the following reasons: first, the DBT we considered is suboptimal as we did not optimize the performance over the choice of sparsity (p, λ). Furthermore, the LRT uses an optimal decision statistic and the exact measurements to perform the test. Currently, we are in pursuit of finding the optimal sparsity that can yield further improvement in performance.

- The issue of space-time sampling is key to the performance of the DBT. Any scenario that exhibits high spatial correlations (e.g., highly reflective targets or more sensors/unit area) can yield further improvement in performance of the DBT. For example, Fig. 2.10 illustrates the performance of the DBT for $\tau = 0$ when the number of sensors is increased to 300. By comparing the perturbation to the fixed initial frame, we only perform spatial smoothing of the sensor location estimates. We observe that the denser sampling of sensors have resulted in better spatial smoothing, which eliminates more false alarms resulting in an improved performance. For example, at a false alarm level $\alpha = 0.01$, the DBT with 100 sensors yields $\beta \approx 0.48$, while the DBT with 300 sensors yields $\beta \approx 0.66$. If the sampling time for the sensors and the computation time of the DBT algorithm is much faster than the target motion, the DBT can yield better performance by taking advantage of more temporal correlations.

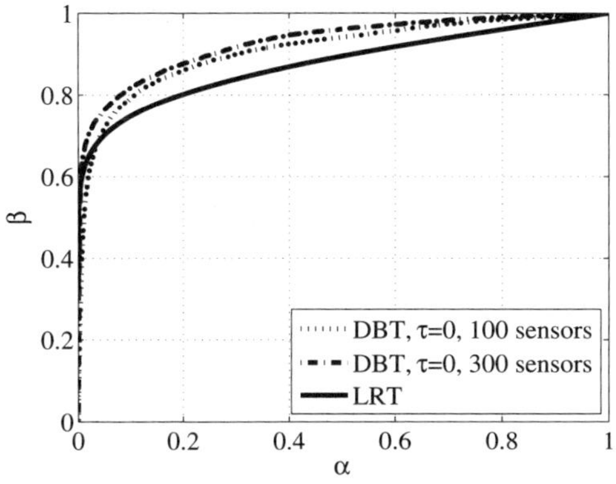

Fig. 2.10. ROC curve for the LRT and the DBT link level tracking algorithm with $\tau = 0$ for different spatial sampling. LRT (solid line), DBT for a moving target using 10×10 grid of sensors (dotted), and DBT using 300 randomly located sensors (dashed dotted).

- The disadvantage of LRT in this setting is that the test is performed independently on each sensor link. Further improvements in the probability of detection can be achieved when the LRT is derived for the full spatio-temporal model.
- In the performance analysis, we assumed steady state operation, i.e., perfect knowledge of the inter-sensor distances are obtained a priori in the absence of target. If such knowledge is unavailable and distances need to be estimated, the LRT tracker must be modified to a generalized likelihood ratio test (GLRT). The DBT can estimate the initial set of distances more accurately from the RSS measurements by taking advantage of spa-

tial correlations and hence can yield a higher probability of detection than the GLRT.

Spatial localization from link level localization

Our objective is to approximately locate the target relative to the location of the sensors. There are a number of ways in which this link level estimate can be translated into estimated target coordinates in space. For example, one could use as an estimate the midpoint of the convex hull generated by the positions of those sensors that detect the target according to the LRT or the DBT. An example of the midpoint tracking algorithm is shown in Fig. 2.11. Another estimate can be found by the intersection of convex regions

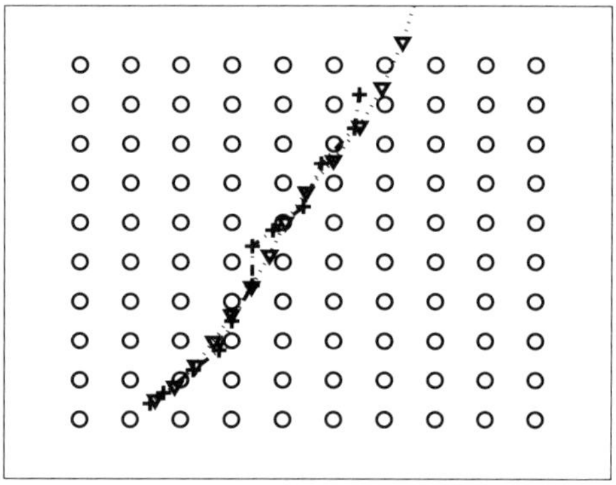

Fig. 2.11. A simple tracking algorithm based on link level tracking. True sensor locations (circle), true trajectory of the target (diamond), estimated trajectory (plus).

corresponding to the sensor links that show the presence of the target through the optimal decision rule. These estimates do not require a physical model of the target trajectory. However, given a target motion model, standard filtering techniques such as the Kalman filter or particle filter (PF) can be used to obtain refined target position estimates from the link level data.

Future work

Given the set of tagged sensors, i.e., sensor links with high output in the LRT, we have reduced the problem to that of binary sensing, where the knowledge of the presence of the target is stored as the decisions made on each of the links. For accurate estimation of targets positions, we can now use the popular particle filtering techniques proposed on binary sensing models to perform multi-target tracking given a small set of anchor nodes. Moreover, most sensor networks are remotely operated and limited in power. We can pose a power constraint by limiting the number of inter-sensor measurements to a small s of the $n(n-1)$ ($s \ll n(n-1)$) total sensor links at each time step. The problem of choosing s from $n(n-1)$ links is a combinatorially hard problem. So we propose a convex relaxation to the problem, which chooses the set of active links by minimizing the predicated mean square error of the state of the target. This approach has been shown to achieve near optimal performance in our earlier work [38].

2.6 Conclusions

In this chapter, we presented the sparsity penalized MDS algorithm for simultaneous localization and tracking. We are interested in tracking a target relative to the sensor coordinates. The subset selection capability of our proposed sparsity constraint allows the algorithm to find only those which have changed their location estimate due to the presence of a target. We use these sensors to perform link level tracking. We formulate a model for the inter-sensor RSS measurements in the presence and absence of a target by conducting actual experiments in free space. Using this model, we illustrated the performance of our algorithm for link level target tracking. Currently, we are in pursuit of optimal sensor scheduling strategies for physical level tracking.

Acknowledgements

We would like to thank Mr. Xing Zhou for his enthusiastic assistance in conducting the experiments discussed in this chapter. This work was partially supported by a grant from the National Science Foundation CCR-0325571.

References

[1] D. L. Alspach and H. W. Sorensen. Nonlinear Bayesian estimation using Gaussian sum approximations. *IEEE Trans. Automat. Contr.*, 82:1032–1063, 1987.

[2] A. Arora, P. Dutta, S. Bapat, V. Kulathumani, H. Zhang, V. Naik, V. Mittal, H. Cao, M. Demirbas, M. Gouda, Y. Choi, T. Herman, and S. Kulkarni. A line in the sand: a wireless sensor network for detection, classification, and tracking. *Computer Networks*, 46(5):605–634, 2004.

[3] Y. Bar-Shalom. *Multitarget Multisensor Tracking: Advanced Applications.* Artech House, 1990.

[4] R. R. Brooks, P. Ramanathan, and A. M. Sayeed. Distributed target classification and tracking in sensor networks. *Proc. IEEE*, 91(8):1163–1171, 2003.

[5] S. Capkun, L. Buttyan, and J. Hubaux. SECTOR: Secure tracking of node encounters in multi-hop wireless networks. In *Proc. 1st ACM Workshop on Security of Ad Hoc and Sensor Networks (SASN)*, pages 21–32, 2003.

[6] V. Cevher and J. H. McClellan. Sensor array calibration via tracking with the extended Kalman filter. In *Proc. IEEE Intl. Conf. Acoust., Speech, Signal Processing*, volume 5, pages 2817–2820, 2001.

[7] J. Chamberland and V. V. Veeravalli. Decentralized detection in sensor networks. *IEEE Trans. Signal Processing*, 51(2):407–416, 2003.

[8] W. Cleveland. Robust locally weighted regression and smoothing scatterplots. *J. Am. Statist. Assoc.*, 74(368):829–836, 1979.

[9] J. Costa, N. Patwari, and A. O. Hero III. Distributed multidimensional scaling with adaptive weighting for node localization in sensor networks. *ACM J. Sensor Networking*, 2(1):39–64, 2006.

[10] A. J. Coulson, A. G. Williamson, and R. G. Vaughan. A statistical basis for lognormal shadowing effects in multipath fading channels. *IEEE Trans. on Veh. Tech.*, 46(4):494–502, 1998.

[11] T. Cox and M. Cox. *Multidimensional Scaling.* Chapman & Hall, London, 1994.

[12] M. L. Davidson. *Multidimensional scaling.* Wiley, New York, NY, 1983.

[13] P. M. Djuric, J. H. Kotecha, J. Zhang, Y. Huang, T. Ghirmai, M. F. Bugallo, and J. Miguez. Particle filtering. *IEEE Signal Processing Magazine*, 20(5):19–38, 2003.

[14] L. Doherty, K. S. Pister, and L. E. Ghaoui. Convex position estimation in wireless sensor networks. In *Proc. Twentieth Annual Joint Conference of the IEEE Computer and Communications Societies (INFOCOM)*, volume 3, pages 1655–1663, 2001.

[15] D. L. Donoho, M. Elad, and V. Temlyakov. Stable recovery of sparse overcomplete representations in the presence of noise. *IEEE Trans. on Inform. Theory*, 52(1):6–18, 2006.

[16] C. Estan, S. Savage, and G. Varghese. Automatically inferring patterns of resource consumption in network traffic. In *ACM SIGCOMM*, pages 137–148, 2003.

[17] J. C. Gower and G .B. Dijksterhuis. *Procrustes Problems.* Oxford University Press, 2004.

[18] M. J. Greenacre. *Theory and Applications of Correspondence Analysis.* Academic Press Inc., London, UK, 1984.

[19] C. J. Gregory, R. R. Carthy, and L. G. Pearlstine. Survey and monitoring of species at risk at camp blanding training site, northeastern florida. *Southeastern Naturalist*, 5(3):473–498, 2006.

[20] P. Groenen. *The majorization approach to multidimensional scaling: some problems and extensions.* DSWO Press, 1993.

[21] A. O. Hero III. Geometric entropy minimization (GEM) for anomaly detection and localization. In *Proc. Advances in Neural Information Processing Systems (NIPS)*, 2006.
[22] Y. Hu, A. Perrig, and D. Johnson. Packet leashes: A defense against wormhole attacks in wireless ad hoc networks. In *Proc. Twenty-Second Annual Joint Conference of the IEEE Computer and Communications Societies (INFOCOM)*, volume 3, pages 1976–1986, 2003.
[23] X. Ji and H. Zha. Sensor positioning in wirless ad-hoc sensor networks with multidimensional scaling. In *Proc. IEEE Infocom*, pages 2652–2661, 2004.
[24] S. J. Julier and J. K. Uhlman. A new extension of the kalman filter to nonlinear systems. In *Proc. AeroSense: Eleventh Intl. Symp. on Aerospace/Defense Sensing, Simulations and Control Multi Sensor Fusion, Tracking and Resource Management II*, volume 3068, pages 182–193, 1997.
[25] W. Kim, K. Mechitov, J. Y. Choi, and S. Ham. On target tracking with binary proximity sensors. In *Proc. Fourth Intl Symposium Information Processing in Sensor Networks*, pages 301–308, 2005.
[26] J. Kumagai and S. Cherry. Sensors and sensibility. *IEEE Spectrum*, 41 (7):22–28, 2004.
[27] K. Lange, D. R. Hunter, and I. Yang. Optimization transfer using surrogate objective functions. *Journal of Computational and Graphical Statistics*, 9(1):1–20, 2000.
[28] M. S. Lee and Y. H. Kim. An efficient multitarget tracking for car applications. *IEEE Trans. Industrial Electronics*, 50(2):397–399, 2003.
[29] M. Leoncini, G. Resta, and P. Santi. Analysis of a wirless sensor dropping problem for wide-area environmental monitoring. In *Fourth Intl. Symp. on Inform. Processing Sensor Networks (IPSN)*, pages 239–245, 2005.
[30] R. L. Moses, D. Krishanmurthy, and R. Patterson. A self-localization method for wireless sensor networks. *EURASIP J. Applied Signal Processing*, 4:348–358, 2003.
[31] R. Nagpal, H. Shrobe, and J. Bachrach. Organizing a global coordinate system from local information on an ad hoc sensor network. In *Proc. 2nd Intl Workshop on Inform. Processing Sensor Networks (IPSN), Lecture Notes in Computer Science*, volume 2634, pages 333–348, 2003.
[32] D. Niculescu and B. Nath. Ad hoc positioning systems. In *Proc. IEEE Global Communications Conference (GLOBECOM)*, volume 5, pages 2926–2931, 2001.
[33] R. Niu, P. K. Varshney, and Q. Cheng. Distributed detection in a large wireless sensor network. *Information Fusion*, 7:380–394, 2006.
[34] P. Pathirana, N. Bulusu, S. Jha, and A. Savkin. Node localization using mobile robots in delay-tolerant sensor networks. *IEEE Trans. Mobile Computing*, 4(3):285–296, 2005.

[35] N. Patwari, A. O. Hero III, M. Perkins, N. S. Correal, and R. J. O'Dea. Relative location estimation in wireless sensor networks. *IEEE Trans. Signal Processing*, 51(8):2137–2148, 2003.
[36] N. Patwari, A. O. Hero III, and A. Pacholski. Manifold learning visualization of network traffic data. In *Proc. Workshop on Mining Network Data, Philadelphia, PA*, pages 191–196, 2005.
[37] H. V. Poor. *An Introduction to Signal Detection and Estimation.* Springer-Verlag, New York, N.Y, 1988.
[38] R. Rangarajan, R. Raich, and A. O. Hero III. Single-stage waveform selection for adaptive resource constrained state estimation. In *Proc. IEEE Intl. Conf. Acoust., Speech, Signal Processing*, volume 3, pages 672–675, 2006.
[39] A. Savvides, H. Park, and M. B. Srivastava. The bits and flops of the n-hop multilateration primitive for node localization problems. In *Proc. Intl Workshop on Sensor Nets & Apps.*, pages 112–121, 2002.
[40] Y. Shang, W. Rumi, Y. Zhang, and M. P. Fromherz. Localization from mere connectivity. In *Proc. 4th ACM Intl. Symp. on Mobile ad hoc networking and computing*, pages 201–212, 2003.
[41] N. Shrivastava, R. Mudumbai, U. Madhow, and S. Suri. Target tracking with binary proximity sensors: fundamental limits, minimal descriptions, and algorithms. In *Proc. ACM 4th Intl. Conf. Embedded networked sensor systems (SenSys)*, pages 251–264, 2006.
[42] C. Taylor, A. Rahimi, J. Bachrach, H. Shrobe, and A. Grue. Simultaneous localization, calibration, and tracking in an ad hoc sensor network. In *Proc. 5th Intl. Conf. Proc. Information Processing in Sensor Networks (IPSN)*, pages 27–33, 2006.
[43] E. A. Wan and R. Van Der Merwe. The unscented kalman filter for nonlinear estimation. In *IEEE Symp. Adaptive Systems for Signal Processing, Communications, and Control*, pages 153–158, 2000.
[44] N. Wang, M. H. Wang, and N. Q. Zhang. Wireless sensors in agriculture and food industry: Recent developments and future perspective. *Computers and electronics in agriculture*, 50(1):1–14, 2006.
[45] W. Wang and B. Bhargava. Visualization of wormholes in sensor networks. In *Proc. ACM Workshop on wireless security*, pages 51–60, 2004.

Appendix: Derivation of sparsity penalized dwMDS

To simplify our derivation, we divide the global cost function into multiple local cost functions as follows:

$$C^{(t)} = \sum_{i=1}^{n} C_i^{(t)} + c^{(t)}, \tag{2.19}$$

where $c^{(t)}$ is a constant independent of the sensor locations $\mathbf{X}$ and the local cost function at each sensor node i is

$$C_i^{(t)} = \sum_{j=1, j\neq i}^{n} \bar{w}_{i,j}^{(t)}(\bar{\delta}_{i,j}^{(t)} - d_{i,j}(\mathbf{X}))^2 + 2\sum_{j=n+1}^{n+m} \bar{w}_{i,j}^{(t)}(\bar{\delta}_{i,j}^{(t)} - d_{i,j}(\mathbf{X}))^2 + r_i\|\bar{\mathbf{x}}_i - \mathbf{x}_i\|^2 + \lambda\|\mathbf{x}_i - \mathbf{x}_i^{(t-1)}\|^p, \quad (2.20)$$

where $\bar{w}_{i,j}^{(t)} = \sum_{l=1}^{M} w_{i,j}^{(t),l}$ and $\bar{\delta}_{i,j}^{(t)} = \sum_{l=1}^{M} w_{i,j}^{(t),l}\delta_{i,j}^{(t),l}/\bar{w}_{i,j}^{(t)}$. The cost function $C_i^{(t)}$ depends only the measurements made by sensor node i and the positions of the neighboring nodes, i.e., nodes for which $w_{i,j}^{(t),l} > 0$; $C_i^{(t)}$ is local to node i [9]. The local cost function in (2.20) can be rewritten as

$$C_i^{(t)}(\mathbf{X}) = c_1^{(t)} + c_2^{(t)}(\mathbf{X}) - c_3^{(t)}(\mathbf{X}) + c_4^{(t)}(\mathbf{X}), \quad (2.21)$$

where

$$\begin{aligned} c_1^{(t)} &= \sum_{j=1, j\neq i}^{n} \bar{w}_{i,j}^{(t)}(\bar{\delta}_{i,j}^{(t)})^2 + 2\sum_{j=n+1}^{n+m} \bar{w}_{i,j}^{(t)}(\bar{\delta}_{i,j}^{(t)})^2 \\ c_2^{(t)}(\mathbf{X}) &= \sum_{j=1, j\neq i}^{n} \bar{w}_{i,j}^{(t)}d_{i,j}^2(\mathbf{X}) + 2\sum_{j=n+1}^{n+m} \bar{w}_{i,j}^{(t)}d_{i,j}^2(\mathbf{X}) + r_i\|\bar{\mathbf{x}}_i - \mathbf{x}_i\|^2 \\ c_3^{(t)}(\mathbf{X}) &= 2\sum_{j=1, j\neq i}^{n} \bar{w}_{i,j}^{(t)}\bar{\delta}_{i,j}^{(t)}d_{i,j}(\mathbf{X}) + 4\sum_{j=n+1}^{n+m} \bar{w}_{i,j}^{(t)}\bar{\delta}_{i,j}^{(t)}d_{i,j}(\mathbf{X}) \\ c_4^{(t)}(\mathbf{X}) &= \lambda\|\mathbf{x}_i - \mathbf{x}_i^{(t-1)}\|^p. \end{aligned} \quad (2.22)$$

The term $c_1^{(t)}$ is independent of $\mathbf{x}_i$. The term $c_2^{(t)}$ is quadratic in $\mathbf{x}_i$. Terms $c_3^{(t)}$ and $c_4^{(t)}$ are neither affine nor quadratic functions of $\mathbf{x}_i$. A majorizing function for the term $c_3^{(t)}$ is motivated by the following Cauchy-Schwarz inequality,

$$d_{i,j}(\mathbf{X}) = \|\mathbf{x}_i - \mathbf{x}_j\| \geq \frac{(\mathbf{x}_i - \mathbf{x}_j)^T(\mathbf{y}_i - \mathbf{y}_j)}{d_{i,j}(\mathbf{Y})}, \qquad \forall \mathbf{Y}, \quad (2.23)$$

where $\mathbf{Y} = [\mathbf{y}_1, \ldots, \mathbf{y}_n]$. For $c_4^{(t)}$, we present a quadratic majorizing function, which can be obtained from the following relation

$$\alpha^{p/2} \leq \alpha_0^{p/2} + \frac{p}{2}(\alpha - \alpha_0)(\alpha_0)^{(\frac{p}{2}-1)}, \qquad \forall \alpha, \alpha_0 > 0. \quad (2.24)$$

The above inequality follows from a linear approximation to the concave function $f(\alpha) = \alpha^{p/2}$ via Taylor series expansion. Choosing $\alpha = \|\mathbf{x}_i - \mathbf{x}_i^{t-1}\|^2$ and $\alpha_0 = \|\mathbf{y}_i - \mathbf{x}_i^{t-1}\|^2$ yields

$$\|\mathbf{x}_i - \mathbf{x}_i^{t-1}\|^p \leq \|\mathbf{y}_i - \mathbf{x}_i^{t-1}\|^p + \frac{p}{2}\frac{\|\mathbf{x}_i - \mathbf{x}_i^{t-1}\|^2 - \|\mathbf{y}_i - \mathbf{x}_i^{t-1}\|^2}{\|\mathbf{y}_i - \mathbf{x}_i^{t-1}\|^{2-p}}, \quad (2.25)$$

the majorizing function for the $c_4^{(t)}$ term. Substituting the inequalities from (2.23) and (2.25) in (2.21), we obtain the majorizing function for the local cost function as

$$\begin{aligned} T_i^{(t)}(\mathbf{X},\mathbf{Y}) = c_1^{(t)} &+ \sum_{j=1,j\neq i}^{n} \bar{w}_{i,j}^{(t)} d_{i,j}^2(\mathbf{X}) + 2\sum_{j=n+1}^{n+m} \bar{w}_{i,j}^{(t)} d_{i,j}^2(\mathbf{X}) + r_i\|\bar{\mathbf{x}}_i - \mathbf{x}_i\|^2 \\ &+2\sum_{j=1,j\neq i}^{n} \bar{w}_{i,j}^{(t)}\bar{\delta}_{i,j}^{(t)} \frac{(\mathbf{x}_i-\mathbf{x}_j)^T(\mathbf{y}_i-\mathbf{y}_j)}{d_{i,j}(\mathbf{Y})} \\ &+4\sum_{j=n+1}^{n+m} \bar{w}_{i,j}^{(t)}\bar{\delta}_{i,j}^{(t)} \frac{(\mathbf{x}_i-\mathbf{x}_j)^T(\mathbf{y}_i-\mathbf{y}_j)}{d_{i,j}(\mathbf{Y})} \\ &+\lambda\|\mathbf{y}_i - \mathbf{x}_i^{(t-1)}\|^p + \frac{\lambda p}{2}\frac{\|\mathbf{x}_i - \mathbf{x}_i^{(t-1)}\|^2 - \|\mathbf{y}_i - \mathbf{x}_i^{(t-1)}\|^2}{\|\mathbf{y}_i - \mathbf{x}_i^{(t-1)}\|^{2-p}}. \end{aligned} \tag{2.26}$$

Since $T_i^{(t)}(\mathbf{X},\mathbf{Y})$ is a majorizing function to $C_i^{(t)}(\mathbf{X})$, it is easy to verify that the function $T^{(t)}(\mathbf{X},\mathbf{Y}) = \sum_{i=1}^n T_i^{(t)}(\mathbf{X},\mathbf{Y})$ is a majorizing function to the global cost function $C^{(t)}(\mathbf{X})$. The partial derivative of $T^{(t)}(\mathbf{X},\mathbf{Y})$ with respect to $\mathbf{x}_i$ is straightforward as all the expressions in (2.26) are linear or quadratic in $\mathbf{x}_i$. The partial derivative of $T^{(t)}(\mathbf{X},\mathbf{Y})$ with respect to $\mathbf{x}_i$ is given by

$$\frac{\partial T^{(t)}(\mathbf{X},\mathbf{Y})}{\partial \mathbf{x}_i} = \frac{\partial T_i^{(t)}(\mathbf{X},\mathbf{Y})}{\partial \mathbf{x}_i} + \sum_{k\neq i}\frac{\partial T_k^{(t)}(\mathbf{X},\mathbf{Y})}{\partial \mathbf{x}_i}, \tag{2.27}$$

where

$$\begin{aligned} \frac{\partial T_i^{(t)}(\mathbf{X},\mathbf{Y})}{\partial \mathbf{x}_i} = 2\sum_{j=1,j\neq i}^{n} &\left(\bar{w}_{i,j}^{(t)}(\mathbf{x}_i-\mathbf{x}_j) - \bar{w}_{i,j}^{(t)}\bar{\delta}_{i,j}^{(t)}\frac{(\mathbf{y}_i-\mathbf{y}_j)}{\|\mathbf{y}_i-\mathbf{y}_j\|}\right) \\ &+4\left(\sum_{j=n+1}^{n+m} \bar{w}_{i,j}^{(t)}(\mathbf{x}_i-\mathbf{x}_j) - \bar{w}_{i,j}^{(t)}\bar{\delta}_{i,j}^{(t)}\frac{(\mathbf{y}_i-\mathbf{y}_j)}{\|\mathbf{y}_i-\mathbf{y}_j\|}\right) \\ &+2r_i(\mathbf{x}_i - \bar{\mathbf{x}}_i) + \lambda p\frac{(\mathbf{x}_i - \mathbf{x}_i^{(t-1)})}{\|\mathbf{y}_i - \mathbf{x}_i^{(t-1)}\|^{2-p}} \end{aligned} \tag{2.28}$$

and

$$\frac{\partial T_k^{(t)}(\mathbf{X},\mathbf{Y})}{\partial \mathbf{x}_i} = 2\left(\bar{w}_{i,k}^{(t)}(\mathbf{x}_i-\mathbf{x}_k) - \bar{w}_{i,k}^{(t)}\bar{\delta}_{i,k}^{(t)}\frac{(\mathbf{y}_i-\mathbf{y}_k)}{\|\mathbf{y}_i-\mathbf{y}_k\|}\right). \tag{2.29}$$

Substituting (2.28) and (2.29) in (2.27) yields,

$$\frac{\partial T^{(t)}(\mathbf{X},\mathbf{Y})}{\partial \mathbf{x}_i} = 4\left(\sum_{j=1,j\neq i}^{n+m} \bar{w}_{i,j}^{(t)}(\mathbf{x}_i-\mathbf{x}_j) - \bar{w}_{i,j}^{(t)}\bar{\delta}_{i,j}^{(t)}\frac{(\mathbf{y}_i-\mathbf{y}_j)}{\|\mathbf{y}_i-\mathbf{y}_j\|}\right)$$
$$+2r_i(\mathbf{x}_i-\bar{\mathbf{x}}_i)+\lambda p\frac{(\mathbf{x}_i-\mathbf{x}_i^{(t-1)})}{\|\mathbf{y}_i-\mathbf{x}_i^{(t-1)}\|^{2-p}}. \quad (2.30)$$

Setting the derivatives to zero yields the following recursive update rule

$$\mathbf{x}_i^k = \frac{1}{a_i}\left(c_i + \left[\mathbf{x}_1^{(k-1)},\ldots,\mathbf{x}_N^{(k-1)}\right]\mathbf{b}_i^{(k-1)}\right), \quad (2.31)$$

where $\mathbf{x}_i^k$ denotes the location of node i at iteration k. Furthermore, $\mathbf{b}_i^k = [b_1^k, b_2^k, \ldots, b_N^k]$ and

$$b_i^k = 4\left(\sum_{j=1,j\neq i}^{n+m}\frac{\bar{w}_{i,j}^{(t)}\bar{\delta}_{i,j}^{(t)}}{\|\mathbf{x}_i^k-\mathbf{x}_j^k\|}\right), \quad (2.32)$$

$$b_j^k = 4\left(\bar{w}_{i,j}^{(t)} - \frac{\bar{w}_{i,j}^{(t)}\bar{\delta}_{i,j}^{(t)}}{\|\mathbf{x}_i^k-\mathbf{x}_j^k\|}\right), \quad j\neq i, \quad (2.33)$$

$$a_i = 4\sum_{j=1,j\neq i}^{n+m}\bar{w}_{i,j}^{(t)} + 2r_i + \frac{\lambda p}{\|\mathbf{x}_i^k-\mathbf{x}_i^{t-1}\|^{2-p}}, \quad (2.34)$$

$$c_i = 2r_i\bar{\mathbf{x}}_i + \frac{\lambda p\mathbf{x}_i^{(t-1)}}{\|\mathbf{x}_i^{k-1}-\mathbf{x}_i^{(t-1)}\|}. \quad (2.35)$$

The dwMDS algorithm in [9] obtains a recursive update for location $\mathbf{x}_i$ by setting the derivatives of the surrogate to the ith local cost function $(T_i^{(t)}(\mathbf{X},\mathbf{Y}))$ to zero. This is equivalent to minimizing the global cost function only under anchor free localization ($m=0$) and no a priori information ($r_i=0$). However, in our algorithm, we use the local cost functions only to derive a majorizing function for the global cost function and not in the minimization. Moreover, the algorithm is still decentralized in its implementation though we minimize the global cost function with respect to the sensor locations $\mathbf{X}$.

Appendix: Optimal likelihood ratio test

To test the presence of a target on a sensor link i,j, we pose the following hypotheses testing problem

$$\begin{aligned}
&\mathrm{H}_0 : P_1,\ldots,P_M \sim \mathcal{N}(\bar{P},\sigma_0^2)\\
&\mathrm{H}_1 : P_1,\ldots,P_M|\hat{P} \sim \mathcal{N}(\hat{P},\sigma_0^2),\ \text{i.i.d}, \quad \hat{P}\sim\mathcal{N}(\bar{P},\sigma_1^2),
\end{aligned}$$

where $P_1,\ldots,P_M$ are the measurements made by a particular link i,j. We leave out the indices i,j in the measurements for brevity. $\bar{P}$ is the mean received power in the sensor link i, j. We assume it can be obtained during the system setup in the absence of targets. Denote the measurements by the M-element vector $\mathbf{p} = [P_1, P_2, \ldots, P_M]^T$. Then the hypotheses can be written as

$$\begin{aligned} \mathrm{H}_0 &: \mathbf{p} \sim \mathcal{N}(\bar{P}\mathbf{1}, \sigma_0^2 \mathbf{I}) \\ \mathrm{H}_1 &: \mathbf{p} \sim \mathcal{N}(\bar{P}\mathbf{1}, \sigma_1^2 \mathbf{1}\mathbf{1}^T + \sigma_0^2 \mathbf{I}). \end{aligned}$$

To construct the LRT, we first compute the log likelihood ratio as

$$\begin{aligned} \Lambda &= \log\left(\frac{f(\mathbf{p}|\mathrm{H}_1)}{f(\mathbf{p}|\mathrm{H}_0)}\right) \\ &= \frac{1}{2}(\mathbf{p} - \bar{P}\mathbf{1})^T(\mathbf{C}_0^{-1} - \mathbf{C}_1^{-1})(\mathbf{p} - \bar{P}\mathbf{1}) + \frac{1}{2}\log\left(\frac{|\mathbf{C}_0|}{|\mathbf{C}_1|}\right), \end{aligned} \tag{2.36}$$

where $\mathbf{C}_0 = \sigma_0^2\mathbf{I}$, $\mathbf{C}_1 = \sigma_1^2\mathbf{1}\mathbf{1}^T + \sigma_0^2\mathbf{I}$ and $|\mathbf{C}|$ denotes the determinant of a matrix $\mathbf{C}$. The eigendecompositions of the covariance matrices $\mathbf{C}_0$ and $\mathbf{C}_1$ can be written as

$$\begin{aligned} \mathbf{C}_0 &= \mathbf{V}_0\mathbf{D}_0\mathbf{V}_0^T, \\ \mathbf{C}_1 &= \mathbf{V}_1\mathbf{D}_1\mathbf{V}_1^T, \end{aligned}$$

where $\mathbf{D}_i$ is a diagonal matrix composed of the eigenvalues $\lambda_1^i, \ldots, \lambda_M^i$ and $\mathbf{V}_i$ is the matrix of corresponding eigenvectors. The eigenvalues of the covariance matrix $\mathbf{C}_1$ are given by $\lambda_1^1 = \sigma_1^2 M + \sigma_0^2$ and $\lambda_i^1 = \sigma_0^2$, $i = 2, \ldots, M$. The corresponding eigenvectors are $\mathbf{v}_1 = \mathbf{1}/\sqrt{M}, \mathbf{v}_2, \ldots, \mathbf{v}_M$, where $\{\mathbf{v}_i\}_{i=1}^M$ are a set of orthogonal unit norm vectors. The eigenvalues of $\mathbf{C}_0$ are all σ_0^2 and it is easy to verify that $\mathbf{v}_1, \ldots, \mathbf{v}_M$ are eigenvectors to $\mathbf{C}_0$, i.e., $\mathbf{V}_0 = \mathbf{V}_1$. Thus

$$\mathbf{C}_0^{-1} - \mathbf{C}_1^{-1} = \mathbf{V}_0 \mathrm{diag}\left(\frac{M\sigma_1^2}{M\sigma_1^2 + \sigma_0^2}, 0, \ldots, 0\right)\mathbf{V}_0^T = \frac{\sigma_1^2 M}{\sigma_1^2 M + \sigma_0^2}\frac{\mathbf{1}\mathbf{1}^T}{M}. \tag{2.37}$$

Substituting (2.37) in (2.36) and collecting constant terms at the right hand side yields the optimal LRT as

$$|\bar{p} - \bar{P}| \underset{\mathrm{H}_0}{\overset{\mathrm{H}_1}{\gtrless}} \gamma, \tag{2.38}$$

where $\bar{p} = \sum_{i=1}^M P_i/M$ is the minimal sufficient statistics of this test. Under H_0, $\bar{p}$ is distributed as $\mathcal{N}(\bar{P}, \sigma_0^2/M)$ and under H_1, $\bar{p}$ is $\mathcal{N}(\bar{P}, \sigma_0^2/M + \sigma_1^2)$. We find γ to satisfy a false alarm of level α, i.e.,

$$P\left(|\bar{p} - \bar{P}| > \gamma|\mathrm{H}_0\right) = 2Q\left(\frac{\sqrt{M}\gamma}{\sigma_0}\right) = \alpha, \tag{2.39}$$

which implies $\gamma = (\sigma_0/\sqrt{M})Q^{-1}(\alpha/2)$. The probability of correct decision, β is then given by

$$\begin{aligned}
\beta &= P\left(|\bar{p} - \bar{P}| > \gamma|\mathrm{H}_1\right) \\
&= 2Q\left(\frac{\gamma}{\sqrt{\sigma_0^2/M + \sigma_1^2}}\right) \\
&= 2Q\left(Q^{-1}(\alpha/2)\sqrt{\frac{\sigma_0^2}{\sigma_0^2 + M\sigma_1^2}}\right). \qquad (2.40)
\end{aligned}$$

Part II

Distributed Computation over Unreliable Communication Networks

3

Error Exponents for Decentralized Detection in Tree Networks *

Wee Peng Tay and John N. Tsitsiklis

Laboratory for Information and Decision Systems
Massachusetts Institute of Technology
Cambridge, MA, USA
{wptay, jnt}@mit.edu

3.1 Introduction

Consider a set of sensors, one of them designated as the fusion center. We are given two hypotheses H_0 and H_1, with associated probability spaces. In this chapter, we consider only simple hypothesis testing, i.e., the probability measures under both hypotheses are known to the network. The goal of the network is to make a decision on the true hypothesis based on information provided by observations made at each sensor node. This is commonly known as the decentralized detection problem. Decentralized detection in sensor networks has attracted a lot of interest in recent years, because of new technologies (especially, the availability of low-cost sensing devices) and numerous potential applications. The decentralized detection problem was first formulated and studied by [3], which considers a "parallel configuration" whereby each sensor makes an observation and sends a quantized version of that observation to a fusion center. The goal is to make a decision on the two possible hypotheses, based on the messages received at the fusion center. The main difference between this scenario and the classical *centralized* decision system is that the fusion center has no access to the raw observation made at each sensor. Rather, a sensor transmits a summary of its observation via a transmission function to the fusion center. (When the outputs of the transmission functions are restricted to a finite alphabet, these are known as quantizers.) The network aims to minimize the probability of error or some other cost function at the fusion center, by choosing optimal transmission functions and fusion rules. Various properties and variants of the decentralized detection problem in a parallel configuration have been extensively studied over the last twenty-five years; examples include the following: [4, 5, 6, 7, 8] study the

* This chapter is an abridged version of [1] and [2], and an overview of the results therein. This research was supported, in part, by the National Science Foundation under contracts ECCS-0701623 and ECS-0426453.

properties of optimal fusion rules and quantizers at sensor nodes; [9] shows the existence of optimal strategies, and proves that likelihood ratio quantizers are optimal for a large class of problems including the decentralized detection problem; and [10, 11, 12, 13, 14] consider constrained decentralized detection. The reader is referred to [15, 16] for a survey of the work done in this area.

We are interested in networks operating in a regime of limited communication capabilities. Our focus on this regime reflects an emphasis on networks consisting of many, small, and inexpensive devices that have limited battery life and power, and cannot afford to communicate frequently or to transmit a lot of data. Indeed, with abundant communication capabilities, the sensors could just share all their measurements, in which case the network aspects become immaterial, and we are faced with much easier, classical, centralized information processing problems.

Suppose we have n sensors dispersed in a large geographical region. If the sensors are organized in a parallel configuration (cf. the left-hand side in Figure 3.1), some of the sensors may have to communicate to a far away receiver. The energy expended for communicating can be reduced significantly if the sensors are organized in an in-tree architecture, as in the right-hand side in Figure 3.1, with sensors sending their messages first to an intermediate aggregator. Moreover, all sensors, except the aggregators, in the right-hand side figure expend approximately the same amount of energy if each uses the same transmission function. This ensures that the lifetimes of the sensors are uniform geographically (aggregators can be special nodes that have a larger energy supply).

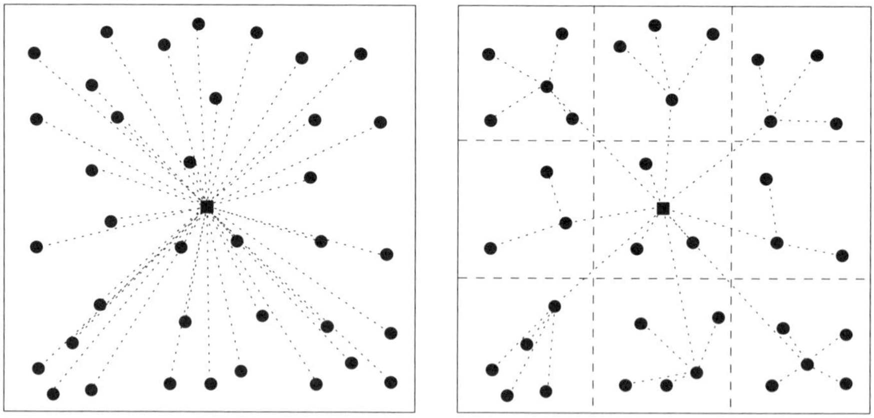

Fig. 3.1. Two alternative architectures of a geographically dispersed sensor network. The architecture on the left is known as a parallel configuration, while that on the right is a tree configuration.

Special cases of the tandem configuration (sensors arranged in a serial network), and some specific tree configurations have been studied in [17], where it is shown that it is optimal for sensor nodes to employ likelihood ratio quantizers. Tree configurations are also discussed in [18, 6, 19, 20, 21, 22, 23]. However, the exact form of optimal strategies in tree configurations is difficult to derive. Nevertheless, to obtain necessary conditions for the optimal transmission functions or fusion rule, one can analyze the problem using a person-by-person (PBP) optimality approach. In such an analysis, all nodes' transmission functions, except for one particular node, are fixed. Then, one can derive the form of the optimal transmission function at that particular node. Under a conditional independence assumption, and when transmission functions are quantizers, typical results show that likelihood ratio quantizers are PBP optimal. However, finding optimal quantizer thresholds requires the solution of a nonlinear system of equations, with as many equations as there are thresholds. Closed form formulae for the optimal thresholds used in the likelihood ratio quantizers at each node are known only for trees with a small number of nodes; e.g. [6] considers configurations with at most four sensor nodes. Therefore, characterizing the overall performance is hard, even for networks of moderate size.

Because of these difficulties, to obtain useful insights into the detection performance of large scale networks, we resort to asymptotics. In the Neyman-Pearson framework, one can focus on minimizing the error exponent.

$$\limsup_{n\to\infty} \frac{1}{n} \log \beta_n, \tag{3.1}$$

where β_n is the Type II error probability at the fusion center and n is the number of sensors, while keeping the Type I error probability less than some given threshold. Suppose that $f_i(n)$, $i = 1, \ldots, N$, are functions taking positive values. Then, it is easy to see that

$$\limsup_{n\to\infty} \frac{1}{n} \log \sum_{i=1}^{N} f_i(n) = \limsup_{n\to\infty} \frac{1}{n} \log \max_{1\le i\le N} f_i(n).$$

Therefore, by minimizing the error exponent (3.1), we also approximately minimize the probability of a "dominant" error event (one can think of a Type II error event as the disjoint union of several smaller error events, one of which has the highest probability of occurrence). In the asymptotic regime of large n, an efficient system design strives to minimize the probability of the most likely of the "rare" error events, and this motivates a design that minimizes (3.1). As we will see, studying the error exponent also makes the problem tractable, and produces elegant results that provide useful insights into the original problem.

The rest of this chapter is organized as follows. We describe the basic model, and introduce some concepts and notations in Section 3.2. In Section 3.3, we characterize the performance of the parallel configuration, then in

Section 3.4, we consider more general tree networks with bounded height. We study the impact of sensor failures and unreliable communications on the detection performance in Section 3.5. Finally, we offer some concluding remarks in Section 3.6.

3.2 The Basic Model

We now introduce the basic model, notations and assumptions made in this chapter. We consider a decentralized binary detection problem involving $n-1$ sensors and a fusion center; we will be interested in the case when n increases to infinity. We are given two probability spaces $(\Omega, \mathcal{F}, \mathbb{P}_0)$ and $(\Omega, \mathcal{F}, \mathbb{P}_1)$, associated with two hypotheses H_0 and H_1. We use E_j to denote the expectation operator with respect to $\mathbb{P}_j$. Each node v observes a random variable X_v taking values in some set $\mathcal{X}$. Under either hypothesis H_j, $j = 0, 1$, the random variables X_v are i.i.d., with marginal distribution $\mathbb{P}_j^X$.

Our main goal is to characterize the optimal performance (over all transmission strategies) of a tree network of sensors, under an appropriate asymptotic performance criterion. We consider tree configurations with a bounded height h. The parallel configuration is a special case of a tree, with height $h = 1$. We first define formally a tree network below.

3.2.1 Tree Networks

We use a directed tree $T_n = (V_n, E_n)$ to represent the sensor network. Here, V_n is the set of nodes, of cardinality n, and E_n is the set of directed arcs of the tree. One of the nodes (the "root") represents the fusion center, and the remaining $n-1$ nodes represent the remaining sensors. We use the special symbol f to denote the root of T_n. The arcs in E_n are oriented so that they all point towards the root or the fusion center.

A node u is a *predecessor* of node v if there exists a directed path from u to v. In this case, we also say that v is a *successor* of u. An *immediate predecessor* of node v is a node u such that $(u, v) \in E_n$. An immediate successor is similarly defined. Let the set of immediate predecessors of v be $C_n(v)$. If v is a leaf node, $C_n(v)$ is naturally defined to be empty. The *length* of a path is defined as the number of arcs in the path. The *height* of the tree T_n is the length of the longest path from a leaf to the root, and will be denoted by h_n.

Since we are interested in asymptotically large values of n, we consider a *sequence* of trees $(T_n)_{n\geq 1}$. While we could think of the sequence as representing the evolution of the network as sensors are added, we do not require the sequence E_n to be an increasing sequence of sets; thus, the addition of a new sensor to T_n may result in some arcs being deleted and some new arcs being added. We define the height of a sequence of trees to be $h = \sup_{n\geq 1} h_n$. We are interested in tree sequences of bounded height, i.e., $h < \infty$. For a tree with height h, we say that a node is at *level* k if it is connected to the fusion center via a path of length $h - k$. Hence the fusion center f is at level h.

Let $l_n(v)$ be the number of leaves of the sub-tree rooted at the node v. Thus, $l_n(f)$ is the total number of leaves.

3.2.2 Strategies

Consider a node v, other than the fusion center f. The node v receives messages $Y_{u,n}$ from its immediate predecessors $u \in C_n(v)$. Because of capacity or other cost constraints, node v can only transmit a summary of its received messages and its own observation X_v, if any. It uses a transmission function γ_v to form its message $Y_{v,n} = \gamma_v(X_v, \{Y_{u,n} : u \in C_n(v)\})$. Let all messages be symbols in a fixed alphabet $\mathcal{T}$. Thus, if the number of immediate predecessors of v is $|C_n(v)| = d$, then the transmission function γ_v maps $\mathcal{X} \times \mathcal{T}^d$ to $\mathcal{T}$. We also assume that for each $d \geq 0$, we are given a set of transmission functions $\Gamma(d)$ that the sensor v can choose from. For convenience, we denote $\Gamma(0)$ by Γ. This is the set of transmission functions available to leaf nodes. We assume that all transmissions are perfectly reliable, unless there is a statement to the contrary.

The role of the fusion center f is to make a decision on the true hypothesis, based on the messages it receives from its immediate predecessors. Suppose that it has d immediate predecessors. Recall that in centralized Neyman-Pearson detection, randomization can reduce the Type II error probability. Hence, we assume that the fusion center has access to a random variable, which is uniformly distributed in $[0, 1]$, and independent of everything else. The fusion center uses a randomized fusion rule $\gamma_f : \mathcal{T}^d \times [0, 1] \mapsto \{0, 1\}$ to make a decision. Let $Y_{f,n}$ be a binary-valued random variable indicating the decision of the fusion center.

A *strategy* $\gamma^{(n)}$ consists of a collection of transmission functions, one for each sensor, and a fusion rule for the fusion center. Strategies in which only the leaves make observations will be of special interest to us. In such a scenario, every other node v simply fuses the messages it has received, and forwards a message $Y_{v,n} = \gamma_v(\{Y_{u,n} : u \in C_n(v)\})$ to its immediate successor. We call a strategy of this type a *relay strategy*. A tree network in which we restrict to relay strategies will be called a *relay tree*. Finally, in a relay tree, nodes other than the root and the leaves will be called *relay nodes*.

3.2.3 Neyman-Pearson Hypothesis Testing

In Neyman-Pearson hypothesis testing, we require that the Type I error probability $\mathbb{P}_0(Y_{f,n} = 1)$ be no more than a given $\alpha \in (0, 1)$. A strategy $\gamma^{(n)}$ is said to be *admissible* if it meets this constraint. We define $\beta^*(T_n)$ as the infimum of $\mathbb{P}_1(Y_{f,n} = 0)$, over all admissible strategies. Similarly, we define $\beta_R^*(T_n)$ as the infimum of $\mathbb{P}_1(Y_{f,n} = 0)$, over all admissible relay strategies. Typically, $\beta^*(T_n)$ or $\beta_R^*(T_n)$ will converge to zero as $n \to \infty$. Our goal is to determine if such convergence takes place exponentially fast, and to characterize the Type II error exponent, defined by

$$g^* = \limsup_{n\to\infty} \frac{1}{n} \log \beta^*(T_n), \qquad g_R^* = \limsup_{n\to\infty} \frac{1}{l_n(f)} \log \beta_R^*(T_n).$$

For a relay tree, g_R^* is defined using $l_n(f)$ instead of n due to the fact that only the leaves make observations. Therefore, g_R^* measures the rate of error decay per observation. In the case of a parallel configuration, we use the special notation g_P^* to denote the error exponent.

For any $\gamma \in \Gamma$, let $\mathbb{P}_j^\gamma = \mathbb{P}_j^X \circ \gamma^{-1}$ be the probability law of $\gamma(X)$, and let the Kullback-Leibler divergences be

$$\bar{x}_{0,\gamma} = \mathrm{E}_0\Big[\log \frac{\mathrm{d}\mathbb{P}_1^\gamma}{\mathrm{d}\mathbb{P}_0^\gamma}\Big], \qquad \bar{x}_{1,\gamma} = \mathrm{E}_1\Big[\log \frac{\mathrm{d}\mathbb{P}_1^\gamma}{\mathrm{d}\mathbb{P}_0^\gamma}\Big].$$

It is well known that $\bar{x}_{0,\gamma} \le 0 \le \bar{x}_{1,\gamma}$ [24]. Moreover, both inequalities are strict as long as the measures $\mathbb{P}_j^\gamma$ are not indistinguishable. We will make the following assumptions throughout this chapter. However, for our results to hold, Assumption 2 can be weakened somewhat; see [25].

Assumption 1 *The measures $\mathbb{P}_0^X$ and $\mathbb{P}_1^X$ are equivalent, i.e., they are absolutely continuous w.r.t. each other. Furthermore, there exists some $\gamma \in \Gamma$ such that $\bar{x}_{0,\gamma} < 0 < \bar{x}_{1,\gamma}$.*

Assumption 2 $E_0\big[\log^2 \frac{\mathrm{d}\mathbb{P}_1^X}{\mathrm{d}\mathbb{P}_0^X}\big] < \infty$.

Suppose that the node v sends a message $Y_{v,n} = y$ to its immediate successor. Let the log-likelihood ratio of the message sent by v be

$$\mathcal{L}_{v,n}(y) = \log \frac{\mathrm{d}\mathbb{P}_{1,n}^{(v)}}{\mathrm{d}\mathbb{P}_{0,n}^{(v)}}(y),$$

where $\mathrm{d}\mathbb{P}_{1,n}^{(v)}/\mathrm{d}\mathbb{P}_{0,n}^{(v)}$ is the Radon-Nikodym derivative of the distribution of $Y_{v,n}$ under H_1 w.r.t. the distribution under H_0. If the transmission alphabet $\mathcal{T}$ is a discrete set, then this is just the ratio

$$\log \frac{\mathbb{P}_1(Y_{v,n} = y)}{\mathbb{P}_0(Y_{v,n} = y)}.$$

We will make extensive use of the following class of transmission functions. Recall that $l_n(v)$ is the number of leaves of the sub-tree rooted at node v.

Definition 1 *A (1-bit) Log-Likelihood Ratio Quantizer (LLRQ) with threshold t for a non-leaf node v, with $|C_n(v)| = d$ immediate predecessors, is a binary-valued function on $\mathcal{T}^d$, defined by*

$$\text{LLRQ}_{d,t}\big(\{y_u : u \in C_n(v)\}\big) = \begin{cases} 0, \text{ if } x \le t, \\ 1, \text{ if } x > t, \end{cases}$$

where

$$x = \frac{1}{l_n(v)} \sum_{u \in C_n(v)} \mathcal{L}_{u,n}(y_u).$$

Note that if a node v uses a LLRQ, it ignores its own observation X_v and acts as a relay. If all non-leaf nodes use a LLRQ, we have a special case of a relay strategy. We assume that LLRQs are available choices of transmission functions for all non-leaf nodes. As we will see, LLRQs will play an important role in our results.

Assumption 3 *For all* $t \in \mathbb{R}$ *and* $d > 0$, $\text{LLRQ}_{d,t} \in \Gamma(d)$.

For simplicity, we define the sum of the log-likelihood ratios of the received messages at node v, as follows:

$$S_n(v) = \sum_{u \in C_n(v)} \mathcal{L}_{v,n}(Y_{v,n}).$$

3.3 The Parallel Configuration

We consider here the special case of a network with a parallel configuration. The following proposition shows that the Type II error probability falls exponentially fast with the number of nodes n. Moreover, an asymptotically optimal strategy consists of using identical transmission functions for each sensor. A proof can be found in [26].

Proposition 1 *If Assumptions 1-2 hold, then*

$$g_p^* = \inf_{\gamma \in \Gamma} \bar{x}_{0,\gamma}.$$

Moreover, the error exponent stays the same if we restrict all sensors to using the same transmission function.

The quantity $-\bar{x}_{0,\gamma}$ may be recognized as the Kullback-Leibler divergence, which measures the "discrimination" between the two probability measures $\mathbb{P}_0^\gamma$ and $\mathbb{P}_1^\gamma$. Hence, the asymptotically optimal strategy is to choose a transmission function that produces the greatest discrimination between the two hypotheses.

3.4 Tree Architectures

In this section, we consider general tree networks with a bounded height. Our objective is to study g^* and g_R^* for different sequences of trees. Since the fusion center of a parallel configuration can simulate a relay network (by carrying out all the operations taking place at each relay node, internally in the fusion center), we have

$$g_P^* \leq g_R^*. \tag{3.2}$$

Recall that $l_n(f)$ is the number of leaves in the network. Let

$$z = \liminf_{n\to\infty} \frac{l_n(f)}{n}$$

be the proportion of nodes that are leaves. By comparing the performance to a centralized system where all raw observations are transmitted directly to the the fusion center, and using a similar argument as above, we obtain the first inequality in the expression below,

$$\bar{x}_0 = \mathrm{E}_0\big[\log \frac{\mathrm{d}\mathbb{P}_1^X}{\mathrm{d}\mathbb{P}_0^X}\big] \leq g^* \leq z g_R^*. \tag{3.3}$$

The second inequality follows because an optimal strategy is at least as good as an optimal relay strategy; the factor z arises because we have normalized g_R^* by $l_n(f)$ instead of n.

In the following, we provide a method to propagate error bounds along a tree network, and derive upper bounds similar to that in Cramér's Theorem for the parallel configuration [27]. We consider specifically a h-uniform tree, defined as follows.

Definition 2 (h-uniform tree) *A tree T_n is said to be h-uniform if the length of every path from a leaf to the root is exactly h. A sequence of trees $(T_n)_{n\geq 1}$ is said to be h-uniform if there exists some $n_0 < \infty$, so that for all $n \geq n_0$, T_n is h-uniform.*

It turns out that it is easier to work with h-uniform trees, and as shown on page 83, a height uniformization procedure can be performed on any given tree. Moreover, the detection performance of this height uniformized tree cannot be better than the original tree.

3.4.1 Error Bounds for h-Uniform Relay Trees

We consider the special case of a 1-bit h-uniform relay tree, in which all relay nodes at level k use a LLRQ with a common threshold t_k. Let $t^{(k)} = (t_1, t_2, \ldots, t_k)$, for $k \geq 1$, and $t^{(0)} = \emptyset$. For $j = 0, 1$, $k \geq 1$, and $\lambda \in \mathbb{R}$, we define recursively

$$\Lambda_{j,0}(\gamma;\lambda) = \Lambda_{j,0}(\gamma,\emptyset;\lambda) = \log \mathrm{E}_j\Big[\Big(\frac{\mathrm{d}\mathbb{P}_1^\gamma}{\mathrm{d}\mathbb{P}_0^\gamma}\Big)^\lambda\Big],$$

$$\Lambda_{j,k}^*(\gamma,t^{(k)}) = \sup_{\lambda\in\mathbb{R}}\big\{\lambda t_k - \Lambda_{j,k-1}(\gamma,t^{(k-1)};\lambda)\big\}, \tag{3.4}$$

$$\Lambda_{j,k}(\gamma,t^{(k)};\lambda) = \max\big\{-\Lambda_{1,k}^*(\gamma,t^{(k)})(j+\lambda), \Lambda_{0,k}^*(\gamma,t^{(k)})(j-1+\lambda)\big\}. \tag{3.5}$$

Here, $\Lambda_{j,k}^*(\gamma,t^{(k)})$ is the Fenchel-Legendre transform of $\Lambda_{j,k-1}$ [27], and can be visualized as in Figure 3.2. We will be interested in the case where

$$\bar{x}_{0,\gamma} < 0 < \bar{x}_{1,\gamma}, \tag{3.6}$$

$$t_1 \in (\bar{x}_{0,\gamma}, \bar{x}_{1,\gamma}), \tag{3.7}$$

$$t_k \in (-\Lambda_{1,k-1}^*(\gamma,t^{(k-1)}), \Lambda_{0,k-1}^*(\gamma,t^{(k-1)})), \text{ for } 1 < k \leq h. \tag{3.8}$$

The reader is referred to [1] for an argument that shows the above requirements on the thresholds t_k to be feasible.

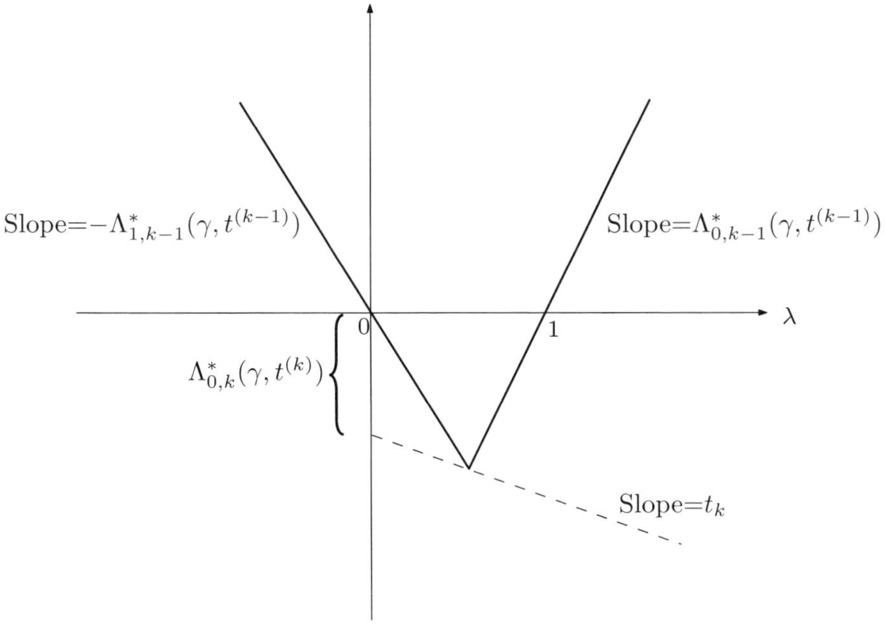

Fig. 3.2. Typical plot of $\Lambda_{0,k-1}(\gamma,t^{(k-1)};\lambda)$, $k \geq 2$.

Proposition 2 below shows that the Type I and II error exponents are essentially upper bounded by $-\Lambda_{0,h}^*(\gamma,t^{(h)})$ and $-\Lambda_{1,h}^*(\gamma,t^{(h)})$ respectively. Note that we recover the classical Chernoff bound when the network has height $h = 1$, i.e., the network is a parallel configuration. Let $p_n(v)$ be the total number of predecessors of v, i.e., the total number of nodes in the sub-tree rooted at v, not counting v itself. Thus, $p_n(f) = n - 1$. Recall that that for a given h-uniform sequence of trees $(T_n)_{n\geq 1}$, there exists a n_0 such that for all $n \geq n_0$, T_n is h-uniform.

Proposition 2 *Fix some $h \geq 1$, and consider an h-uniform sequence of trees. Suppose that Assumptions 1-2 hold. Suppose that, for every n, every leaf node uses the same transmission function $\gamma \in \Gamma$, which satisfies (3.6), and that every level k node ($k \geq 1$) uses a LLRQ with threshold t_k, satisfying (3.7)-(3.8).*

(i) For all nodes v of level $k \geq 1$ and for all $n \geq n_0$,

$$\frac{1}{l_n(v)} \log \mathbb{P}_1\Big(\frac{S_n(v)}{l_n(v)} \leq t_k\Big) \leq -\Lambda^*_{1,k}(\gamma, t^{(k)}) + \frac{p_n(v)}{l_n(v)} - 1,$$
$$\frac{1}{l_n(v)} \log \mathbb{P}_0\Big(\frac{S_n(v)}{l_n(v)} > t_k\Big) \leq -\Lambda^*_{0,k}(\gamma, t^{(k)}) + \frac{p_n(v)}{l_n(v)} - 1.$$

(ii) Suppose that for all $n \geq n_0$ and for all level 1 nodes v, we have $l_n(v) \geq N$. Then, for all $n \geq n_0$, we have

$$\frac{1}{l_n(f)} \log \mathbb{P}_1\Big(\frac{S_n(f)}{l_n(f)} \leq t_h\Big) \leq -\Lambda^*_{1,h}(\gamma, t^{(h)}) + \frac{h}{N},$$
$$\frac{1}{l_n(f)} \log \mathbb{P}_0\Big(\frac{S_n(f)}{l_n(f)} > t_h\Big) \leq -\Lambda^*_{0,h}(\gamma, t^{(h)}) + \frac{h}{N}.$$

3.4.2 Optimal Error Exponent

The following proposition shows that Type II error probabilities decay exponentially (the error exponents are negative). The bounded height assumption is crucial for this result. Indeed, for the case of a tandem configuration, the error probability seems to decay at a sub-exponential rate [28]. The following is proved in [29].

Proposition 3 *Consider a sequence of trees of height h, and suppose Assumptions 1-3 hold. Then,*

$$-\infty < g_P^* \leq g_R^* < 0 \qquad \textit{and} \qquad -\infty < \bar{x}_0 \leq g^* < 0.$$

¿From (3.2), we have $g_P^* \leq g_R^*$, i.e., a relay network performs at best as well as a parallel configuration. Next, we want to know when a relay network has the same error exponent as a parallel configuration. A proof of the following proposition can be found in [29].

Proposition 4 *Consider a sequence of trees of height h in which $z = 1$. Suppose that Assumptions 1-3 hold. Then,*

$$g_P^* = g^* = g_R^*.$$

Furthermore, if the sequence of trees is h-uniform, the optimal error exponent does not change even if we restrict to relay strategies in which every leaf uses the same transmission function and all other nodes use a 1-bit LLRQ with the same threshold.

Proposition 4 is surprising as it establishes that the performance of every network possessing certain qualitative properties is comparable to that of a parallel configuration. This result has important ramifications: suppose that all nodes are restricted to be at most h hops away from the fusion center, then a system designer can reduce the energy consumption (e.g., by employing a h-hop spanning tree that minimizes the overall energy consumption), without losing detection efficiency, as long as the number of nodes n is large, and the proportion of leaf nodes is large. For example, consider the case of nodes uniformly distributed in a square as in Figure 3.1, and suppose that the cost of transmitting a message from one node to another is proportional to the Euclidean distance between the nodes. It is well known that finding a h-hop constrained Minimum Spanning Tree (MST) is NP-hard (see [30]). However, heuristics that achieve a cost of the same order of magnitude as the h-hop constrained MST can be employed to design a suitable network architecture [31]. This involves dividing the square into suitable sub-squares, and it can be verified that with high probability, as $n \to \infty$, the network we obtain has the property that $z = 1$.

If a sequence of trees is h-uniform and $z = 1$, it can be shown that the following simple relay strategy ϵ-achieves the optimal error exponent:

1. all leaf nodes transmit with the same transmission function $\gamma \in \Gamma$, such that $\bar{x}_{0,\gamma} \leq g_P^* + \epsilon/2$;
2. all other nodes use a 1-bit LLRQ with the same threshold $t = \bar{x}_{0,\gamma} + \epsilon/2$.

This is a convenient strategy since only leaf nodes need to make observations, while all the rest of the nodes act as relay nodes. Moreover, transmitting only 1 bit is sufficient, and all relay nodes use the same 1-bit LLRQ. This may be useful in situations where the nodes are simple, low-cost devices.

For a general sequence of trees with height h, we can perform a height uniformization procedure to obtain an h-uniform sequence of trees as follows. We let $A_n \subset V_n$ be the set of nodes whose immediate predecessors include leaves of the tree T_n.

Height Uniformization Procedure. Consider a tree $T_n = (V_n, E_n)$ of height h, and a node v that has at least one leaf as an immediate predecessor (i.e., $v \in A_n$). Let D_n be the set of leaves that are immediate predecessors of v, and whose paths to the fusion center f are of length $k < h$. Add $h - k$

nodes, $\{u_j : j = 1, \ldots, h-k\}$, to V_n; remove the edges (u, v), for all $u \in D_n$; add the edges (u_1, v), and (u_{j+1}, u_j), for $j = 1, \ldots, h-k-1$; add the edges (u, u_{h-k}), for all $u \in D_n$. This procedure is repeated for all $v \in A_n$. The resulting tree is h-uniform. □

It is clear that any strategy on the height uniform tree can be simulated by a strategy on the original tree. Furthermore, it can be shown that the height uniformization procedure preserves the property that $z = 1$ [29]. As the height uniformized tree sequence cannot perform better than the original tree sequence, using the strategy as described above for h-uniform trees is an ϵ-optimal strategy for the original tree sequence.

Next, we want to consider when the sufficient condition $z = 1$ is also a necessary condition for a relay network to have the same asymptotically optimal performance as a parallel configuration. Non-trivial necessary conditions for the equality $g_R^* = g_P^*$ to hold are, in general, difficult to obtain, because they depend on the nature of the transmission functions available to the sensors. Suppose that sensors are allowed to simply forward undistorted all of the messages that they receive, then the equality $g_R^* = g_P^*$ holds trivially. Hence, we need to impose some restrictions on the set of transmission functions available, as in the assumption that follows. Let B_n be the set of nodes all of whose predecessors are leaves.

Assumption 4

(a) There exists a $n_0 \geq 1$ such that for all $n \geq n_0$, we have $l_n(v) > 1$ for all $v \in B_n$.

(b) Let $X_1, X_2, \ldots$ be i.i.d. random variables under either hypothesis H_j, each with distribution $\mathbb{P}_j^X$. For $k > 1$, $\gamma_0 \in \Gamma(k)$, and $\gamma_i \in \Gamma$, $i = 1, \ldots, k$, let $\xi = (\gamma_0, \ldots, \gamma_k)$. Let ν_j^ξ be the distribution of $\gamma_0(\gamma_1(X_1), \ldots, \gamma_k(X_k))$ under hypothesis H_j. We assume that

$$g_P^* = \inf_{\gamma \in \Gamma} \bar{x}_{0,\gamma} < \inf_{\xi \in \Gamma(k) \times \Gamma^k} \frac{1}{k} E_0 \Big[\log \frac{d\nu_1^\xi}{d\nu_0^\xi} \Big], \tag{3.9}$$

for all $k > 1$.

Assumption 4 holds in most cases of interest. There is no loss of generality in assuming part (a), because if in a relay tree we have $l_n(v) = 1$ for some $v \in B_n$, we can remove the predecessor of v, and treat v as a leaf sensor. As for part (b), it is easy to see that the L.H.S. of (3.9) is always less than or equal to the R.H.S., hence we have only excluded those cases where (3.9) holds with equality. We are essentially assuming that when the messages $\gamma_1(X_1), \ldots, \gamma_k(X_k)$ are summarized (or quantized) by γ_0, there is some loss of information, as measured by the associated Kullback-Leibler divergences.

Proposition 5 *Suppose that Assumptions 1-4 hold. Then, $g_R^* = g_P^*$ iff $z = 1$.*

3.5 Unreliable Networks

So far, we have assumed that all nodes are error-free, and all communications are reliable. We next study the impact of failure-prone sensors and unreliable communications on the detection performance.

3.5.1 Sensor Failures

We model the case of sensor failures by using a random number of nodes. Variants of the decentralized detection problem with a random number of nodes distributed in a parallel configuration have been studied in [32, 33, 34]. In [32] and [34], the authors consider the case of spatially correlated signals, and analyze the detection performance of a simple but suboptimal strategy. In [33], the objective is not to find an optimal transmission strategy. Rather, the authors assume that nodes in a parallel configuration make i.i.d. observations under either hypothesis, quantize their observations using some known quantizer that is identical for all nodes, and use a special multiple access protocol called Type-Based Random Access (TBRA) (in all problems so far, we have implicitly assumed some sort of orthogonal multiple access protocol in which messages from different nodes do not corrupt one another, whereas in TBRA, messages are combined additively over the transmission medium). In this section, our goal is to characterize the asymptotically optimal performance for tree networks with bounded height, and to develop an optimal transmission strategy, assuming i.i.d. observations and the usual orthogonal multiple access protocol. The results in this section are a summary of those in [2].

To model sensor failures, we construct a random tree as in a Galton-Watson process, but with a limited 'time span' of h (which corresponds to our tree having a height h). We start with the fusion center f, and let the number of immediate predecessors of f be a random variable $N_f = |C(f)|$, with distribution law μ_h. Then, we let each node v in the random set $C(f)$ have $N_v = |C(v)|$ immediate predecessors, where N_v has marginal law μ_{h-1}. We continue this process until the level 0 nodes are reached. Hence, each level k node v (with $k \geq 1$) has N_v immediate predecessors, where N_v is a random variable with law μ_k. Furthermore, we also assume that all these random variables are independent, and independent of the hypothesis. We call such a random tree a GW-tree.

We want to model a dense network, therefore we consider the case when each of the laws μ_k, $k = 1, \ldots, h$, has asymptotically large mean. Let λ_k be the mean of μ_k. We let $\lambda^* = \min_{1 \leq k \leq h} \lambda_k$ increase to infinity, by allowing the laws μ_k to vary accordingly. However, we require that the distributions satisfy the following assumption.

Assumption 5 *Let $\tilde{N}_k$ be random variables with distribution μ_k, $k = 1, \ldots, h$. We have*

$$E[\tilde{N}_k^2] = (1 + o(1))\lambda_k^2, \tag{3.10}$$

where $o(1)$ stands for a term that goes to 0 as $\lambda_k \to \infty$.

It is easy to check that both the Poisson distribution and Binomial distribution satisfy the above assumption. Under Assumption 5, Chebychev's inequality shows that the distribution of $\tilde{N}_k$ is clustered around its mean.

Lemma 1 *For all $\eta > 0$, $\mathbb{P}(|\tilde{N}_k/\lambda_k - 1| > \eta) \to 0$ as $\lambda_k \to \infty$.*

Let β^*_{GW} be the infimum of the Type II error probability, with the minimization taken over all strategies (to be more precise, we have to consider a family of strategies, see [2]), subject to the constraint that the Type I error probability is not more than $\alpha \in (0, 1)$. Our goal is to characterize the optimal error exponent

$$\limsup_{\lambda^* \to \infty} \frac{1}{\lambda(h)} \log \beta^*_{GW},$$

where $\lambda(h) = \prod_{k=1}^{h} \lambda_k$ is the expected number of nodes.

Let the log-likelihood ratio of the received messages at v be S_v. Motivated by the ϵ-optimal strategies for non-random tree networks, it is natural to define the following class of transmission policies.

Definition 3 *A transmission function for a level k node v is called a Mean-normalized Log-Likelihood Ratio (MLLR) quantizer at level k with threshold t if*

$$Y_v = \begin{cases} 0, & \text{if } S_v/\lambda(k) \le t, \\ 1, & \text{otherwise.} \end{cases}$$

Again, we assume that MLLR quantizers are valid quantizers for each node.

Assumption 6 *Every node of level $k \geq 1$ has access to MLLR quantizers.*

In the case where there are no sensor failures, i.e., $N_v = \lambda_k$ a.s. for all non-leaf nodes v, Proposition 4 shows that the Type II error probability decays exponentially fast with $\lambda(h)$, at rate g^*_P. The proposition below shows that this remains true for a GW-tree.

Proposition 6 *Suppose that Assumptions 1, 2, 5 and 6 hold. Then, for all $\alpha \in (0, 1)$, the optimal error exponent of a GW-tree of height h is given by*

$$\lim_{\lambda^* \to \infty} \frac{1}{\lambda(h)} \log \beta^*_{GW} = g^*_P. \tag{3.11}$$

Furthermore, for any $\epsilon \in (0, -g^*_P)$, and any large enough λ^*, the following strategy satisfies the Type I error probability constraint, and its error exponent is bounded above by $g^*_P + \epsilon$:

(i) *each leaf uses the same transmission function* $\gamma \in \Gamma$, *with* $\bar{x}_{0,\gamma} \leq g_P^* + \epsilon/2 < 0$; *and*

(ii) *for* $k \geq 1$, *every level* k *node uses a MLLR quantizer with threshold* $t_k = \bar{x}_{0,\gamma} + \epsilon/2^{h-k+1}$.

Consider the scenario where each node in the network may fail, independently, with some probability p. A network of n failure-prone sensors cannot be better (on the average) than a network of np failure-proof sensors. However, in the limit of large n, Proposition 6 shows that the asymptotically optimal performance of both networks is essentially the same.

3.5.2 Unreliable Communications

We now consider the case where each sensor in a tree of height h is constrained to sending one bit to its immediate successor, and the channel between any two nodes is a binary symmetric channel (BSC) with known crossover probability $\eta \in (0, 1/2)$. Suppose that $(T_n)_{n\geq 1}$ is a h-uniform tree sequence. For simplicity, we discuss only relay strategies in this section. For every non-leaf node v, we assume that $|C_n(v)| \geq c_n$, for some $c_n \to \infty$ as $n \to \infty$. This models a dense sensor network with bounded height h. We again consider the problem of minimizing the Type II error exponent, with the Type I error constrained to be no larger than $\alpha \in (0, 1)$. However, in this case, the appropriate error exponent to consider, as we will see later in Proposition 8, is

$$\limsup_{n\to\infty} \frac{1}{|C_n(f)|} \log \beta^*(T_n),$$

where we have normalized the error exponent by $|C_n(f)|$, the number of immediate predecessors of the fusion center, instead of the total number of nodes n.

Consider an immediate predecessor u of the node v. The node u transmits a 1-bit message $Y_{u,n} = y_u$ over the BSC to node v. Let $\bar{Y}_{u,n} = \bar{y}_u$ be the received message at node v. We now define LLRQs as in Definition 1, but with respect to the received messages $\bar{Y}_{u,n}$, i.e., the likelihood ratios are replaced with the Radon-Nikodym derivative of the distribution of $\bar{Y}_{u,n}$ under H_1 with respect to that under H_0.

Let us first consider the simple case when $h = 1$, i.e., the parallel configuration. For each $\gamma \in \Gamma$ and $j = 0, 1$, we define the probability measures $\mathbb{Q}_j^\gamma$ on the space $\{0, 1\}$ as follows. Let

$$\mathbb{Q}_j^\gamma(\{0\}) = (1-\eta)\mathbb{P}_j^\gamma(\{0\}) + \eta\mathbb{P}_j^\gamma(\{1\}),$$
$$\mathbb{Q}_j^\gamma(\{1\}) = (1-\eta)\mathbb{P}_j^\gamma(\{1\}) + \eta\mathbb{P}_j^\gamma(\{0\}).$$

Then, the following proposition is a consequence of Proposition 1.

Proposition 7 *Suppose that Assumptions 1-3 hold. For $h = 1$, and all $\alpha \in (0,1)$, the optimal error exponent is*

$$\lim_{n\to\infty} \frac{1}{|C_n(f)|} \log \beta^*(T_n) = \inf_{\gamma\in\Gamma} E^{\mathbb{Q}_0^\gamma}\big[\log \frac{d\mathbb{Q}_1^\gamma}{d\mathbb{Q}_0^\gamma}\big],$$

where $E^{\mathbb{Q}_0^\gamma}$ is the expectation operator under $\mathbb{Q}_0^\gamma$. Furthermore, there is no loss in optimality if we restrict all the leaf nodes in $C_n(f)$ to using the same transmission function $\gamma \in \Gamma$.

The optimal error exponent in the case $h \geq 2$ is markedly different from that in Proposition 7. Let Bern(η) denote the Bernoulli distribution on $\{0,1\}$ that takes the value 1 with probability η. The following proposition is proved in [2].

Proposition 8 *Suppose that Assumptions 1-3 hold. For $h \geq 2$, and for all $\alpha \in (0,1)$, the optimal error exponent is*

$$\begin{aligned}\lim_{n\to\infty} \frac{1}{|C_n(f)|} \log \beta^*(T_n) &= -\big(\eta \log \frac{\eta}{1-\eta} + (1-\eta)\log\frac{1-\eta}{\eta}\big)\\ &= -D(\eta) < 0, \end{aligned} \tag{3.12}$$

where $D(\eta)$ is the Kullback-Leibler divergence function of Bern($1-\eta$) w.r.t. Bern(η).

Again, it can be shown that a strategy that ϵ-achieves the optimal error exponent is the following:

(i) All leaves use the same transmission function $\gamma \in \Gamma$, where γ is chosen so that $\mathbb{P}_0(\gamma(X) = 0) \neq \mathbb{P}_1(\gamma(X) = 0)$.
(ii) Every level 1 node uses a LLR with threshold 0.
(iii) All other nodes use the majority rule: send a 1 if and only if more than half of the received messages are equal to 1.
(iv) The fusion center uses a LLR with threshold $t = -D(\eta) + \epsilon$.

Compared to the result in Proposition 3, the above proposition shows that the detection performance of a h-uniform relay tree network of height at least 2, in the presence of unreliable communications, is significantly worse than that of a similar network with reliable communications. Indeed, in the case of unreliable communications, the error probability decays exponentially fast with $|C_n(f)|$, instead of n.

3.6 Conclusions

We have considered Neyman-Pearson decentralized detection in sensor networks with tree architectures of bounded height. Although the problem of finding exact optimal strategies, and hence of characterizing the optimal detection performance, for a fixed number of nodes n is computationally intractable, the asymptotically optimal performance is surprisingly the same as the well-known detection performance of the parallel configuration, under most practical cases of interest. Indeed, when the leaf nodes dominate, a tree network with bounded height has the same error exponent as a parallel configuration. Under a mild condition, the property that leaf nodes dominate the network is also shown to be a necessary condition for achieving the same optimal error exponent as the parallel configuration.

We have only considered the Neyman-Pearson criterion in this chapter. A similar analysis is possible within a Bayesian framework. It is shown in [35] that error probabilities also decay exponentially fast in a Bayesian setting. However, the same performance as the parallel configuration can no longer be achieved.

We also discussed the impact of sensor failures and unreliable communications on detection performance. Our results provide a useful insight into the performance of height uniform tree sequences. Suppose that all nodes can only send 1-bit messages, and every node, except the leaf nodes, has the same number of immediate predecessors c_n. A network can operate in two modes, in a 'sensor failure' mode and in an 'unreliable communications' mode. In the 'sensor failure' mode, if a sensor determines that its message cannot be received reliably by its intended recipient, it remains silent instead of transmitting. In the 'unreliable communications' mode, it transmits its 1-bit message regardless. Our results indicate that in the 'sensor failure' mode, the error probability decays exponentially with c_n^h, whereas in the 'unreliable communications' mode, it decays exponentially with c_n. Therefore, it is better for the network to operate in the 'sensor failure' mode, when the height of the tree network is greater than one.

Several issues remain outstanding, and are areas for further research. Our results are valid in the large n regime; however, a significantly larger number of nodes may be needed before a relay network can approximate, in a certain sense, the performance of a parallel configuration. Another issue is that although the error exponents are the same, the ratio $\beta^*(T_n)/\beta_P^*$, where β_P^* is the optimal error probability of the parallel configuration, could be diverging to infinity as n increases. Therefore, it is of interest to study the *exact asymptotics* of this problem. Finally, the case where sensor observations are correlated remains a difficult problem. For the case of correlated sensor observations in parallel configurations, the reader is referred to [36, 37] for recent results.

References

[1] W.-P. Tay, J. N. Tsitsiklis, and M. Z. Win, "Data fusion trees for detection: Does architecture matter?" in *Proc. Allerton Conf. on Communi-*

cation, Control, and Computing, Monticello, IL, Sep. 2006.
[2] ——, "Detection in dense wireless sensor networks," in *Proc. IEEE Wireless Commun. and Networking Conf.*, Hong Kong, Mar. 2007.
[3] R. R. Tenney and N. R. Sandell, "Detection with distributed sensors," *IEEE Trans. Aerosp. Electron. Syst.*, vol. 17, pp. 501–510, 1981.
[4] Z. Chair and P. K. Varshney, "Optimal data fusion in multiple sensor detection systems," *IEEE Trans. Aerosp. Electron. Syst.*, vol. 22, pp. 98–101, 1986.
[5] G. Polychronopoulos and J. N. Tsitsiklis, "Explicit solutions for some simple decentralized detection problems," *IEEE Trans. Aerosp. Electron. Syst.*, vol. 26, pp. 282–292, 1990.
[6] A. R. Reibman and L. W. Nolte, "Design and performance comparison of distributed detection networks," *IEEE Trans. Aerosp. Electron. Syst.*, vol. 23, pp. 789–797, 1987.
[7] P. Willett and D. Warren, "The suboptimality of randomized tests in distributed and quantized detection systems," *IEEE Transactions on Information Theory*, vol. 38, pp. 355–361, Mar. 1992.
[8] W. W. Irving and J. N. Tsitsiklis, "Some properties of optimal thresholds in decentralized detection," *IEEE Trans. Autom. Control*, vol. 39, pp. 835–838, 1994.
[9] J. N. Tsitsiklis, "Extremal properties of likelihood-ratio quantizers," *IEEE Trans. Commun.*, vol. 41, pp. 550–558, 1993.
[10] C. Rago, P. Willett, and Y. Bar-Shalom, "Censoring sensors: A low-communication-rate scheme for distributed detection," *IEEE Trans. Aerosp. Electron. Syst.*, vol. 32, no. 2, pp. 554–568, 1996.
[11] S. Appadwedula, V. V. Veeravalli, and D. Jones, "Energy-efficient detection in sensor networks," *IEEE J. Sel. Areas Commun.*, vol. 23, no. 4, pp. 693–702, 2005.
[12] J.-F. Chamberland and V. V. Veeravalli, "Decentralized detection in sensor networks," *IEEE Trans. Signal Process.*, vol. 51, no. 2, pp. 407–416, 2003.
[13] ——, "Asymptotic results for decentralized detection in power constrained wireless sensor networks," *IEEE J. Select. Areas Commun., Special Issue on Wireless Sensor Networks*, vol. 22, no. 6, pp. 1007–1015, 2004.
[14] W.-P. Tay, J. N. Tsitsiklis, and M. Z. Win, "Censoring sensors: Asymptotics and the value of cooperation," in *Proc. Conf. on Inform. Sci. and Sys.*, Princeton, NJ, Mar. 2006, pp. 62–67.
[15] J. N. Tsitsiklis, "Decentralized detection," *Advances in Statistical Signal Processing*, vol. 2, pp. 297–344, 1993.
[16] R. Viswanathan and P. K. Varshney, "Distributed detection with multiple sensors: part I - fundamentals," *Proc. IEEE*, vol. 85, pp. 54–63, 1997.
[17] L. K. Ekchian and R. R. Tenney, "Detection networks," in *Proc. 21st IEEE Conf. Decision Control*, 1982, pp. 686–691.

[18] R. Viswanathan, S. C. A. Thomopoulos, and R. Tumuluri, "Optimal serial distributed decision fusion," *IEEE Trans. Aerosp. Electron. Syst.*, vol. 24, no. 4, pp. 366–376, 1988.
[19] Z. B. Tang, K. R. Pattipati, and D. L. Kleinman, "Optimization of detection networks: part I- tandem structures," *Systems, Man and Cybernetics, IEEE Transactions on*, vol. 21, no. 5, pp. 1044–1059, 1991.
[20] ——, "Optimization of detection networks: part II- tree structures," *Systems, Man and Cybernetics, IEEE Transactions on*, vol. 23, no. 1, pp. 211–221, 1993.
[21] J. D. Papastavrou and M. Athans, "On optimal distributed decision architectures in a hypothesis testing environment," *IEEE Trans. Autom. Control*, vol. 37, no. 8, pp. 1154–1169, 1992.
[22] A. Pete, K. Pattipati, and D. Kleinman, "Optimization of detection networks with multiple event structures," *IEEE Trans. Autom. Control*, vol. 39, no. 8, pp. 1702–1707, 1994.
[23] S. Alhakeem and P. K. Varshney, "A unified approach to the design of decentralized detection systems," *IEEE Trans. Aerosp. Electron. Syst.*, vol. 31, no. 1, pp. 9–20, 1995.
[24] T. A. Cover and J. A. Thomas, *Elements of Information Theory*, 1st ed. New York, NY: John Wiley & Sons, Inc., 1991.
[25] P.-N. Chen and A. Papamarcou, "New asymptotic results in parallel distributed detection," *IEEE Trans. Inf. Theory*, vol. 39, no. 6, pp. 1847–1863, Nov. 1993.
[26] J. N. Tsitsiklis, "Decentralized detection by a large number of sensors," *Math. Control, Signals, Syst.*, vol. 1, pp. 167–182, 1988.
[27] A. Dembo and O. Zeitouni, *Large Deviations Techniques and Applications.* New York, NY: Springer-Verlag, 1998.
[28] W.-P. Tay, J. N. Tsitsiklis, and M. Z. Win, "On the sub-exponential decay of detection probabilities in long tandems," in *Proc. IEEE Int. Conf. Acoustics, Speech, and Signal Processing*, Honolulu, HI, Apr. 2007.
[29] ——, "Data fusion trees for detection: Does architecture matter?" *IEEE Trans. Inf. Theory*, 2006, submitted for publication.
[30] M. R. Garey and D. S. Johnson, *A guide to the theory of NP-Completeness.* New York, NY: W. H. Freeman, 1979.
[31] A. E. F. Clementi, M. D. Ianni, M. A., L. M., R. G., and S. R., "Divide and conquer is almost optimal for the bounded-hop MST problem on random Euclidean instances," in *Proc. Structural Information and Communication Complexity*, Mont Saint-Michel, France, May 2005, pp. 89–98.
[32] R. Niu and P. K. Varshney, "Distributed detection and fusion in a large wireless sensor network of random size," *EURASIP Journal on Wireless Communications and Networking*, vol. 2005, no. 4, pp. 462–472, 2005.
[33] L. T. Animashree Anandkumar, "A large deviation analysis of detection over multi-access channels with random number of sensors," in *Acous-*

tics, Speech and Signal Processing, 2006. ICASSP 2006 Proceedings. 2006 IEEE International Conference on, vol. 4, 2006.

[34] T. Q. S. Quek, D. Dardari, and M. Z. Win, "Energy efficiency of dense wireless sensor networks: to cooperate or not to cooperate," *IEEE J. Sel. Areas Commun.*, vol. 25, pp. 459–469, 2007.

[35] W.-P. Tay, J. N. Tsitsiklis, and M. Z. Win, "Bayesian detection in bounded height tree networks," in *Proc. of Data Compression Conf.*, Snowbird, UT, Mar. 2007.

[36] J.-F. Chamberland and V. V. Veeravalli, "How dense should a sensor network be for detection with correlated observations?" *IEEE Transactions on Information Theory*, vol. 52, no. 11, pp. 5099–5106, Nov. 2006.

[37] W. Li and H. Dai, "Distributed detection in large-scale sensor networks with correlated sensor observations," in *Proc. 43rd Allerton Annual Conference on Communication, Control, and Computing*, Sep. 2005.

4

Function Computation in Wireless Sensor Networks*

Lei Ying, R. Srikant, and Geir E. Dullerud

Coordinated Science Lab
University of Illinois at Urbana-Champaign
Urbana, IL 61801

4.1 Introduction

With the wide availability of inexpensive wireless technology and sensing hardware, wireless sensor networks are expected to become commonplace because of their broad range of potential applications. A wireless sensor network consists of sensors that have sensing, computation and wireless communication capabilities. Each sensor monitors the environment surrounding it, collects and processes data, and when appropriate transmits information so as to cooperatively achieve a global detection objective. One important feature of wireless sensor networks is that the network is often designed for a specific purpose, and the sensors are required to collaborate to achieve a global objective. This is one fundamental distinction between wireless networks used for communication and wireless networks used for sensing. In wireless communication networks, the protocols are designed so that they are not application-specific, and therefore the network can support a constantly evolving set of applications. Contrasting this, in sensor networks, the architecture and protocols can be designed for each specific application, exploiting its structure, to reduce the energy usage within the network. Note that the objectives of sensor networks are to retrieve useful information from sensor measurements, so many of these objectives can be regarded as a function computation of sensor measurements. For example, counting the number of intruders or hot-spots is equivalent to computing a "sum" function, and detecting an abnormal event could be same as computing a threshold function. Recently there has been a lot of interest in function computation in wireless sensor networks. For example, in [5], the authors have designed a block coding scheme to compress the amount of information to be transmitted in a sensor network computing some functions. In [8], [14] and [15], energy consumption of function computation is studied for large scale sensor networks, where the energy consumption of

* The research was supported by a Vodafone Fellowship and NSF Grant CNS 05-19535.

each bit is assumed to be same. Also, several recent works [2], [1] and [16] have discussed using silence to convey information in sensor networks to save energy, and indicate the potential benefit of non-traditional communication in sensor networks.

In this chapter, we consider two different scenarios. In Section 4.2 we consider a multi-hop network with noisy communication channels where the measurement of each sensor consists of one bit; the goal is for the fusion center to compute symmetric functions — those functions determined by the sum of the observed bits. To achieve this, we would like to design a distributed algorithm while minimizing the total transmission energy consumed by the network. Specifically, distributed symmetric function computation with binary data, which is also called a counting problem in this chapter, is as follows: each node is in either state "1" or "0", and the fusion center's goal is to count the number of sensors in state "1". Since the wireless channels are unreliable, we adopt the multi-user diversity idea introduced in [4], where the network is divided into small cells, and the sensors in the same cell cooperate to estimate the number of "1"s in the cell. Since energy consumption is our major concern, we also use data-aggregation to further reduce the energy consumption. Note that only the total number of "1"s instead of the individual measurements of each sensor are needed at the fusion center. Our algorithm forms a tree rooted at the fusion center, and the cell-counting results are aggregated and transmitted along the rooted-tree in a multi-hop fashion. Assume that each sensor uses r^α units of energy to transmit each bit, where r is the transmission range of the sensor. We first show that the transmission energy consumption is at least $\Omega\left(\left(\sqrt{\frac{\log n}{n}}\right)^\alpha\right)$, and then propose a distributed function computation algorithm whose energy consumption is $O\left(n(\log\log n)\left(\sqrt{\frac{\log n}{n}}\right)^\alpha\right)$. The contents of Section 4.2 are a more detailed version of the results presented in [14].

In Section 4.3, we consider a different scenario where wireless channels are assumed to be reliable, but sensors are densely deployed so that the sensor measurements are highly correlated. Due to the broadcast nature of wireless channels, a transmission from one sensor can be heard by other sensors in its neighborhood. We consider a collocated network [5] where a transmission can be heard by all sensors in the network, and investigate the average energy consumption of real-time communication where the sensors send the measurements at each time slot without accumulating block measurements. We show that the exploitation of correlation can lead to a further reduction of energy consumption, and propose a methodology to analyze and design minimum cost communication and computation in wireless sensor networks. The results in Section 4.3 are somewhat preliminary since the proposed algorithm could be computationally very expensive to implement. However, our hope is that, by making a connection to stochastic control and dynamic programming

(DP), one may be able to use results from the vast literature on approximate solutions to DP problems to ease the computational burden.

4.2 Distributed Function Computation in Noisy Wireless Sensor Networks

4.2.1 Notation

The following notation is used throughout this chapter. Given a sequence of random variables $X(n)$ indexed by n, and positive function $f(n)$, we will say that

(i) $X(n) = O(f(n))$, when there exists a positive constant c such that

$$\lim_{n\to\infty} \Pr(X(n) \leq cf(n)) = 1 \text{ holds.}$$

(ii) $X(n) = \Omega(f(n))$ when there exists a positive constant c such that

$$\lim_{n\to\infty} \Pr(X(n) \geq cf(n)) = 1 \text{ holds.}$$

(iii) $X(n) = \Theta(f(n))$ when both $X(n) = \Omega(f(n))$ and $X(n) = O(f(n))$ hold.

Note that the above definitions also apply in the obvious way to deterministic functions.

4.2.2 Model

We consider a random network of n sensors that are uniformly and independently distributed on a unit square. Upon the occurrence of a certain event, sensor k records b_k, where b_k can take a value either "1" or "0." The sensors have the capability to transmit this data over noisy wireless channels, and based on the data transmitted by the sensors in the network, a fusion center tries to evaluate some symmetric function $f(b_1, \ldots, b_n)$, i.e., a function which has the property that

$$f(b_1, \ldots, b_n) = f(\sigma(b_1, \ldots, b_n)),$$

for any permutation σ. Symmetric functions form a large class of functions, which includes almost all statistical functions like max, min, mean, etc. A key property of a symmetric function is that the function value only depends on the frequency-histogram; for the binary measurement case, the function only depends on the total number of "1"s in the network. So in this section, we will design algorithms to count the number of "1"s in the sensors' measurements, i.e., to compute

$$\sum_{i=1}^{n} b_i.$$

Since counting and computation are equivalent for symmetric functions, we will interchangeably use the terms counting and computation in this section.

Let S_i denote the location of sensor i and $|S_i - S_j|$ denote the distance from sensor i to sensor j. We use the protocol model in [7] for wireless interference with some additional assumptions.

(1) All nodes use the same transmission radius r, and the power required to transmit one bit is r^α.
(2) A transmission from sensor i can be received at sensor j only if $|S_i - S_j| \leq r$ and $|S_k - S_j| \geq (1+\Delta)r$ for each sensor $k \neq i$ which transmits at the same time, where Δ is a protocol-specified guard-zone to prevent interference.
(3) A binary modulation scheme is used so that each transmission is either 1 or 0.
(4) Even if a transmission is received at the receiver, there is some probability $p < 1/2$ with which the received bit is flipped, i.e., the channel is a binary symmetric channel with error probability p.

Note that this model only holds when the near-field effects are negligible, which is assumed in this section.

By a counting algorithm, we mean a set of protocols (which may depend on n) to convey the appropriate information from the sensors to the fusion center and a protocol at the fusion center to use the received information to compute the number of "1"s in the network. Given an algorithm for counting, we define the energy required by the algorithm to be the maximum energy required for the computation over all possible values of the measurements. Our goal is to characterize the minimum energy required subject to the constraint that the probability of error in the computation in a random network with n sensors goes to zero as $n \to \infty$. We only consider the transmission energy used for counting, and assume other energy expenditure, such as the energy used for computation, receiving, coordination etc, is negligible.

4.2.3 A Trivial Lower Bound on the Energy Consumption

We note that, for accurate function computation, the network has to be connected. Let $\mathcal{C}(n,r)$ denote the event that a random network with n nodes and common transmission radius r is connected. Now, we introduce the following two connectivity results.

Lemma 1. *Suppose that* $r \leq \frac{1}{6}\sqrt{\frac{\log n}{2n}}$. *Then given* $\epsilon > 0$, *there exists* n_0 *such for any* $n \geq n_0$, *we have*

$$\Pr\left(\mathcal{C}(n,r)\right) \leq \epsilon.$$

Proof. We divide the unit square into small cells with side length $\sqrt{\log n/(2n)}$, and each cell is further divided into nine min-cells as in Figure 4.1. Since

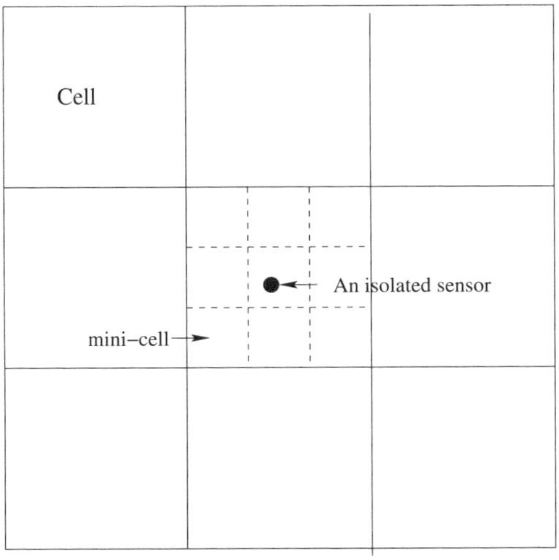

Fig. 4.1. An Example of An Isolated Sensor

$r \leq \frac{1}{6}\sqrt{\frac{\log n}{2n}}$, it is easy to see that the sensors in the central mini-cell of a cell can only communicate with the sensors in the same cell. If a cell contains only one sensor and the sensor is positioned in the central mini-cell, then the sensor is said to be isolated and the cell is said to be an isolated cell. Recall the n nodes are uniformly, independently partitioned in the unit square. To obtain a bound on $\Pr(\mathcal{C}(n,r))$, we first pretend that the number of sensors in the unit square is distributed according to a spatial Poisson process of intensity n. In other words, the number of sensors in an area $\mathcal{A}$ is a Poisson random variable with mean $n\mathcal{A}$. Further, the numbers of sensors in non-overlapping areas are independent. Thus, the number of sensors in a cell is a Poisson random variable with mean $\log n/2$ and the probability that a cell is isolated is

$$\frac{1}{9} \times \frac{\log n}{2\sqrt{n}},$$

where $\log n/(2\sqrt{n})$ is the probability that the number of sensors in the cell is one, and $1/9$ is the probability that the sensor happens to be in the central mini-cell. Note that the probability that the network is connected increases with the number of sensors in the unit square, which we denote by N_p. Thus, we have

$$\begin{aligned}
\Pr(\mathcal{C}(n,r)) &\leq \Pr(\mathcal{C}(N_p,r)|N_p \geq n) \\
&\leq \frac{\Pr(\mathcal{C}(N_p,r))}{\Pr(N_p \geq n)} \\
&\leq 2\Pr(\mathcal{C}(N_p,r)) \\
&\leq 2\left(1 - \frac{1}{18}\frac{\log n}{\sqrt{n}}\right)^{\frac{n}{2\log n}},
\end{aligned}$$

where the third inequality holds since $\Pr(N_p \geq n) \geq 1/2$. It is easy to see that

$$\left(1 - \frac{1}{18}\frac{\log n}{\sqrt{n}}\right)^{\frac{n}{2\log n}}$$

goes to zero when n goes to infinity, so lemma holds.

We would like to comment that in [6], it has been shown that the network is asymptotically disconnected if $r \leq \sqrt{(\log n + c(n))/(\pi n)}$ and $\limsup_n c(n) < \infty$. This lower bound on r is tighter than the one in Lemma 1. However, Lemma 1 captures the lower bound up to the right order, and we have presented the proof is much easier and the result is sufficient for the purpose of the next lemma. To complement the necessary condition for connectivity in Lemma 1, we next present a sufficient condition for connectivity.

Lemma 2. *Suppose that the unit square is partitioned into* $(\lfloor\frac{n}{c_1 \log n}\rfloor)^2$ *square cells, and further let* n_i *denote the number of sensors in cell* i*. Then, for large enough* n*,*

$$Pr\left(\frac{c_1 \log n}{2} \leq n_i \leq 2c_1 \log n \;\; \forall i\right) \geq 1 - \frac{2n^{\left(1-\frac{c_1}{8}\right)}}{c_1 \log n}, \tag{4.1}$$

which implies that if $r \geq 2\sqrt{2}\lfloor\frac{n}{c_1 \log n}\rfloor$*, then*

$$\Pr\left(\mathcal{C}(n,r)\right) \geq 1 - \frac{2n^{\left(1-\frac{c_1}{8}\right)}}{c_1 \log n}. \tag{4.2}$$

Thus, if $c_1 > 8$*,*

$$\lim_{n\to\infty} \Pr(\mathcal{C}(n,r)) = 1.$$

Proof. Consider cell i. Note that the probability that a particular sensor is positioned in the cell is given by $(c_1 \log n)/n$. From the Chernoff bound [11], we have that

$$\Pr(n_i \geq 2c_1 \log n) \leq e^{-c_1 \log n/3}$$

and

$$\Pr\left(n_i \leq \frac{c_1}{2} \log n\right) \leq e^{-c_1 \log n/8},$$

which implies that

$$\Pr\left(\frac{c_1}{2} \log n \leq n_i \leq 2c_1 \log n\right) \geq 1 - 2e^{-c_1 \log n/8}.$$

So from the union bound, we have

$$\Pr\left(\frac{c_1}{2} \log n \leq n_i \leq 2c_1 \log n \;\; \forall i\right) \geq 1 - 2\frac{n}{c_1 \log n} e^{-c_1 \log n/8} = 1 - \frac{2n^{\left(1-\frac{c_1}{8}\right)}}{c_1 \log n}.$$

Assume that the common transmission radius $r \geq 2\sqrt{2}\lfloor\frac{n}{c_1 \log n}\rfloor$, then nodes in neighboring cells can communicate with each other. Thus, the network is connected if there is at least one node in each cell, and inequality (4.2) holds.

Note that a result slightly weaker than the one in Lemma 2 was first obtained in [12].

It is obvious that, in the worst-case when all sensors have a "1," each sensor has to broadcast its value once. Thus, we have the following trivial lower bound on the energy consumption.

Lemma 3 (A Trivial Lower Bound). *The minimum total transmission energy required to count is*

$$\Omega\left(n\left(\sqrt{\frac{\log n}{n}}\right)^{\alpha}\right) \tag{4.3}$$

.

Proof. Connectivity of the network is a necessary condition for correct counting. To guarantee connectivity, it has been shown in Lemma 1 that the transmission range of the sensors should be chosen as $\Omega\left(\sqrt{\frac{\log n}{n}}\right)$. Thus, the energy used per sensor transmission is $\Omega\left(\left(\sqrt{\frac{\log n}{n}}\right)^{\alpha}\right)$. There are n sensors in the network, each of which must make at least one transmission; thus, the total transmission energy required is $\Omega\left(n\left(\sqrt{\frac{\log n}{n}}\right)^{\alpha}\right)$.

4.2.4 An Upper Bound on the Energy Consumption

In this subsection, we propose a counting algorithm whose energy consumption is only a factor of $\log\log n$ more than the lower bound. We first present two well-known results for the reader's convenience. First, we study the error probability win a binary symmetric channel when repetition coding is used. Consider a binary symmetry channel with error probability p where each bit is transmitted m times, and the receiver decodes the data using majority rule. Then we have following well-known bound [3] on the error probability, where the proof is provided for completeness.

Lemma 4. *Suppose one bit of data is transmitted m times over a binary symmetric channel with error probability p, and the receiver decodes the bit using majority rule. Then, the probability of decoding error is no greater than*

$$(4p(1-p))^{\frac{1}{2}m}.$$

Proof. Define m independent binary random variables $\{I_i\}$, where $I_i = 0$ with probability p and $I_i = 1$ with probability $1-p$. Using the Chernoff bound, we have

$$\Pr\left(\sum_{i=1}^{m} I_i < \frac{m}{2}\right) \le e^{\frac{m}{2}\log(4(1-p)p)} = (4p(1-p))^{\frac{1}{2}m}.$$

We also need the following coding theorem [3] for discrete memoryless channels for our analysis.

Theorem 1 (Gallager's Coding Theorem) *For any discrete memoryless channel with capacity C, any positive integer N, and any positive $R < C$, there exist block codes with $M = 2^{NR}$ codewords of length N for which the decoding error probability of each codeword is less than $4e^{-NE_r(R)}$, where $E_r(R)$ is a non-increasing function of R.*

Now, we consider the counting problem in detail. We first define the routing strategy. To transmit sensor information to the fusion center, we divide the unit square into a regular lattice of B cells, and fix the transmission radius to be

$$r = \sqrt{\frac{8}{B}}, \tag{4.4}$$

which guarantees that a sensor can reach any other sensors within adjacent (common edge or corner) cells. We then adopt the hierarchical routing architecture of [5].

Routing Strategy: For each cell, we choose one sensor as the cell-center. Designating the fusion center as the root, we form a rooted tree as in Figure 4.2, whose vertices include all the cell-centers, and whose links can only be between cell-centers of adjacent (common edge or corner) cells. Sensors

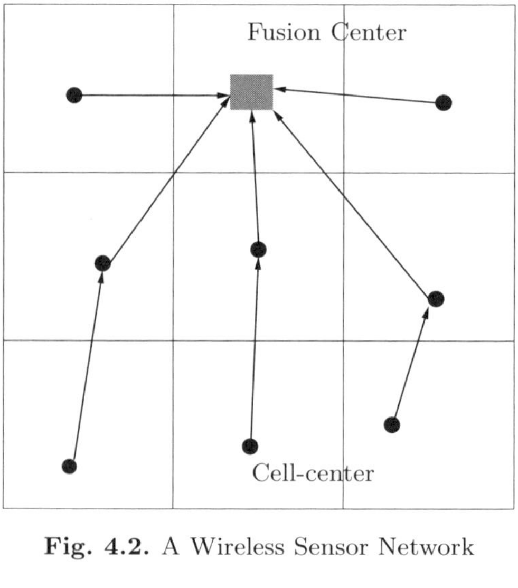

Fig. 4.2. A Wireless Sensor Network

first transmit data to the cell-centers, and then the data are aggregated and transmitted along the rooted-tree to the fusion center.

We let $P(i)$ denote the parent of cell-center i, $C(i)$ denote the set of the children of cell-center i in the rooted tree, $H_{\max}$ denote the depth of the tree, and $H(i)$ denote the depth of the cell-center i in the tree ($H(\text{fusion center}) = 0$).

Note that the network is connected and the routing strategy is feasible if there is at least one sensor in each cell. It is easy to see that

$$E[\text{Number of sensors in each cell}] = \frac{n}{B}.$$

In [9, 13, 12], it has been shown that the number of sensors in each cell is n/B with high probability when $B = O\left(\frac{n}{\log n}\right)$. Further, from Lemma 2, we have that if

$$B = \left(\left\lfloor \frac{n}{c_1 \log n} \right\rfloor\right)^2 \tag{4.5}$$

and $c_1 > 8$, then $\max_i n_i = O(\log n)$ and $\min_i n_i = \Omega(\log n)$ both hold. Throughout this section, B is chosen as in (4.5) with $c_1 = 8$, and

$$r = 8\sqrt{\frac{\log n}{n}}.$$

Thus, the probability that the routing strategy is feasible approaches 1 as n goes to infinity. In the following sections, we propose a distributed algorithm which works when each cell has at least one sensor. We assume that the algorithms report an error if the assumption does not hold.

Note that the wireless transmissions in neighboring cells will interfere with each other, so we adopt the cell scheduling scheme used in [7, 5].

Cell Scheduling: Without loss of generality, we assume that $\Delta = 0.05$. We group every 5×5 cells into a super-cell, and index the cells within each super-cell from 1 to 25 as in Figure 4.3. We divide each time slot into 25 mini-slots, and at mini-slot i, the mini-cells with index i are chosen to be active, for example, all mini-cells with index 1 (as in Figure 4.3) are active in the first mini-slot of every time slot. When a cell is active, one sensor in the cell could be selected to transmit. In our algorithms, transmissions will occur only within a cell or between neighboring cells. Thus, it is easy to verify that there is only one transmitter within a distance $1.05r$ for each receiver, and simultaneous transmissions do not interfere with each other under the cell scheduling.

Now given the routing strategy and cell scheduling algorithm, we will define protocols for intra-cell and inter-cell information processing and data aggregation. The protocols consist of two distinct parts:

(1) Intra-Cell-Protocol: The information within cells is aggregated at the respective cell-centers.
(2) Inter-Cell-Protocol: The information aggregated by cell-centers is transmitted, and aggregated further, along the rooted tree to the fusion center.

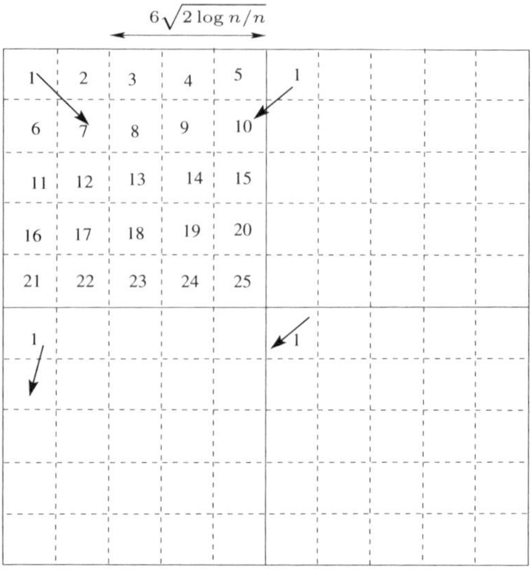

Fig. 4.3. Cell Scheduling ($\Delta = 0.05$))

We use the idea in [4] to design an algorithm for which the energy consumed is $\Theta\left(n\left(\log\log n\right)\left(\sqrt{\frac{\log n}{n}}\right)^{\alpha}\right)$. In wireless sensor networks, transmissions by a sensor can be heard by any sensor within its transmission range. Suppose there are $\tilde{n}$ sensors in sensor k's transmission range, then there are $\tilde{n}$ independent receptions for each bit sent by sensor k. The main idea in [4] is to use the reception diversity to obtain a good estimate of the bit transmitted by sensor k. But it requires additional transmissions among sensors; for example, it takes $\tilde{n}$ more transmissions for $\tilde{n}$ sensors to report the bits they received from sensor k. We will show how to use in-network processing to reduce the number of transmissions required to exploit the reception diversity.

Recall that b_k is the bit sensor k has. For cell i, define Δ_i as the set of indices of the sensors in cell i, and γ_i as the counting of cell-center i, so

$$\gamma_i = \sum_{k \in \Delta_i} b_k$$

if the counting is correct. For easy reference, we also define $\lambda = -\log(4p(1-p))$.

Counting-Algorithm-I:

When the a cell is active, the sensors first transmit and process measurements according to Intra-cell-protocol-I, and then the counting results are aggregated and transmitted according to Inter-cell-protocol-I.

Intra-Cell-Protocol-I (At cell i):

(i) The sensors in cell i take turns to transmit their bits. When it is the turn of sensor k, it broadcasts its bit $\lceil \frac{4}{\lambda} (\log \log n) \rceil$ times. Then, all other sensors in the cell will receive $\lceil \frac{4}{\lambda} (\log \log n) \rceil$ bits from sensor k. Sensor j ($j \neq k$) sets α_{jk} to be the majority of the bits received from sensor k, and sets A_j to be

$$A_j = b_j + \sum_{k \in \Delta_i, k \neq j} \alpha_{jk}$$

after all sensors broadcast their bits.

(ii) Select $\left\lceil \frac{n_i}{\log \log n} \right\rceil$ sensors in the cell. Each selected sensor j represents A_j using $\lceil \log_2 n_i \rceil$ bits, codes it using a block code with rate R_1 such that $E_r(R_1)/R_1 \geq 1$, and then broadcasts A_j once.

(iii) Suppose $\tilde{A}_j$ is the output of the binary symmetric channel between the cell-center and sensor j with input A_j. Cell-center i sets γ_i to be any mode of sequence $\{\tilde{A}_j\}$.

Inter-Cell-Protocol-I:

Define η_i to be the aggregated information of the subtree rooted at cell-center i. When cell-center i is scheduled, cell-center i sets

$$\eta_i = \gamma_i + \sum_{j \in C(i)} \tilde{\eta}_j,$$

where $\tilde{\eta}_j$ is the output of the channel between cell center j and cell center i with input η_j. Since $0 \leq \eta_i \leq n$, note that η_i can be represented using $\lceil \log_2 n \rceil$ bits. If i is the fusion center, then $\gamma_c = \eta_i$. Otherwise, it transmits η_i to cell-center $P(i)$ using a block code with rate R_2 such that $E_r(R_2)/R_2 > 1$.

We now analyze the energy requirement of Counting-Algorithm-I. First, in Lemma 5, we show that under Intra-Cell-Protocol-I,

$$\Pr\left(\text{All } \gamma_i \text{ are correct } \;\middle|\; \frac{c_1}{2} \leq \frac{n_i}{\log n} \leq 2c_1 \ \forall i \right) \geq 1 - \frac{1}{c_1 \log n}.$$

Then, in Lemma 6 and Theorem 2, we show that

$$\Pr\left(\gamma_c \text{ is correct } | \gamma_i \text{ is correct } \forall i\right) \geq 1 - \frac{4}{c_1 \log n}.$$

Finally, Theorem 2 quantifies the energy requirement of Counting-Algorithm-I.

Lemma 5. *Suppose $\frac{c_1}{2} \log n \leq n_i \leq 2c_1 \log n$ for all i. Then, by executing Intra-Cell-Protocol-I, the cell-centers can obtain γ_i with*

$$Pr\left(\gamma_i = \sum_{k \in \Delta_i} b_k \ \forall i \;\middle|\; 2c_1 \geq \frac{n_i}{\log n} \geq \frac{c_1}{2} \ \forall i \right) \geq 1 - \frac{1}{c_1 \log n} \tag{4.6}$$

and the number of transmissions required in each cell is $\Theta\left((\log n)\left(\log \log n\right)\right)$.

Proof. In the following analysis, we assume $\frac{c_1}{2}\log n \le n_i \le 2c_1 \log n$ holds for all i. First, the number of transmissions in each cell under Intra-Cell-Protocol-I is

$$n_i \left\lceil \frac{4}{\lambda}(\log\log n) \right\rceil + \left\lceil \frac{n_i}{\log\log n} \right\rceil \lceil \log_2 n_i \rceil = \Theta((\log n)(\log\log n)).$$

Next we investigate the probability that γ_i is correct, i.e., $\gamma_i = \sum_{k\in\Delta_i} b_k$. From Lemma 4, we have

$$\Pr(\alpha_{jk} = b_k) \ge 1 - (4p(1-p))^{\frac{2\log\log n}{\lambda}}.$$

Note that A_j is correct if α_{jk} is correct for all $k \in \Delta_i$. From the union bound, we have

$$\Pr\left(A_j = \sum_{k\in\Delta_i} b_k\right) \ge 1 - n_i(4p(1-p))^{\frac{2\log\log n}{\lambda}} \ge 1 - \frac{2c_1}{\log n}.$$

Consider step (ii) of Intra-Cell-Protocol-I, from Theorem 1,

$$\Pr\left(\tilde{A}_j = A_j\right) \ge 1 - 4e^{-\frac{E_r(R_1)}{R_1}\log_2 n_i} \ge 1 - 4e^{-\log\log n},$$

where the last inequality holds because $n_i \ge \frac{c_1}{2}\log n$. Thus,

$$\Pr\left(\tilde{A}_j = \sum_{k\in\Delta_i} b_k\right) \ge 1 - \frac{2c_1+4}{\log n}.$$

Note that $\{\alpha_{jk}\}$ are i.i.d. for all $j \in \Delta_i$, so $\{A_j\}$ are identical and $\{\tilde{A}_j\}$ are i.i.d.. Now define i.i.d. random variables $\{I_j\}$ such that $I_j = 1$ if $\tilde{A}_j = \sum_{k\in\Delta_i} b_k$, and $I_j = 0$ if $\tilde{A}_j \ne \sum_{k\in\Delta_i} b_k$. Since γ_i is the mode of $\{\tilde{A}_j\}$, from Lemma 4, we have

$$\begin{aligned}
\Pr\left(\gamma_i \ne \sum_{k\in\Delta_i} b_k\right) &\le \Pr\left(\sum_j I_j < \frac{1}{2}n_i\right) \\
&\le \left(4\left(\frac{4c_1+4}{\log n}\right)\left(1 - \frac{4c_1+4}{\log n}\right)\right)^{\frac{n_i}{2\log\log n}} \\
&\le e^{-(\log\log n - \log(16c_1+16))\frac{n_i}{2\log\log n}} \\
&\le e^{-\log n}.
\end{aligned}$$

There are at most $\frac{n}{c_1\log n}$ cells in the network, so

$$\Pr\left(\gamma_i = \sum_{k\in\Delta_i} b_k \;\forall i\right) \ge 1 - \frac{n}{c_1\log n}e^{-\log n} = 1 - \frac{1}{c_1\log n},$$

and the lemma holds.

Now, suppose that all γ_i are correct. Since η_i can be represented using $\lceil \log_2 n \rceil$ bits, each cell-center has $\lceil \log_2 n \rceil$ bits to transmit under Inter-Cell-Protocol-I.

Lemma 6. *Suppose all cell-centers have the correct γ_i, then under Inter-Cell-Protocol-I, the probability that the fusion center obtains the correct γ_c is bounded as follows:*

$$Pr\left(\gamma_c = \sum_k b_k \,\middle|\, \gamma_i = \sum_{k \in \Delta_i} b_k \ \forall i\right) \geq 1 - \frac{4}{c_1 \log n}, \tag{4.7}$$

and the number of transmissions required is $\Theta(n)$.

Proof. Suppose all cell-centers have the correct γ_i, then $\gamma_c = \sum_k b_k$ if all η_i's are correctly received. From Theorem 1, there exists a block code satisfying the conditions given in step (i) of Inter-Cell-Protocol-I. Thus, for a given i,

$$\begin{aligned} \Pr\left(\eta_i \text{ is correctly received}\right) &\geq 1 - 4e^{-\frac{E_r(R_2)}{R_2}\log_2 n} \\ &\geq 1 - 4e^{-\log n}, \end{aligned}$$

and from the union bound,

$$\begin{aligned} &\Pr\left(\gamma_c = \sum_k b_k \,\middle|\, \gamma_i = \sum_{k \in \Delta_i} b_k \ \forall i\right) \\ &= \Pr\left(\text{All } \eta_i\text{'s are correctly received} \,\middle|\, \gamma_i = \sum_{k \in \Delta_i} b_k \ \forall i\right) \\ &\geq 1 - \frac{4n}{c_1 \log n} e^{-\log n} \\ &= 1 - \frac{4}{c_1 \log n}. \end{aligned}$$

¿From Lemma 5 and Lemma 6, we have shown that, under Counting-Algorithm-I, the number of sensors in state "1" can be counted accurately with high probability when the number of sensors is large enough. Using these lemmas, we have following theorem, which provides an upper bound on the energy requirement to solve our counting problem.

Theorem 2 *The number of sensors in state "1" can be counted accurately with high probability by total transmission energy consumption*

$$O\left(n(\log\log n)\left(\sqrt{\frac{\log n}{n}}\right)^{\alpha}\right),$$

and Counting-Algorithm-I is an asymptotically correct algorithm that achieves this energy consumption. Specifically, the probability of computation error at the fusion center is upper bounded by $\frac{7}{c_1 \log n}$.

Proof. Recall that

$$c_1 > \max\left\{8, \frac{4}{\lambda}\right\}.$$

From inequalities (4.1), (4.6) and (4.7), we have

$$\Pr\left(\gamma_c = \sum_k b_k\right) \geq 1 - \frac{7}{c_1 \log n},$$

which converges to one when n goes to infinity. So Counting-Algorithm-I is asymptotically correct.

Further, from Lemma 5 and Lemma 6, the number of transmissions under Counting-Algorithm-I is $\Theta(n(\log\log n))$. Since the common transmission range is $\sqrt{\frac{8c_1 \log n}{n}}$, the total energy consumption is

$$\Theta\left(n(\log\log n)\left(\sqrt{\frac{\log n}{n}}\right)^{\alpha}\right). \tag{4.8}$$

The theorem holds because there may exist other algorithms that consume less energy.

A simple lower bound has been obtained in Lemma 3. Comparing it with the upper bound in Theorem 2, we can see that the upper bound differs by a factor of *only* $(\log\log n)$ from the lower bound. But it is still not clear how good our bound is. A more general computational problem than ours, i.e., one of knowing all the bits in the network, is considered for a broadcast network in [4]. The number of transmissions required there is also shown to be $O\left(n\left(\log\log n\right)\right)$. This suggests that one may be able to improve our upper bound on the energy usage since counting is easier than detecting all the bits in the network. On the other hand, parity computation which is a simpler problem than counting is also studied in [4], but the number of transmissions needed is again $O\left(n\left(\log\log n\right)\right)$, the same complexity as Counting-Algorithm-I. To the best of our knowledge, this is the best upper bound in the literature for parity computation in broadcast networks. Further, our network with its multihop architecture also requires more transmissions for the data from the sensors to reach the fusion center. This suggests that our upper bound on energy usage is quite good.

In this section, we presented the simplest case, where each sensor has only one binary measurement to report, to demonstrate that energy savings can be achieved by sensor collaboration and data aggregation. More general cases can be found in [14] and [15]. In [14], besides the simplest case, we also studied the case where each sensor has N binary measurements to report and the symmetric function needs to be computed for each measurement. We showed that the total transmission energy consumption can be reduced to $O\left(n\left(\max\left\{1, \frac{\log\log n}{N}\right\}\right)\left(\sqrt{\frac{\log n}{n}}\right)^{\alpha}\right)$ per measurement. When

$N = \Omega(\log\log n)$, the energy consumption is $\Theta\left(n\left(\sqrt{\frac{\log n}{n}}\right)^{\alpha}\right)$ per measurement, which is a tight bound. We also considered the case that we only want to know roughly, i.e., how many sensors have "1." The answer can be obtained with the transmission energy consumption $\Theta\left(n\left(\sqrt{\frac{\log n}{n}}\right)^{\alpha}\right)$. All these results can be extended the cases where the sensor measurements taken value from $\{0, \ldots, m-1\}$, and the details are presented in [15].

4.3 Minimum Cost Real-time Function Computation in Wireless Sensor Networks

In the previous section, we investigated function computation in wireless sensor networks, where we did not make any assumption on data correlation. However, in the case where sensors are densely deployed, the sensor measurements could be highly correlated. This additional energy savings can be achieved by exploiting such a data correlation. Further, in the previous problem, we assumed that sensors use the same amount of energy to transmit a "0" or "1." In this section, we will consider the case where the sensor data is correlated, communication costs can be different, and transmission channels need not not be binary. For example, silence, which is a signal with zero cost, could be used to convey information. Then to reduce the energy consumption, we should use low-cost signals as frequently as possible. In the following simple example, we illustrate that energy savings can be achieved by exploiting the data correlation.

4.3.1 A Simple Example

Consider a network with three sensors each with a binary measurement value. The observations are random variables and their joint distribution is given in in Table 4.1.

Event	Probability
0 0 1	$\frac{1}{6}$
0 1 0	$\frac{1}{6}$
0 1 1	$\frac{1}{6}$
1 0 1	$\frac{1}{3}$
1 1 0	$\frac{1}{6}$

Table 4.1. Joint Distribution of Sensor Observations

An event in the table is a particular set of possible observations. For example, the first row indicates that the probability of sensor 1 observing a "0," sensor 2 observing a "0", and sensor 3 observing a "1" is 1/6. Events not in the table are unlikely to occur and hence have zero probability.

Assume that the sensor can use two transmission signals $\{S, P\}$ to convey the information, where S indicates silence, with cost zero; and P is some pulse with energy cost E. Now we consider following three different cases.

(1) Now suppose the sensors are deployed far away from each other, so no sensor can hear the others' transmissions. Each sensor can use S to represent either 0 or 1. Let F_i denote the encoding scheme. It is easy to see the minimum cost scheme is as in Table 4.2, and the expected cost is $4E/3$. Note that the minimum cost scheme for sensor i is obtained by computing the probability of seeing 0 and the probability of seeing 1 at sensor i, and using S to represent the bit with higher probability.

Sensor	Encoding Scheme
Sensor 1	$F_1(0) = S$ and $F_1(1) = P$
Sensor 2	$F_2(0) = S$ and $F_2(1) = P$
Sensor 3	$F_3(0) = P$ and $F_3(1) = S$

Table 4.2. Case 1 Encoding Scheme: Isolated Sensors

(2) Suppose that the sensors do not know each others' observations, but can hear each other. After one sensor transmits its information, all other sensors can update the probability of the various possible events, and use this information to reduce their transmission costs. The sensors cannot simultaneously transmit, but transmit in sequential order. In this collocated case, the encoding scheme of sensor 2 depends on what it hears from sensor 1, and the encoding scheme of sensor 2 depends on what it hears from sensors 1 and 2. The optimal transmission scheme is given in Table 4.3. It is easy to see that each event is associated with a se of distinct transmission signals and the expected cost is E.

Sensor	Encoding Scheme
Sensor 1	$F_1(0) = S$ and $F_1(1) = P$
Sensor 2	$F_2(S,0) = P$, $F_2(S,1) = S$, $F_2(P,0) = S$, and $F_2(P,1) = P$
Sensor 3	$F_3(S,P,1) = S$, $F_3(S,S,0) = P$, $F_3(S,S,1) = S$ $F_3(P,S,1) = S$, and $F_3(P,P,0) = S$

Table 4.3. Case 2 Encoding Scheme: Collocated Network

We can see that, even for this simple example, significant energy savings (25%) can be achieved. Next, we propose a stochastic control approach to obtain the minimum cost transmission scheme, for a general collocated network.

4.3.2 Model

Consider a wireless sensor network consisting of n sensors, and each sensor has a measurement X_i where X_i takes values from $\mathcal{X}$, and $\mathcal{X}$ is the space of observations, and $|\mathcal{X}| = m$. The network has a special node called the fusion center whose goal is to compute some function $G(\mathbf{X})$ based on sensor observations $\mathbf{X}$. We consider a collocated network [5] where a transmission can be heard by all sensors in the network, and only one transmission is allowed at one time. Let $\mathbf{X}_i^j$ denote $(X_i, \ldots, X_j)$, x_i denote a realization of X_i, $\mathbf{x}_i^j$ denote a realization of $\mathbf{X}_i^j$, and $\Pr\left(\mathbf{x}_i^j\right)$ be the abbreviation of $\Pr\left(\mathbf{X}_i^j = \mathbf{x}_i^j\right)$.

In this section, we assume that sensors sequentially take turns to transmit, and the order is fixed such that sensor 1 transmits first, then sensor 2 and so on, as in Figure 4.4. Each sensor only transmits once, after which the fusion

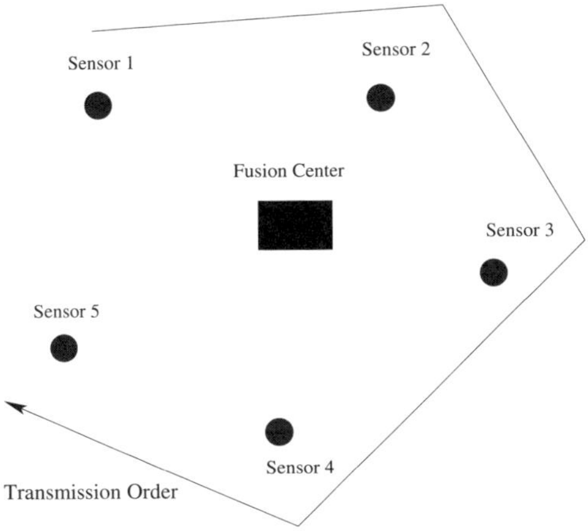

Fig. 4.4. The Order of Sensor Transmissions

center needs to compute $G(\mathbf{X})$. Note that the fusion center does not need to recover $\mathbf{X}$, it only needs to know enough information to compute the function value. Let C_i denote the signal node i transmits, and c_i be a realization of C_i. We assume $c_i \in \mathcal{S}$, where $\mathcal{S}$ is the set of signals and $|\mathcal{S}| = m$. For simplicity, we let c_i denote both the signal and the cost of the signal. Furthermore, it is easy to see that C_i is a function of the transmissions before node i and the measurement X_i of node i. Accordingly, we define an encoding function F_i such that

$$C_i = F_i(\mathbf{C}_1^{i-1}, X_i).$$

Further define $\mathbf{F}_1^i$ such that

$$\mathbf{F}_1^i(\mathbf{X}_1^i) = (F_1(X_1), F_2(F_1(X_1), X_2), \ldots, F_i(\mathbf{F}_1^{i-1}(\mathbf{X}_1^{i-1}), X_i)),$$

and let $\mathbf{F}$ denote $\mathbf{F}_1^n$. An encoding scheme $\mathbf{F}$ is said to be feasible if

$$\mathbf{F}(\mathbf{x}) \neq \mathbf{F}(\mathbf{y}),$$

for every $\mathbf{x}$ and $\mathbf{y}$ such that $\Pr(\mathbf{x}) > 0$, $\Pr(\mathbf{y}) > 0$ and $G(\mathbf{x}) \neq G(\mathbf{y})$. Let $\mathcal{F}$ denote the set of feasible encoding functions, and $\sum \mathbf{F}(\mathbf{X}) = \sum_{i=1}^n F_i(\mathbf{F}_1^{i-1}(\mathbf{X}_1^{i-1}), X_i)$. Our goal is find $\mathbf{F}^*$ such that

$$\begin{aligned} \mathbf{F}^* \quad &= \quad \arg\min_{\mathbf{F}} E\left[\sum \mathbf{F}(\mathbf{X})\right] \\ &\text{subject to: } \ \mathbf{F} \in \mathcal{F}. \end{aligned} \tag{4.9}$$

4.3.3 Stochastic Control Approach

In this subsection, we propose a stochastic control approach to solve the minimum cost problem (4.9). We use superscript -1 to indicate decoding functions, for example, $\mathbf{F}^{-1}$ is the decoding function of $\mathbf{F}$ such that $\mathbf{F}^{-1}(\mathbf{c}) \in (2^{\mathcal{X}})^n$ is the preimage of $\mathbf{c}$. Note that decoding functions and known to all sensors and the fusion center. The decoding process is as follows: When the fusion center receives $\mathbf{c}_1^i$, it can decode $\mathbf{c}_1^i$ to obtain $F_i^{-1}(\mathbf{c}_1^i)$, which is a subset of $\mathcal{X}$. After all sensors transmit, the fusion center decodes the received information and obtains

$$\mathcal{Y} = (F_1^{-1}(c_1), \ldots, F_n^{-1}(\mathbf{c}_1^n)).$$

Then the fusion center takes $G(\mathcal{Y})$ as the function computation result, which is unique and correct if the encoding function $\mathbf{F}$ is feasible. Similarly, sensor i can compute $(\mathbf{F}_1^{i-1})^{-1}(\mathbf{c}_1^{i-1})$ after receiving $\mathbf{c}_1^{i-1}$. Note that $(\mathbf{F}_1^{i-1})^{-1}(\mathbf{c}_1^{i-1}) \in (2^{\mathcal{X}})^{i-1}$. Let $\mathfrak{A}_i$ be a random variable taken values from $2^{\mathcal{X}}$, and $\mathfrak{a}_i$ be a realization of $\mathfrak{A}_i$. Note that $\mathfrak{a}_i$ is a subset of $\mathcal{X}$. We can first design an encoding scheme $\tilde{\mathbf{F}}$ based on $\mathfrak{A}_1^{i-1}$ and X_i, and then obtain $\mathbf{F}$ as follows:

(1) First design $c_i = \tilde{F}_i(\mathfrak{a}_1^{i-1}, x_i)$ for each $(\mathfrak{a}_1^{i-1}, x_i)$ such that $\Pr(\mathbf{x}_1^{i-1}) > 0$ for all $\mathbf{x}_1^{i-1} \in \mathfrak{a}_1^{i-1}$, and $\Pr(x_i|\mathfrak{a}_1^{i-1}) > 0$.
(2) Let

$$F_i(\mathbf{C}_1^{i-1}, X_i) = \tilde{F}_i((\tilde{\mathbf{F}}_1^{i-1})^{-1}(\mathbf{C}_1^{i-1}), X_i). \tag{4.10}$$

Note that we impose conditions on $\mathfrak{a}_1^{i-1}$ and x_i in step (1) since we only need to design coding scheme for events with positive probability. The relationship between F_i and $\tilde{F}_i$ is illustrated in Figure 4.5.

We next model the coding process as a controlled Markov chain $\{\mathfrak{Z}_i\}$, where $\mathfrak{Z}_i = (\mathfrak{A}_1, \ldots, \mathfrak{A}_{i-1}, \mathcal{X}, \ldots, \mathcal{X})$. Note $\mathfrak{Z}_i$ represents the information obtained at the fusion center after first i sensors transmit. Next let $\mathfrak{z}_i$ be a realization of $\mathfrak{Z}_i$. The transition probability of the controlled Markov chain under strategy $\tilde{\mathbf{F}}$ is defined to be

$$\begin{aligned} \Pr\left(\mathfrak{z}_{i+1} \middle| \mathfrak{z}_i, \tilde{F}_i\right) &= \Pr\left((\mathfrak{a}_1, \ldots, \mathfrak{a}_i, \mathcal{X}, \ldots, \mathcal{X}) | (\mathfrak{a}_1, \ldots, \mathfrak{a}_{i-1}, \mathcal{X}, \ldots, \mathcal{X}), \tilde{F}_i\right) \\ &= \Pr\left(\mathfrak{a}_i | \, \mathfrak{a}_1^{i-1}\right) \end{aligned}$$

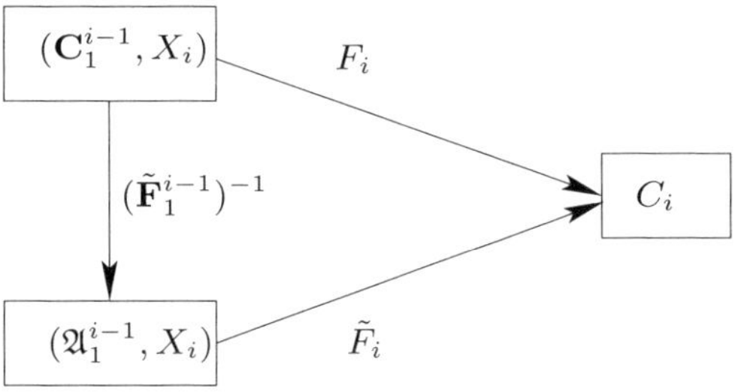

Fig. 4.5. The Relation between F_i and $\tilde{F}_i$

if the following conditions hold:

(1) $\Pr(\mathbf{x}_1^{i-1}) > 0$ for every $\mathbf{x}_1^{i-1} \in \mathfrak{a}_1^{i-1}$;
(2) $\mathfrak{a}_i = \tilde{F}_i^{-1}(\mathfrak{a}_1^{i-1}, c)$ for some $c \in \mathcal{S}$.

Otherwise $\Pr\left(\mathfrak{z}_{i+1} \middle| \mathfrak{z}_i, \tilde{F}_i\right) = 0$. Note that we impose the first condition since we only design coding scheme for events with positive probability, and impose the second condition to guarantee that $\mathfrak{a}_i$ will be conveyed by some signal. The cost of decision $\tilde{F}_i$ given $\mathfrak{z}_{i-1}$ is

$$C\left(\mathfrak{z}_i, \tilde{F}_i\right) = E\left[\tilde{F}_i\left(\mathfrak{a}_1^{i-1}, X_i\right)\right].$$

Since the function $G(\mathbf{X})$ needs to be computed accurately at the end of sensor transmissions, we impose a cost on $\mathfrak{Z}_n$ such that

$$C^p(\mathfrak{z}_n) = \begin{cases} \infty, & \text{if } \exists \mathbf{x}, \mathbf{y} \in \mathfrak{z}_n \text{ s.t. } G(\mathbf{x}) \neq G(\mathbf{y}); \\ 0, & \text{otherwise.} \end{cases}$$

Thus, from the definition of cost (4.11), we can see that problem (4.9) is equivalent to the following stochastic control problem

$$\min_{\tilde{\mathbf{F}}} E\left[C^p(\mathfrak{Z}_n) + \sum_{i=1}^{n} C(\mathfrak{Z}_i, \tilde{F}_i) \middle| \mathfrak{Z}_1\right]. \tag{4.11}$$

So the minimum cost scheme $\tilde{\mathbf{F}}^*$ can be obtained by solving this standard stochastic control problem, and $\mathbf{F}^*$ can be further obtained from equation (4.10).

We would like to comment that the goal of this section is to model the minimum cost real-time function computation as a standard stochastic control problem, thus providing a methodology to design the minimum cost transmission scheme. Note that $\mathfrak{a}_i \in 2^{\mathcal{X}}$, so $\mathfrak{z}$ can take $2^{i|\mathcal{X}|}$ different values in general and the complexity of stochastic control problem (4.11) increases exponentially both in n and $|\mathcal{X}|$. So to make this approach applicable to large n or large $\mathcal{X}$, we need to further reduce the complexity, which is still an open problem for general cases. However, for a special case where $\Pr(\mathbf{x}) > 0$ for any $\mathbf{x} \in \mathcal{X}^n$ and $G(\mathbf{x}) \neq G(\mathbf{y})$ if $\mathbf{x} \neq \mathbf{y}$, a simple $\mathbf{F}^*$ can be obtained from (4.11).

Theorem 3 *Suppose that* $\Pr(\mathbf{x}) > 0$ *for any* $\mathbf{x} \in \mathcal{X}^n$, $G(\mathbf{x}) \neq G(\mathbf{y})$ *if* $\mathbf{x} \neq \mathbf{y}$, *and* $c_i > c_j$ *if* $i > j$. *Given* $\mathbf{x}_1^{i-1}$, *we order* x_i *according to* $\Pr(x_i|\mathbf{x}_1^{i-1})$, *from the largest to the smallest, and let* $I(x_i|\mathbf{x}_1^{i-1})$ *denote the rank of* x_i. *Then the minimum cost scheme is*

$$F_i^*(x_i, \mathbf{c}_1^{i-1}) = c_{I(x_i|(F^{*i-1}_1)^{-1}(\mathbf{c}_1^{i-1}))}. \tag{4.12}$$

Proof. Since $\Pr(\mathbf{x}) > 0$ for any $\mathbf{x} \in \mathcal{X}^n$ and and $G(\mathbf{x}) \neq G(\mathbf{y})$ if $\mathbf{x} \neq \mathbf{y}$, it is easy to see that $C^p(\mathfrak{z}_n) = 0$ if $\mathfrak{z}_n \in \mathcal{X}^n$, and $C^p(\mathfrak{z}_n) = \infty$ otherwise. Thus $\tilde{\mathbf{F}}^*$ needs to satisfy

$$\tilde{\mathbf{F}}^*(\mathbf{x}) \neq \tilde{\mathbf{F}}^*(\mathbf{y}) \text{ if } \mathbf{x} \neq \mathbf{y}, \tag{4.13}$$

which implies that

$$\tilde{F}_i^*(x_i, \mathbf{x}_1^{i-1}) \neq \tilde{F}_i^*(y_i, \mathbf{x}_1^{i-1}) \text{ if } x_i \neq y_i. \tag{4.14}$$

Now let $\tilde{\mathcal{F}}^*$ denote the set of $\tilde{\mathbf{F}}$ satisfying (4.13), and $\tilde{\mathcal{F}}_i^*$ denote the set of $\tilde{F}_i$ satisfying (4.14), then problem (4.11) can be re-written as

$$\min_{\tilde{\mathbf{F}}} E\left[C^p(\mathfrak{Z}_n) + \sum_{i=1}^n C(\mathfrak{Z}_i, \tilde{F}_i)\middle| \mathfrak{Z}_1\right] = \min_{\tilde{\mathbf{F}} \in \tilde{\mathcal{F}}^*} E\left[\sum_{i=1}^n C(X_i, \tilde{F}_i)\right]$$
$$= \sum_{i=1}^n \sum_{\mathbf{x}_1^{i-1} \in \mathcal{X}^{i-1}} \left(\min_{\tilde{F}_i \in \tilde{\mathcal{F}}_i^*} \sum_{x_i \in \mathcal{X}} \tilde{F}_i(x_i, \mathbf{x}_1^{i-1}) \Pr\left(x_i|\mathbf{x}_1^{i-1}\right)\right) \Pr\left(\mathbf{x}_1^{i-1}\right).$$

Given $\mathbf{x}_1^{i-1}$, it is easy to see that the optimal $\tilde{F}_i$ is to assign the lower cost signal to the event with higher probability, so (4.12) is the optimal scheme.

4.4 Conclusions

In Section 4.2, we investigated counting problems in multi-hop networks with noisy communication channels. We considered the case where each sensor has a single measurement, and showed by construction that feasible algorithms exist whose energy consumption is $O\left(n(\log\log n)\left(\sqrt{\frac{\log n}{n}}\right)^{\alpha}\right)$. In Section 4.3, we investigated minimum cost real-time function computation in single-hop networks, and showed that the problem can be solved using a stochastic control approach. Function computation is a fundamental problem in sensor networks since the objectives of sensor networks are to retrieve useful information from sensor measurements, and these objectives can be thought as function computation problems. We demonstrated in this chapter how to design energy-efficient algorithms by exploiting the structure of functions and by optimizing algorithms for the specific objectives of sensor networks.

References

[1] A. Dhulipala, C. Fragouli and A. Orlitsky Silence based communication for sensor networks. In *IEEE International Symposium on Information Theory,* July 2006.

[2] C. Fragouli and A. Orlitsky Silence is golden and time is money: power-aware communication for sensor networks, C. Fragouli and A. Orlitsky. In *the Allerton Conference,* September 2005.

[3] R. G. Gallager. Information Theory and Reliable Communication. John Wiley & Sons, New York, 1968.

[4] R. G. Gallager. Finding parity in a simple broadcast network. In *IEEE Transactions on Information Theory,* vol. 34, pp 176-180, 1988.

[5] A. Giridhar and P. R. Kumar. Computing and communicating functions over sensor networks. In *IEEE Journal on Selected Areas in Communications,* pp. 755–764, vol. 23, no. 4, April 2005.

[6] P. Gupta and P. Kumar. Critical power for asymptotic connectivity in wireless network. In *Stochastic Analysis, Control, Optimization and Applicaitons: a Volume in Honor of W.H.Fleming,* W. McEneaney, G. Yin and Q. Zhang, Eds., 1998

[7] P. Gupta and P. Kumar. The capacity of wireless networks. In *IEEE transactions of Information Theory,* vol. 46, no.2, pp. 388-404, 2000.

[8] N. Khude, A. Kumar and A. Karnik. Time and Energy Complexity of Distributed Computation in Wireless Sensor Networks. In *Proceedings of the IEEE Infocom,* 2005.

[9] S. R. Kulkarni and P. Viswanath. A Deterministic Approach to Throughput Scaling in Wireless Networks. In *IEEE Trans. on Information Theory,* Vol. 50, No.6, pp. 1041-1049, June 2004.

[10] E. Kushilevitz and Y. Mansour. Computation in Noisy Radio Networks In *Proceedings of the ninth annual ACM-SIAM symposium on Discrete algorithms,* pp. 236-243, 1998.

[11] M. Mitzenmacher and E. Upfal. *Probability and Computing: Randomized Algorithms and Probabilistic Analysis.* Cambridge, 2005.

[12] S. Toumpis and A. J. Goldsmith Large wireless network under fading, mobility, and delay constraints. In *Procedings of IEEE INFOCOM,* 2004.

[13] F. Xue and P. R. Kumar. The number of neighbors needed for connectivity of wireless networks. Wireless Networks, pp. 169–181, vol.10, no. 2, March 2004.

[14] L. Ying, R. Srikant and G. E. Dullerud. Distributed Symmetric Function Computation in Noisy Wireless Sensor Networks with Binary Data. In *Proc. 4rd Intl. Symposium on Modelling and Optimization in Mobile, Ad Hoc, and Wireless Networks,* April 2006.

[15] L. Ying, R. Srikant and G. E. Dullerud. Distributed Symmetric Function Computation in Noisy Wireless Sensor Networks. Preprint. Available at http://www.ifp.uiuc.edu/~lying/Publicaitons/DSFComputation_J.pdf

[16] Y. Zhu and R. Sivakumar Challenges: communication through silence in wireless sensor networks. In *Proceedings of the 11th annual international conference on Mobile computing and networking.*

5

Network Coding for Distributed Storage in Wireless Networks

Alexandros G. Dimakis and Kannan Ramchandran

Department of Electrical Engineering and Computer Science,
University of California, Berkeley, CA 94704.
{adim,kannanr} @eecs.berkeley.edu

5.1 Introduction

We will address some of the problems related to storing information in multiple storage devices that are individually unreliable, and connected in a network. As an application consider a sensor network deployment in a remote and inaccessible environment where sensor nodes are taking measurements (possibly after processing) and storing data in the network, over long time periods. A data collector may appear at any location in the network and try to retrieve as much useful data as possible. Another scenario is a sensor network deployed in a time-critical or emergency situation (e.g. fire, flood, earthquake). Here, the focus is on maximizing the amount of sensed data than can be retrieved from a rapidly failing network. In both scenarios, many storage nodes are expected to fail and redundancy is necessary to guarantee the required reliability. This redundancy in the information representation can be introduced either through replication or through erasure coding. It is well known that information representations that use erasure codes require far less redundancy to provide the same level of reliability [42] and have been used in numerous applications (e.g. Reed-Solomon codes [37]). After extensive studies, essentially optimal erasure codes exist today, with linear encoding and decoding complexity [28, 40]. However, when coding is performed in an unstructured (and possibly dynamic) network, new issues arise that have not been addressed in classical coding theory. Specifically:

- *Communication* between storage and data nodes comes with a cost, since energy is a precious resource in sensor networks. Therefore, the code should be constructed with the minimal possible communication between nodes. This means that sparsity in the generator matrix of the code is critical for such applications.
- The information is sensed in multiple distributed locations and global *coordination* is difficult to achieve. Hence the code construction should be distributed and based on local knowledge.

- The sensor network will be deployed in a dynamic environment and the the encoded storage might need to *evolve over time*, to reflect such dynamic changes. For example when storage nodes are failing, new encoded packets need to be generated from existing encoded packets, naturally leading to network coding schemes.

In this chapter we discuss these issues and various related schemes that have been proposed in the recent literature including our own contributions. In summary, we will be interested in distributed, scalable and energy-efficient algorithms to generate and dynamically maintain encoded information representations in networks.

5.1.1 The Distributed Networked Storage Problem

We will be using the abstractions of a *data node* which is a *source* of information that must be stored, and a *storage node* which corresponds to a storage device with limited memory and communication capabilities. A physical sensor mote can have both sensing capabilities and sufficient memory, and hence can be both a data node and a storage node of our abstract model. This separation is useful because it simplifies the presentation and can be easily mapped back to actual devices.

The classical distributed storage problem consists of having multiple (distributed) storage nodes (e.g. hard disks) for storing data which is initially located at *one single data node* (see Figure 5.1).

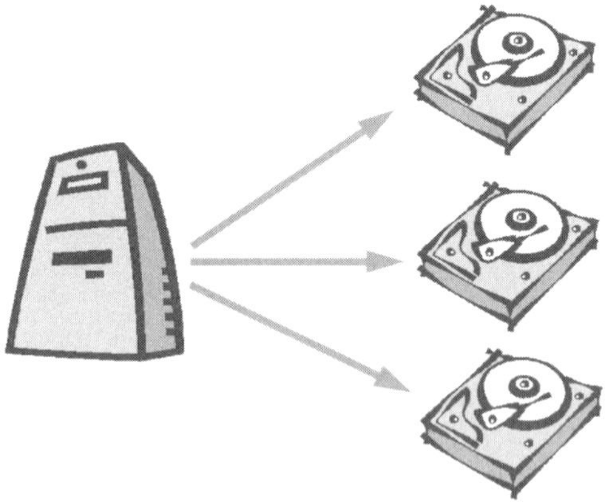

Fig. 5.1. The classical distributed storage setup. Data which is initially centralized is encoded and stored in distributed storage nodes.

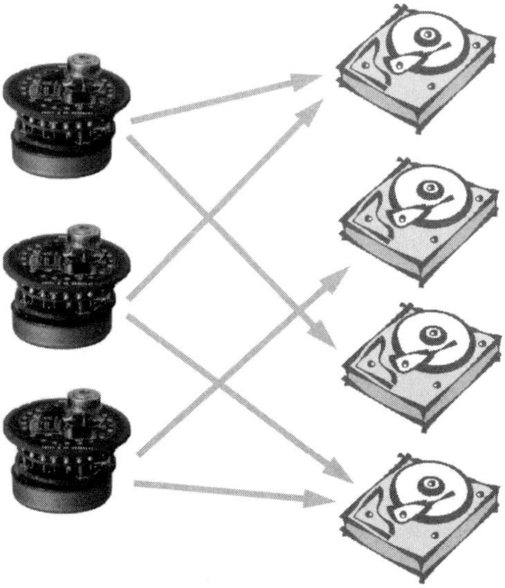

Fig. 5.2. Distributed networked storage. Initially distributed data stored in multiple storage nodes.

The *distributed networked storage* problem arises when both the data sources and the storage nodes are distributed (see Figure 5.2) and hence we have multiple data and storage nodes.

We make the following assumptions:

- We assume that there are k data-generating nodes and without loss of generality we will assume that each data node generates one data packet containing the information of interest.
- Further, assume we have $n \geq k$ storage nodes that will be used as storage and relay devices. Sensor nodes have limited memory, and we model that by assuming that each node can store only one data packet (or a combination having the same number of bits as a data packet). This is a key requirement to the scalability of the network. A data packet contains measurements over a time interval and can have significant size.
- The ratio $k/n = R$ (code rate) is assumed fixed as k and n scale. For example, we can assume that some fixed ratio (for example 10%) of nodes in a sensor network are generating data. These assumptions are only to simplify the presentation, and in practice the k data nodes and n storage nodes can be any arbitrary (possibly overlapping) subsets of nodes in a larger network.
- We are interested in schemes that require no routing tables, centralized processing or global knowledge or coordination of any sort. We rely on a packet routing layer that can route packets to *uniformly random locations*

in the network. Constructing such *random sampling algorithms* which are distributed and localized is key for the construction of codes in networks.

5.1.2 Outline

The chapter is structured as follows: in Section 5.2 we briefly present some properties of linear coding and network coding . Section 5.3 presents decentralized erasure codes and fountain codes and describes how their properties can be useful for storage in sensor networks. We also present randomized algorithms to construct such encodings with minimal coordination and knowledge. Section 5.4 discusses how encoded representations can be used for answering various queries other than data recovery. Finally, Section 5.5 discusses distributed network algorithms for randomly sampling sensor nodes with minimal communication; a mechanism that is necessary for constructing the codes described in this chapter.

5.2 Background

We will start with a brief survey of linear erasure codes and network coding. One important point is how the algebraic properties of the generator matrix of a code correspond to requirements from the network algorithm used to construct and maintain the encoded representation.

5.2.1 Linear Erasure Codes

Nearly all the coding schemes proposed in the literature are linear codes over finite fields. Erasure coding is a generalization of replication that divides the initial data object into k packets (or blocks) which are then used to generate n encoded packets of the same size. Good erasure codes have the property that *any* k out of the n encoded packets suffice to recover the original k data packets. In erasure coding we assume that the only types of errors that can happen are erasures of packets (due to failure of the corresponding storage node), but the packets that survive are always correct. Note also that we will be dealing with erasures of packets, not bits within a packet. Good erasure codes can yield much higher reliability compared to replication schemes for the same number of storage nodes. The most common erasure codes are Reed-Solomon codes, which are very widely employed in numerous applications like computer network distributed storage systems, and redundant disk arrays. Low-density parity-check (LDPC) codes and more recently Fountain codes [27] were proposed as alternatives with randomized construction and faster encoding and decoding times.

A toy example of a linear code over $GF(2^8)$ is given in Figure 5.3. In the example there are two data nodes X_1 and X_2 and three storage nodes

Y_1, Y_2, Y_3. We assume the data nodes have gathered a number of measurements. In the example we choose $u = 8$ bits to represent each number in our field which corresponds to $GF(2^8)$. The bits of the data measurements are divided into blocks of u bits which correspond to elements in $GF(2^8)$ (for example $X_1(1) = 002, X_1(2) = 080, X_1(3) = 220$). The data packet X_1 is routed to storage nodes Y_1, Y_3 and X_2 to Y_2, Y_3. Once a storage node receives one or more data packets, it must select coefficients f_i to multiply the received packets and subsequently add them to construct one encoded packet. A desired property is that the selection of the coefficients is done without any coordination, i.e. each storage node selects them uniformly and independently in $GF(2^8)$. Each coefficient then multiplies each block independently, multiple blocks are added (under the arithmetic of the Galois Field) and the results are cascaded into a new block packet Y_i that has exactly the same size as all the data packets. For example Y_3 has stored a packet that corresponds to $2X_1 + 1X_2$. Using this notation we mean that $Y_3(i) = 2X_1(i) + 1X_2(i)$ for $i = 1, 2, 3$. Each storage node will also store the coefficients f_i that it selected. This introduces an overhead storage that can be made arbitrarily small by coding over larger blocks [18, 11].

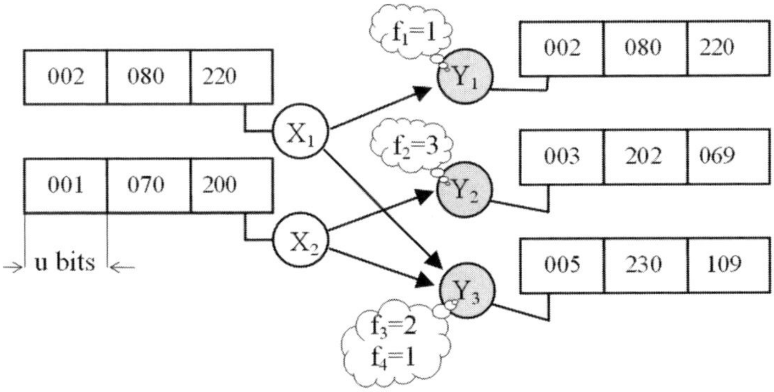

Fig. 5.3. A simple example of a linear code over $GF(2^8)$. Here $k = 2$, $n = 3, k/n = 2/3$, $q = 256$. The primitive polynomial of the field is $D^8 + D^4 + D^3 + D^2 + 1$. Arithmetic is done by representing numbers as binary coefficients of polynomials and doing polynomial operations modulo the primitive polynomial. For example, $70 \times 3 \rightarrow (D^6 + D^2 + D) \times (D + 1) = D^7 + D^6 + D^3 + D \rightarrow 202$.

Notice that in Figure 5.3 any two out of the three encoding packets can be used to reconstruct the original data.

In general, linear codes can be represented using their *generator matrix* in the form

$$s = mG, \tag{5.1}$$

where s is an $1 \times n$ encoded vector that is stored, m is $1 \times k$ data vector and G is a $k \times n$ matrix with elements selected from a field $GF(q)$. For the example in Figure 5.3,

$$G = \begin{pmatrix} 1 & 0 & 2 \\ 0 & 3 & 1 \end{pmatrix}. \tag{5.2}$$

To reconstruct m the receiver must invert a $k \times k$ sub-matrix G' of G. The key property required for successful decoding (that is also true for the example) is that any sub-matrix selection of G' forms a *full rank matrix*. When this is the case, decoding corresponds to solving a system of linear equations over $GF(q)$ (using for example, Gaussian elimination).

A matrix that has the property that *all* the square sub-matrices G' are full rank corresponds to an Maximum Distance Separable (MDS) code and such combinatorial constructions are quite difficult to achieve.

Reed-Solomon codes [37] construct such matrices by exploiting properties of polynomials over finite fields. The key idea is that any k interpolation points suffice to recover the coefficients of a degree $k-1$ polynomial. The smallest field size q for which MDS codes exist is unknown, and related to the MDS conjecture of algebraic coding theory:
(MDS Conjecture) Let G be a $k \times n$ matrix over $GF(q)$ such that every square sub-matrix G' is nonsingular. Then $q+1 \geq n+k$.

A relaxation of this requirement (nearly-MDS) is that *almost all* the square sub-matrices G' are full rank, or equivalently that a randomly selected G' will be full rank with high probability. A *random linear code* over $GF(q)$ is the code generated by a matrix G that has each entry selected uniformly and independently from the finite field. It is well known [1] that the probability of a randomly selected G' being full rank can made arbitrarily close to one, by selecting a sufficiently large field size q.

5.2.2 Network Coding

Network coding is an exciting new paradigm for communication in networks where data packets are treated as entities which can be algebraically combined rather than simply routed and stored. The first major result [2] was a generalization of the max-flow min-cut theorem for multicasting. If there is one single source and multiple receivers, each receiver cannot hope to have throughput higher than the minimum cut separating it from the source, even if it was the only node being served. The theorem of Ahlswede et al. [2] states that if coding in the intermediate nodes of the network is allowed, *all the receivers* can have throughput equal to the minimum of the min-cuts separating each one from the source. In other words all the receivers can have the same throughput as the one with the weakest connection to the source, without limiting each other. It is easy to construct examples where such throughput cannot be achieved by simply routing packets from the source to the receivers.

Subsequently it was shown that linear coding suffices to achieve the multicast capacity [25, 24] and that random linear coding at intermediate nodes will suffice with high probability [18] for sufficiently large field size.

While most of the initial research on network coding focused on multicasting throughput, the fundamental idea of coding in intermediate nodes in networks has been shown to have advantages in other scenarios such as minimizing network resources [29], network diagnosis [44] and communication in wireless networks [23, 32, 45]. In this chapter we investigate applications of network coding for information storage, see [15] for a general introduction and other applications.

5.3 Coding for Networked Storage

5.3.1 Decentralized Erasure Codes

When trying to store linear combinations of data as information representations in sensor networks, new issues arise that make the existing codes unsuitable. Both random linear codes and Reed-Solomon codes have generator matrices that are *dense*, i.e. almost all the entries of G are non-zero. That means that every data node needs to send its packet to almost all n storage nodes to create the code generating $\Theta(n^2)$ communicating pairs (since $k = Rn$). A second desirable property is that the code can be created without coordination, and more specifically that each data node is choosing where to route its packet independently and also that the storage nodes are selecting their coefficients independently. Algebraically, this corresponds to having a code where every row of the generator matrix is created independently and is sparse. A code with this row independence property is called *decentralized* [11] and this property leads to stateless randomized network algorithms to generate the encoded information.

Randomized linear codes select every entry of G independently and therefore are decentralized [1]. However, they are not sparse and require significant communication to construct them. Algebraically the question is how sparse can a matrix with independent rows be made, and still have the property that square sub-matrices are full rank with high probability.

Decentralized erasure codes [10, 11] answer exactly this question: each data node routes its packet to $d(k) = c \ln k$ storage nodes. Each storage node selects random and independent coefficients f_i and stores a linear combination of the received packets. The main result of [10] is that $d(k) = c \ln k$ where $c > 5\frac{n}{k}$ is sufficient (and optimal up to constants) to ensure that randomly selected submatrices will be full rank with high probability. Decentralized erasure codes therefore have minimal data node degree and logarithmically many nonzero elements in every row.

Any erasure code can be decoded using Gaussian elimination in $O(k^3)$, but one can use the sparsity of the linear equations and have faster decoding.

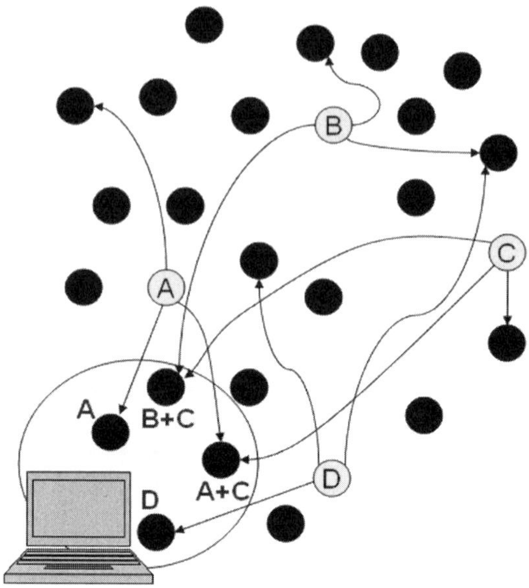

Fig. 5.4. Example of using linear codes for distributed storage. In this example there are $k = 4$ data nodes measuring information that is distributed and $n = 23$ storage nodes. We would like to diffuse the data to the storage nodes so that by accessing any 4 storage nodes it is possible to retrieve the data. Each data node is pre-routing to 3 randomly selected storage nodes. Each storage node has memory to store only one data packet so the ones who receive more than one packet store a linear combination of what they have received. The data collector in the example can recover the data by having access to (A, B+C, A+C, D).

Using the Wiedemann algorithm [43] one can decode decentralized erasure codes in $O(k^2 \log(k))$ time on average.

Randomized Network Algorithm

There is a very simple, robust randomized algorithm to construct a decentralized erasure code in a network: Each data node picks one out of the n storage nodes randomly and routes its packet to a randomly selected storage node. By repeating this process $d(k) = c \ln(k)$ times, we construct the decentralized erasure code. Note that we require a network layer mechanism that can route packets to randomly selected storage nodes in the network. Having a simple distributed mechanism that can perform this task with localized knowledge is key for many randomized algorithms and we discuss this issue in Section 5.5. Each storage node multiplies (over the finite field) whatever it happens to receive with coefficients selected uniformly and independently in $F(q)$ and stores the result and the coefficients. A schematic representation of this is given in Figure 5.5.

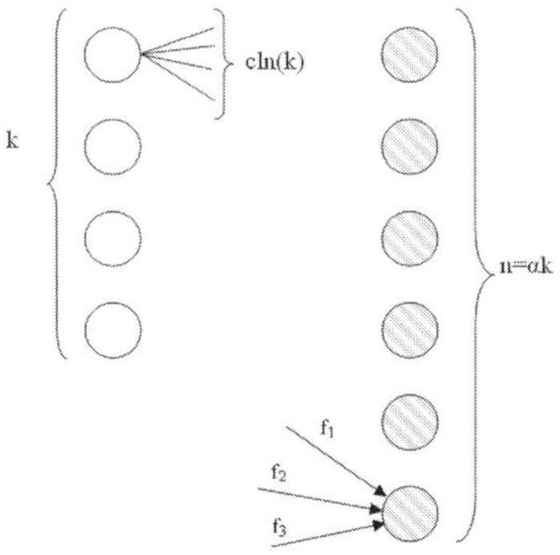

Fig. 5.5. Decentralized erasure codes construction. There are $d(k) = c\ln(k)$ edges starting from each data node and landing independently and uniformly on the storage nodes.

Storage Overhead

In addition to storing the linear combination of the received data packets, each storage node must also store the randomly selected coefficients f_i. The number of coefficients can be bounded by the number of balls that land into a bin when throwing $ck\ln(k)$ balls into n bins. It is standard problem in probabilistic analysis of algorithms [31] that the maximum load (the maximum number of coefficients a storage node will have to store) is $O(\log(k))$ with probability at least $1 - o(1)$. The total number of overhead bits to store the coefficients and data packet IDs is $O(\log(k)(\log(q) + \log(k)))$ which can be easily made negligible by picking larger data packet sizes. Notice that if we denote by $u = \log_2(q)$ the number of bits required to store each f_i, one can reduce the probability of error exponentially in the overhead bits.

Connections to network coding

An equivalent way of thinking of the distributed networked storage problem is that of a random bipartite graph connecting the k data nodes with the n storage nodes and then adding a data collector for every possible subset of size k of the n storage nodes. Then the problem of multicasting the k data packets to all the $\binom{n}{k}$ data collectors is equivalent to making sure that every collection of k storage nodes can reconstruct the original packets. This connection of storage and multicasting was proposed independently in [10, 21].

It has been shown that random linear network codes [25, 18] are sufficient for multicasting problems as long as the underlying network can support the required throughput. Decentralized erasure codes can therefore be seen as random linear network codes [18] on the (random) bipartite graph connecting the data and the storage nodes, where each edge corresponds to one routed packet. One key property is that in distributed storage, the communication graph does not correspond to any physical links but to virtual routing selections that are made by the randomized algorithm. Therefore this graph is not given, but can be *explicitly designed to minimize communication cost.* Essentially, we are trying to make this random bipartite graph as sparse as possible, while keeping the flow high enough and also allowing each data node to act independently. All the good codes described in previous sections have the property that they have very few edges ($o(n^2)$) connecting the data nodes and the storage nodes but can still guarantee very good connectivity between the any two subsets. Such bipartite graphs are called expanders [4] and are fundamental combinatorial objects for coding theory. It is easy to show that if one requires all $\binom{n}{k}$ data collectors to have k-connectivity with the data nodes, the corresponding bipartite graph needs to be dense. It is the probabilistic relaxation (a random data collector will have k-connectivity with high probability) that makes sparsity possible. This concept leads to *probabilistic expanders* that are formally defined and used for error correction in [8].

5.3.2 Fountain Codes

Fountain codes [27, 40] are linear codes over $GF(2)$ with sparse generator matrices and fast encoding and decoding algorithms. In particular, for LT codes [27], each encoded packet is created by first selecting a degree d from a carefully designed degree distribution (called the *robust soliton* [27]), and then taking the bitwise XOR of d randomly selected data packets. Therefore, fountain codes have the *rateless property*: every encoded packet is generated independently and there exists no predetermined rate since they can potentially generate an unbounded number of encoded packets. This corresponds to having every *column* of the generator matrix being independent and sparse (with logarithmic average degree similar to the decentralized codes). The degree distribution of the encoded packets is carefully designed so that a data collector who collects $k+\epsilon$ random packets (where the overhead ϵ is asymptotically vanishing for large k) can decode with a fast back-substitution algorithm which is special case of belief propagation [28] . Raptor codes [40] manage to reduce the degrees from logarithmic to constant by using an appropriate precode.

In this context, one can think of the decentralized property as being the "transpose" of the rateless property. This is because in decentralized codes, it is the rows of the generator matrix that are independent and this corresponds to having each data node acting independently. For sensor network applications, one implicit assumption is that it is easier for a data node to send its data to $d(k)$ randomly selected storage nodes than it is for a storage node to find and request packets from $d'(k)$ data nodes. This is true for many practical scenarios in which there are fewer data nodes that might also be duty-cycled or failing.

5.3.3 Partial data recovery

So far we have been addressing the problem of recovering all k data packets by querying k storage nodes. For this scenario, fountain codes are harder to create in networks, since creating the robust soliton degree distribution at storage nodes requires data node coordination. They however have the advantage of smaller field size (only binary operations) and lower computational complexity at the decoder ($O(k \log k)$ for LT codes versus $O(k^2 \log k)$ for decentralized codes). The pre-coding idea of Raptor codes cannot be easily performed over a network because it requires centralized processing.

Fountain codes can be used for partial recovery problems, where one is interested in querying fewer than k nodes and recover partial information. Creating a fountain code over a network where the data nodes are randomly located on a grid has been addressed in [12]. In this chapter there is no precode, and the user is interested in recovering $(1-\delta)k$ data packets by querying $(1+\epsilon)k$ storage nodes. Random walks [26] can be used to create fountain encoded packets in sensor networks, to guarantee the persistence and reliability of cached data.

Sanghavi [38] investigated the optimal degree distribution for fountain codes when one is interested in recovering $(1-\delta)k$ data packets. Upper bounds on the performance of any degree distribution and lower bounds achieved by optimized distributions for any δ are presented in [38].

5.4 Information representations for query processing

The standard approach in query processing is to flood queries to all nodes, and to construct a spanning tree by having each node maintain a routing table of their parents. This is the approach currently used in both TinyDB and Cougar [30]. Flooding can be pruned by constructing an analog to indexes in the network, and an efficient indexing scheme is the Geographic Hash Table (GHT), which maps IDs and nodes to a metric space [36]. These approaches yield different tradeoffs between reliability over network changes, latency and communication cost. Coding can be used to add storage redundancy in any existing query processing scheme when high reliability or low latency is required.

A common type of query that can often appear in sensor network applications is an aggregate query, involving the average of a sensed quantity in a subset of the nodes. The first problem that arises is the in-network computation of such averages. The simplest algorithm is the construction of a tree-structure that averages the data of interest. Ganesan et al. [16] propose the DIMENSIONS system which uses wavelets to efficiently summarize and store sensor data in a natural hierarchical structure. When high fault-tolerance is required, or the complexity required to form a tree is too high, gossip and consensus algorithms constitute very simple distributed and robust alternatives [6, 13, 39]. Gao et al. [20] exploit the principle of fractionally cascaded information to provide efficient algorithms and theoretical bounds for answering range queries.

5.4.1 Exploiting Data Sparsity

For sensor network applications, the sensed data could be highly structured and this structure can be exploited to improve the performance. One approach for exploiting the structure of the sensed data can be used if we assume that the data is sparse in some basis that is known to the data collector. Recent results (see for example [14]) show that the actual storage devices can be ignorant of the sparse basis and simply make random projections that can be used to reconstruct by solving a linear program. Rabbat et al. [34] showed how gossip algorithms can be used to construct such random projections in a sensor network. Data collectors who obtain access to enough such projections by querying storage nodes can reconstruct the field by exploiting the underlying sparsity. Further, Wang et al. [41] showed how *sparse* random projections can be used and further guarantee a refinable approximation that improves as more sensors are queried.

5.4.2 Distributed Source Coding

If the statistical correlation structure of the data is known (or can be learned), distributed compression can be used to minimize the redundant information without having to collect the data in one location. Distributed Source Coding Using Syndromes (DISCUS) [33] is a practical means of achieving this. The data nodes form the syndromes of the data packets they observe under suitable linear codes. These syndromes are treated as the data which the nodes pre-route to form the decentralized erasure codewords at the storage nodes. The data collector reconstructs the syndromes by gathering the packets from k storage nodes. Using DISCUS decoding, the collector can recover the original data from the syndromes. The correlation statistics, which is required by DISCUS can be learned by observing previous data at the collection point. The data nodes only need to know the rates at which they will compress their packets. This can be either communicated to them or learned adaptively in a distributed network protocol. The syndromes can be considerably shorter than the original data packets if the data observed by the different nodes are significantly correlated as is usually the case in sensor networks. Note that this approach is separating the source coding problem from the storage problem and this may not be optimal in general as shown in [35]. See also [46] for practical constructions.

5.4.3 Growth codes

In sensor network applications involving catastrophic or emergency scenarios such as floods, fires, earthquakes etc., the queries need to be adjusted to network dynamics. The setup is a rapidly failing sensor network where some nodes are sensing information that needs to reach the data collectors as soon as possible. Kamra et al. [22] show how fountain codes can be used for such applications and how the degree distribution needs to evolve over time to maximize the number of immediately recoverable data packets. Specifically, the authors design a dynamically varying degree distribution for partial network recovery to adapt to the data collector having received some data packets already and maximize the probability that the next packet is useful immediately. Growth codes initially create uncoded packets (since a data collector will have received nothing at the time and only degree one packets can be immediately useful). The degree distribution switches to pairwise XORs when the probability that a data collector already has a randomly selected packet becomes larger than the probability that the XOR cannot be decoded immediately.

5.4.4 The Repair problem

If the network is going to store data over long periods of time, with many nodes being duty cycled, it might be useful to monitor the storage nodes and actively refresh redundancy. When the number of storage nodes that actively respond falls below a threshold, fresh nodes (which might have been sleeping until that time) can be deployed to replace the failed ones and prolong data lifetime. However, the problem of creating new encoded packets in response to failures arises. When using replication, a new copy can be made from any other, but if the existing storage nodes are storing linear combinations of data, the problem of creating *new encoded packets from encoded packets* needs to be addressed. Regenerating codes [9] minimize the communication required to generate encoded packets from an existing encoded representation, to repair a failing network. In the same work, the minimal bandwidth required to repair any encoded storage scheme is computed explicitly using a flow formulation. Pyramid codes [19] are practical code constructions which trade space for partial recovery and efficient repair. While the repair problem has not been addressed in detail for sensor networks, simulations for distributed peer-to-peer storage systems suggest significant bandwidth savings over existing repair schemes.

5.5 Network Algorithms for Random Sampling

A very useful primitive operation in sensor networks is being able to find a (uniform) random node with small communication cost. All the coding constructions we have mentioned require such a mechanism, to route packets from data nodes to storage nodes. In addition, other applications like querying average sensor battery life, estimating the number of functional nodes and others could benefit from such a uniform sampling scheme [5].

If only local information is available at each node, it is not clear how to perform such a sampling without global knowledge. A simple idea is to accomplish this by performing a random walk in the network. We show that this is very costly for most relevant network topologies like grids and random geometric graphs. We further show that a simple greedy forwarding algorithm can approximately sample from random nodes using only local information and minimal communication.

5.5.1 Random Walks on Sensor Networks

Assume one wants to find a random node on a network by performing a random walk on the network nodes with properly adjusted transmission probabilities so that the invariant distribution is uniform over nodes. Clearly, after a few steps, one will be close to where the walk started and we need an estimate of how many hops are required before we have reached a truly random node. The number of steps required before the sampling distribution is reasonably close to uniform is measured by the *mixing time* of this Markov chain. To simplify the presentation assume we are dealing with a finite, irreducible and reversible Markov chain[1]. For two probability distributions θ_1, θ_2 defined on a finite space I, define the variation distance to be

$$\Delta(\theta_1, \theta_2) = \frac{1}{2} \sum_{i \in I} |\theta_1(i) - \theta_2(i)|. \tag{5.3}$$

At time step t, assume that the Markov chain has some probability distribution of being at state j after t steps starting from state i: $P_{ij}(t)$. Since the MC is irreducible and aperiodic, it will have a limiting invariant distribution $\pi(j)$ and we know that $P_{ij}(t) \stackrel{t\to\infty}{\to} \pi(j)$. We are interested, assuming the worst case starting state i, to bound the distance of the distribution at time t to the invariant distribution. Define this distance to be

$$d(t) = \max_i \Delta(P_{ij}(t), \pi(j)). \tag{5.4}$$

We define the mixing time (or "variation threshold time" [3]) to be

$$\tau = \min_t \{t \geq 0 : d(t) \leq \frac{1}{2e}\}. \tag{5.5}$$

It is therefore the first time where the Markov chain distribution becomes $1/2e$ close to the invariant, assuming the worst case starting state. The selection of

[1] Note that random walks on connected graphs are always in this class of Markov chains since self-loops can be added to cancel any periodic behavior.

the constant $1/2e$ is for algebraic convenience and any constant smaller than $1/2$ would work.

Therefore, if we use a random walk to sample from a sensor network, we need to perform $\Theta(\tau)$ steps before we are ϵ-close to the uniform distribution. The grid is the most simple model for a wireless sensor network topology, where n nodes are placed on a rectangular 4-connected square of edge length $\sqrt{n} \times \sqrt{n}$.

Unfortunately, even the 2-dimensional torus grid (which mixes faster than the regular grid) has a mixing time $\tau_{\text{Grid}} = \Theta(n)$ [3]. Therefore, one needs to perform $\Theta(n)$ random walk hops (visit approximately all the nodes in the network) to sample one random node.

Random Geometric Graphs

A random geometric graph $G(n, r)$ is formed as follows: place n nodes uniformly and independently in the unit square and connect nodes which are within distance r of each other (see Figure 5.5.1 for an example). Note that to simplify the analysis, some results rely on the assumption that the nodes are placed on the surface of a unit torus. Random geometric graphs have been established as standard models for wireless network topologies following the fundamental work of Gupta and Kumar [17] which shows that in order to have good connectivity and minimize interference, the transmission radius $r(n)$ has to scale like $\Theta(\sqrt{\frac{\log n}{n}})$.

Boyd et al. [7] investigate the question of the mixing time on $G(n, r)$ and establish that both the natural random walk (selecting each edge uniformly at random) and the fastest mixing reversible random walk (selecting the transition probabilities to minimize the mixing time) mix in:

$$\tau_{\text{RGG}} = \Theta(\frac{1}{r(n)^2}), \tag{5.6}$$

and for the critical radius $r(n) = \sqrt{\frac{\log(n)}{n}}$ we obtain

$$\tau_{\text{RGG}} = \Theta(\frac{n}{\log n}). \tag{5.7}$$

This slow mixing suggests that in random geometric graphs, it requires a very large number of hops to sample a random node.

5.5.2 Random Geographic Routing

We will now present a simple scheme that can sample an approximately uniform random node using only $O(1/r)$ hops which is equal to the diameter of the network and therefore order optimal. The key requirement is that we assume that all the nodes know their locations and the locations of their one-hop

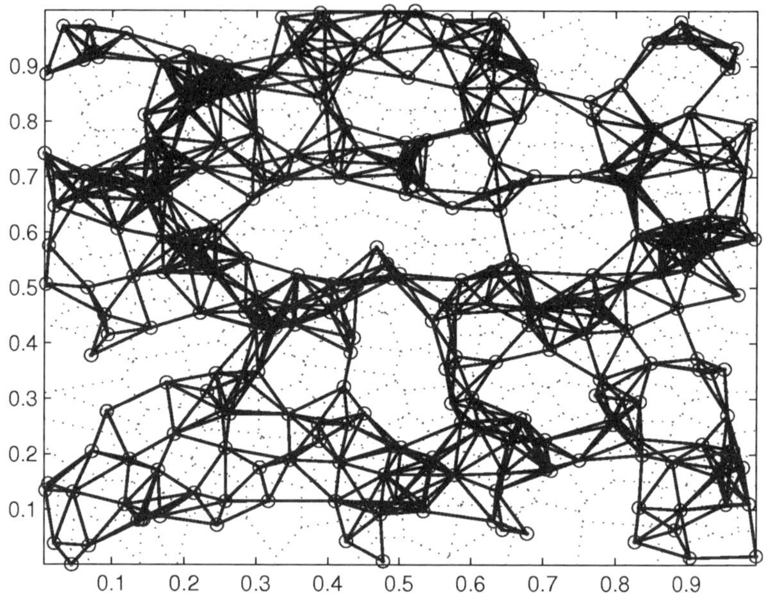

Fig. 5.6. Illustration of a Random Geometric Graph: solid lines represent graph connectivity, and dotted lines show the Voronoi regions associated with each node.

neighbors. Assume that some initial starting node s wants to find a random node in the network. The idea is to select a uniform geographic location in the unit square (called the target) and use greedy geographic routing towards that random target. Random geographic routing was proposed in [10] and independently in [5]. In [5] the authors further propose a rejection sampling scheme that can give sampling distributions that can be made very close to uniform. The performance of random geographic routing on random geometric graphs was analyzed in [13].

More formally, the random geographic routing scheme to sample a random node is the following:

1. Node s chooses a point uniformly in the unit square. Call this the target t. Node s forms the tuple $m_s = (l(s), t)$ where $l(s)$ is the geographic location of node s.
2. Node s sends m_s to its one-hop neighbor closest to t, if any exists. If node r receives a packet m_s, it sends m_s to its one-hop neighbor closest to t. Random geographic routing terminates when a node receives the packet and has no one-hop neighbor closer to the random target. Let v be that node, the output node sampled by the algorithm.

See [13] for a detailed analysis of the rejection sampling overhead for random geometric graphs and an application of random geographic routing for improving the convergence of gossip algorithms.

5.6 Conclusions

In this chapter we showed how erasure and network coding techniques can be used for storage in wireless sensor networks. The main conclusion is that sensor network applications introduce many novel challenges, mainly related to the distributed nature, as well as the communication and coordination constraints that naturally arise.

While some of these challenges have been addressed in the surveyed literature, numerous open problems remain. For example, the questions of combining the erasure encoding with multiresolution and distributed compression architectures, as well as faster encoding and decoding algorithms are among the issues that need to be addressed in future work. Distributed and scalable algorithms naturally fit with the randomized linear network coding theory and we believe that such ideas will be useful for practical applications.

References

[1] S. Acedanski, S. Deb, M. Médard, and R. Koetter. How good is random linear coding based distributed networked storage. In *NetCod*, 2005.
[2] R. Ahlswede, N. Cai, S.-Y. R. Li, and R. W. Yeung. Network information flow. *IEEE Trans. on Information Theory*, 46:1204–1216, 2000.
[3] D. Aldous and J. Fill. *Reversible Markov Chains and Random Walks on Graphs.* in preparation, 2005. (chapters available online), 2005.
[4] N. Alon and J. Spencer. *The Probabilistic Method.* Wiley Interscience, New York, 2000.
[5] B.A. Bash, J.W. Byers, and J. Considine. Approximately uniform random sampling in sensor networks. In *Proc. of the 1st Workshop on Data Management in Sensor Networks (DMSN '04)*, August 2004.
[6] S. Boyd, A. Ghosh, B. Prabhakar, and D. Shah. Gossip algorithms: Design, analysis and applications. In *Proceedings of the 24th Conference of the IEEE Communications Society (INFOCOM 2005)*, 2005.
[7] S. Boyd, A. Ghosh, B. Prabhakar, and D. Shah. Mixing times for random walks on geometric random graphs. In *Proceedings of SIAM ANALCO*, 2005.
[8] C. Daskalakis, A.G. Dimakis, R. M. Karp, and M. J. Wainwright. Probabilistic analysis of linear programming decoding. In *ACM-SIAM Symposium on Discrete Algorithms (SODA)*, January 2007.

[9] A.G. Dimakis, P.B. Godfrey, M.J. Wainwright, and K. Ramchandran. Network coding for distributed storage systems. In *Proceedings of IEEE Infocom*, 2007.

[10] A.G. Dimakis, V. Prabhakaran, and K. Ramchandran. Ubiquitous Acess to Distributed Data in Large-Scale Sensor Networks through Decentralized Erasure Codes. In *IEEE/ACM Int. Symposium on Information Processing in Sensor Networks (IPSN)*, April 2005.

[11] A.G. Dimakis, V. Prabhakaran, and K. Ramchandran. Decentralized erasure codes for distributed networked storage. In *IEEE Transactions on Information Theory*, June 2006.

[12] A.G. Dimakis, V. Prabhakaran, and K. Ramchandran. Distributed fountain codes for networked storage. In *Proceedings of IEEE ICASSP*, 2006.

[13] A.G. Dimakis, A.D. Sarwate, and M.J. Wainwright. Geographic gossip: Efficient aggregation for sensor networks. In *IEEE/ACM Int. Symposium on Information Processing in Sensor Networks (IPSN)*, 2006.

[14] D. Donoho. Compressed sensing. *IEEE Trans. Info Theory*, 52(4):1289–1306, April 2006.

[15] C. Fragouli, J.Y. Le Boudec, and J. Widmer. Network coding: an instant primer. *ACM SIGCOMM Computer Comm. Review*, 2006.

[16] D. Ganesan, D. Estrin, and J. Heidemann. Dimensions: Why do we need a new data handling architecture for sensor networks? In *Proceedings of ACM Computer Communication Rev., Volume 33, Number 1*, 2003.

[17] P. Gupta and P.R. Kumar. The capacity of wireless networks. *IEEE Transactions on Information Theory*, 46(2):388–404, March 2000.

[18] T. Ho, M. Médard, R. Koetter, D. Karger, M. Effros, J. Shi, and B. Leong. A random linear network coding approach to multicast. *IEEE Transactions on Information Theory*, October 2006.

[19] C. Huang, M. Chen, and J. Li. Pyramid codes: Flexible schemes to trade space for access efficiency in reliable data storage systems. In *IEEE International Symposium on Network Computing and Applications (NCA 2007)*, July 2007.

[20] J.Gao, L.J. Guibas, J. Hershberger, and L. Zhang. Fractionally cascaded information in a sensor network. In *IEEE/ACM Int. Symposium on Information Processing in Sensor Networks (IPSN)*, 2004.

[21] A. Jiang. Network coding for joint storage and transmission with minimum cost. In *International Symposium on Information Theory (ISIT)*, July 2006.

[22] A. Kamra, J. Feldman, V. Misra, and D. Rubenstein. Growth codes: Maximizing sensor network data persistence. *Proc. of ACM SIGCOMM*, 2006.

[23] S. Katti, H. Rahul, W. Hu, D. Katabi, M. Medard, and Jon Crowcroft. XORs in the air: Practical wireless network coding. *Proc. of ACM SIGCOMM*, 2006.

[24] R. Koetter and M. Médard. An algebraic approach to network coding. *Transactions on Networking*, October 2003.

[25] S.-Y. R. Li, R. W. Yeung, and N. Cai. Linear network coding. *IEEE Trans. on Information Theory*, 49:371–381, February 2003.
[26] Y. Lin, B. Liang, and B. Li. Data persistence in large-scale sensor networks with decentralized fountain codes. In *Proceedings of IEEE Infocom*, 2007.
[27] M. Luby. LT codes. *Proc. IEEE Foundations of Computer Science (FOCS)*, 2002.
[28] M. Luby, M. Mitzenmacher, M.A. Shokrollahi, and D. Spielman. Improved low-density parity check codes using irregular graphs. *IEEE Trans. Info. Theory*, 47:585–598, February 2001.
[29] D.S. Lun, N. Ratnakar, M. Médard, R. Koetter, D.R. Karger, T. Ho, E. Ahmed, and F. Zhao. Minimum-cost multicast over coded packet networks. *IEEE Transactions on Information Theory*, June 2006.
[30] S. Madden and J. Gehrke. Query processing in sensor networks. In *Pervasive Computing, 3(1)*, 2004.
[31] R. Motwani and P. Raghavan. *Randomized Algorithms.* Cambridge University Press, Cambridge, UK, 1995.
[32] D. Petrović, K. Ramchandran, and J. Rabaey. Overcoming untuned radios in wireless networks with network coding. *IEEE Transactions on Information Theory*, June 2006.
[33] S. S. Pradhan and K. Ramchandran. Distributed source coding using syndromes (DISCUS): Design and construction. *IEEE Trans. Info. Theory*, 49(3):626–643, 2003.
[34] M. Rabbat, J. Haupt, A. Singh, and R. Nowak. Decentralized compression and predistribution via randomized gossiping. In *IEEE/ACM Int. Symposium on Information Processing in Sensor Networks (IPSN)*, 2006.
[35] A. Ramamoorthy, K. Jain, P.A. Chou, and M. Effros. Separating distributed source coding from network coding. In *42nd Allerton Conference on Communication, Control and Computing*, 2004.
[36] S. Ratnasamy, B. Karp, L. Yin, F. Yu, D. Estrin, R. Govindan, and S. Shenker. GHT: A Geographic Hash Table for Data Centric Storage. In *Proceedings of the 1st ACM int. workshop on Wireless sensor networks and applications*, 2002.
[37] I.S. Reed and G. Solomon. Polynomial codes over certain finite fields. In *Journal of the SIAM*, 1960.
[38] Sujay Sanghavi. Intermediate performance of rateless codes. *Information Theory and Applications (ITA)*, 2007.
[39] O. Savas, M. Alanyali, and V. Saligrama. Randomized sequential algorithms for data aggregation in wireless sensor networks. In *Conference on Information Sciences and Systems (CISS)*, 2006.
[40] A. Shokrollahi. Raptor codes. *IEEE Trans. on Information Theory*, June 2006.
[41] W. Wang, M. Garofalakis, and K. Ramchandran. Distributed sparse random projections for refinable approximation. In *IEEE/ACM Int. Symposium on Information Processing in Sensor Networks (IPSN)*, 2007.

[42] Hakim Weatherspoon and John D. Kubiatowicz. Erasure coding vs. replication: a quantitiative comparison. In *Proc. IPTPS*, 2002.
[43] D. H. Wiedemann. Solving sparse linear equations over finite fields. In *IEEE Transactions on Information Theory*, 1986.
[44] C. Wu and B. Li. Echelon: Peer-to-peer network diagnosis with network coding. *Fourteenth IEEE International Workshop on Quality of Service (IWQoS)*, 2006.
[45] Y. Wu. On constructive multi-source network coding. In *International Symposium on Information Theory (ISIT)*, July 2006.
[46] Y. Wu, V. Stankovic, Z. Xiong, and S.-Y. Kung. on practical design for joint distributed source and network coding. In *Proc. 1st Workshop on Network Coding, Theory, and Applications (NetCod)*, April 2005.

Part III

Rate Constrained Field Reconstruction

6

Distributed Field Estimation with One–bit Sensors

Ye Wang, Nan Ma, Manqi Zhao, Prakash Ishwar, and Venkatesh Saligrama

Boston University, Boston, MA 02215 {yw,nanma,mqzhao,pi,srv}@bu.edu

6.1 Introduction

We study the problem of reconstructing a temporal sequence of unknown spatial data fields in a bounded geographical region of interest at a data fusion center from finite bit–rate messages generated by a dense noncooperative network of noisy low–resolution sensors (at known locations) that are statistically identical (exchangeable) with respect to the sensing operation. The interchangeability assumption reflects the property of an unsorted collection of inexpensive mass–produced sensors that behave in a statistically identical fashion. We view each data field as an unknown deterministic function over the geographical space of interest and make only the minimal assumption that they have a known bounded maximum dynamic range. The sensor observations are corrupted by bounded, zero–mean additive noise which is independent across sensors with arbitrary dependencies across field snapshots and has an arbitrary but unknown distribution but a known maximum dynamic range. The sensors are equipped with binary analog–to–digital converters (ADCs) (comparators) with random thresholds that are independent across sensors with arbitrary dependencies across snapshots and are uniformly distributed over a known dynamic range. These modeling assumptions partially account for certain real–world scenarios that include (i) the unavailability of good initial statistical models for data fields in yet to be well studied natural phenomena, (ii) unknown additive sensing/observation noise sources, (iii) additive model perturbation errors, (iv) substantial variation of preset comparator thresholds accompanying the mass–manufacture of low–precision sensors, (v) significant temperature fluctuations across snapshots affecting hardware characteristics, and (vi) the use of intentional dither signals for randomized scalar quantization.

In this work, a data fusion center is any point of data aggregation and/or processing in the sensor network and can be real or virtual. For instance, sensors can be dynamically organized into clusters with different sensors assuming the role of a fusion center at different times [1]. This work does not explicitly address physical–layer network data transport issues. In particular, we do not consider joint source–channel coding strategies. Instead, to conform with the existing base of digital communication architectures, the effective communication links are abstracted into a network of reliable but finite–rate bit–pipes from each sensor to the data fusion center. In practice, sensor data can be moved to the fusion center through a variety of physical–layer transport mechanisms, example, a stationary base–station with directional antenna arrays, a mobile data collector, and passive sensor querying mechanisms involving, for instance, laser–beams and modulating mirrors [2]. We acknowledge that separating the distributed field reconstruction problem into efficient data acquisition and efficient data transport parts through a finite–rate reliable bit–pipe abstraction may be suboptimal [3, p. 449], [4, 5]. For instance, in some scenarios multihop communication is not needed and the characteristics of the field, the communication channel, and the distortion–metric are "matched" to one another. In such a scenario, uncoded "analog" transmission can offer huge performance gains if the synchronization of sensor transmissions can be orchestrated at the physical layer to achieve beamforming gains and the network channel state information is available to the transmitting sensors [4]. For the joint source–channel aspects of this and related problems, see [6, 7, 8, 9]. For networking issues such as sensor scheduling, quality of service, and energy efficiency also see [10] and references therein.

Building upon prior results in [11, 12], and [13], we develop a simple coding and field reconstruction scheme based on noisy one–bit samples of noisy observations and characterize the associated scaling behavior of the MSE of field reconstruction with sensor density in terms of the local and global moduli of continuity of the underlying sequence of fields for fixed, positive, and equal sensor coding rates (bits per sensor per snapshot). These achievable results reveal that for bounded, zero–mean additive observation noise of unknown distribution, the MSE at every point of continuity of every field can be made to go to zero as sensor density increases while simultaneously sending the per–sensor bitrate and any sensing–related network rate overheads (e.g., sensor addresses) to zero. The rate of decay of field reconstruction MSE at a given location is related to the local modulus of continuity of the field at the given location and time. This is possible if the sensor placement and sampling schedule satisfy a certain uniformity property and if the field estimate at any given spatial location is formed using the observations from increasingly many sensors that are located within a vanishingly smaller neighborhood of the location. Specializing these results to the case of spatially constant fields yields an achievable MSE decay rate of $O(1/N)$ where N is the sensor network size.[1] A Cramér–Rao lower–bound on the MSE for parameter estimation establishes that the $O(1/N)$ MSE scaling behavior is order–optimal. Since in our problem

[1] Landau's asymptotic notation: $f(N) = O(g(N)) \Leftrightarrow \limsup_{N\to\infty} |f(N)/g(N)| < \infty$; $f(N) = \Omega(g(N)) \Leftrightarrow g(N) = O(f(N))$; $f(N) = \Theta(g(N)) \Leftrightarrow f(N) = O(g(N))$ and $g(N) = O(f(N))$.

formulation, the per–sensor bitrate is held fixed and equal across sensors, in a scaling sense, the MSE decreases inversely with the total network rate.

These results are consistent with the information–theoretic, total network rate versus MSE scaling results for the CEO problem which was first introduced in [14] and thereafter studied extensively in the information theory literature (see [15, 16] and references therein). However, it should be noted that information–theoretic rate–distortion studies of this and related distributed field reconstruction (multiterminal source coding) problems typically consider stationary ergodic stochastic fields with complete knowledge of the field and observation–noise statistics, block vector–quantization (VQ) and binning operations, and time–averaged (as opposed to worst–case) expected distortion criteria. In VQ, sensors are allowed to collect long blocks of real–valued field samples (of infinite resolution) from multiple field snapshots before a discrete, finite bit–rate VQ operation. The fields are often assumed to be spatially constant and independent and identically distributed (iid) across time (frequently Gaussian) and the observation noise is often assumed to be additive with a known distribution (frequently Gaussian) as in the CEO problem. It should also be noted that the MSE scaling results for the CEO problem in [15] are with respect to the total network rate where the number of agents (or sensors) has already been sent to infinity while maintaining the total network rate a finite value. Recent information–theoretic results for stationary fields under zero observation noise have been developed in [17, 18].

Previous estimation–theoretic studies of one–bit distributed field reconstruction have focused on reconstructing a single field snapshot and have either (i) assumed zero observation noise [11, 12], or (ii) assumed a spatially constant field (equivalent to scalar parameter estimation) with a one–bit communication as opposed to a one–bit sensing constraint. Our proposed system integrates the desirable field sensing and reconstruction properties of these apparently different one–bit field estimation schemes and establishes the statistical and performance equivalence of these approaches. An important hardware implication of this chapter is that noisy op–amps (noisy one–bit ADCs) are adequate for high–resolution distributed field reconstruction. This should be be contrasted with the framework in [13] which implicitly requires sensors to have the ability to quantize their observations to an arbitrarily high bit resolution. A side contribution of this chapter is the holistic treatment of the general distributed field–reconstruction problem in terms of (i) the field characteristics, (ii) sensor placement characteristics, (iii) sensor observation, quantization, and coding constraints with associated sensing hardware implications, (iv) transmission and sensing–related network overhead rates, and (v) reconstruction and performance criteria. We have attempted to explicitly indicate and keep track of what information is known/available/used and what is not.

The randomized scalar quantization model for the sensor comparators not only captures poor sensing capabilities but is also an enabling factor in the high–fidelity reconstruction of signals from quantized noisy observations. As shown in [19] in an information–theoretic setting and alluded to in [12], using *identical* deterministic scalar–quantization (SQ) in all sensors results in the MSE performance being fundamentally limited by the precision of SQ, irrespective of increasing sensor density, even in the absence of sensor observation noise.[2] However, our results further clarify that having "diversity" in the scalar quantizers, achieved, for example, through the means of an intentional random dither, noisy threshold, or other mechanisms, can achieve MSE performance that tends to zero as the density of sensors goes to infinity. Randomization enables high–precision signal reconstruction because zero–mean positive and negative fluctuations around a signal value can be reliably "averaged out" when there are enough independent noisy observations of the signal value. This observation is also corroborated by the findings reported in the following related studies [12, 11, 13, 20, 21].

The rest of this chapter is organized as follows. The main problem description with all technical modeling assumptions is presented in Sect. 6.2. The main technical results of this chapter are summarized and discussed in Sect. 6.3. Sect. 6.4 describes the proposed constructive distributed coding and field reconstruction scheme and the analysis of MSE performance. In Sect. 6.5, we discuss the close connections between the work in [12], [13], and the present work, and establish the fundamental statistical and performance equivalence of the core techniques in these studies. Finally, in Sect. 6.6 we present concluding remarks and comment on ongoing work and future research directions.

6.2 Problem Setup

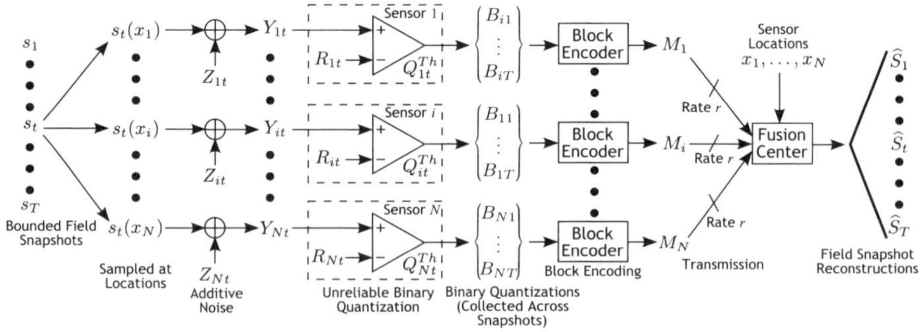

Fig. 6.1. *Block diagram of a distributed field reconstruction sensor–network using randomized* 1*–bit SQ with block–coding.* Sensor i quantizes its noisy observations, $Y_{i1}, \ldots, Y_{iT}$, to the binary values $B_{i1}, \ldots, B_{iT}$. The sensor then generates the message $M_i \in \{1, \ldots, 2^{rT}\}$ based on these quantized values. These messages $\{M_i\}$ are then relayed to the fusion center where the field estimates $\widehat{S}_t$ are produced.

[2] However, VQ with binning does not suffer from this limitation as shown in [17, 18].

6.2.1 Field Model

We consider a sequence of T discrete–time snapshots of a spatio–temporal field. If the spatio–temporal field is temporally bandlimited then the field values at intermediate time points can be interpolated from the estimates at discrete time snapshots if the temporal sampling rate is (strictly) higher than the temporal Nyquist rate of the field. The associated MSE will be no larger than the maximum MSE of the estimates across the discrete–time snapshots times a proportionality constant. Each snapshot is modeled as a continuous. More generally, our results can be extended to arbitrary, amplitude–bounded, measurable functions. For such functions the pointwise MSE bounds given in Sect. 6.3.1 still hold. The estimates at the points of continuity will have MSE tending to 0 as the network size scales. However, the points of discontinuity may have a finite, but non–zero MSE floor. bounded function,

$$s_t : G \rightarrow \mathbb{R} : \ \forall x \in G, \ \forall t \in \{1, \dots, T\}, \ |s_t(x)| \leq a < +\infty,$$

where $G \subseteq \mathbb{R}^d$ is a known geographical region of interest in d–dimensional real space and a is a known bound on the maximum field dynamic range. Although the results of this chapter hold for any G which is bounded and is the closure of its nonempty interior, for simplicity and clarity of exposition, we will assume $G = [0, 1]^d$, the d–dimensional unit–hypercube, in the sequel. Distances are measured with respect to a norm[3] $\|\cdot\|$, which for this work will be assumed to be the Euclidean 2-norm. Since the fields are continuous functions on the compact set G, they are in fact uniformly continuous on G [22].

Results on the fidelity of the field reconstruction will be described in terms of the local and global moduli of continuity associated with the field:

Definition 1. (Local modulus of continuity) *The local modulus of continuity* $\omega_t : [0, \infty) \times G \rightarrow [0, \infty)$ *of the function* $s_t(x)$ *at the point* $x \in G$ *is defined as*

$$\omega_t(\delta, x) \triangleq \sup_{\{x' \in G : \|x - x'\| \leq \delta\}} |s_t(x) - s_t(x')|.$$

Note that for all $x \in G$, $\omega_t(\delta, x)$ *is a nondecreasing function of* δ *and that it* $\longrightarrow 0$ *as* $\delta \longrightarrow 0$ *since* $s_t(x)$ *is continuous at each point* x *in* G.

Definition 2. (Global modulus of continuity) *The global modulus of continuity* $\widetilde{\omega}_t : [0, \infty) \rightarrow [0, \infty)$ *of the function* $s_t(x)$ *is defined as*

$$\widetilde{\omega}_t(\delta) \triangleq \sup_{x \in G} \omega_t(\delta, x).$$

Again note that $\widetilde{\omega}_t(\delta)$ *is a nondecreasing function of* δ *and that it* $\longrightarrow 0$ *as* $\delta \longrightarrow 0$ *since* $s_t(x)$ *is uniformly continuous over* G.

[3] For asymptotic results in which distance $\longrightarrow 0$, any norm on $\mathbb{R}^d$ would suffice since all norms on any finite–dimensional Banach space are equivalent [22, Theorem 23.6, p. 177].

The global and local moduli of continuity of a spatial field respectively reflect the degree of global and local spatial smoothness of the field with smaller values, for a fixed value of δ, corresponding to greater smoothness. For example, for a spatially constant field, that is, for all $x \in G$, $s_t(x) = s_t$ (a constant), we have $\widetilde{\omega}_t(\delta) = 0$ for all $\delta \geq 0$. For $d = 1$ and fields with a uniformly bounded derivative, that is, for all $x \in G$, $\sup_{x \in G} |d(s_t(x))/dx| = \Delta < +\infty$, $\widetilde{\omega}_t(\delta) \leq \Delta \cdot \delta$. More generally, for a Lipschitz–γ spatial function (see [11]) $s_t(x)$, we have $\widetilde{\omega}_t(\delta) \propto \delta^{\gamma}$. Closed–form analytical expressions of moduli of continuity may not be available for arbitrary fields but bounds often are. Sometimes bounds that are tight in the limit as $\delta \longrightarrow 0$ are also available. From Definitions 1, 2, and the boundedness of the field dynamic range, it also follows that for all $\delta \geq 0$, for all $x \in G$, and for all $t \in \{1, \ldots, T\}$, we have

$$0 \leq \omega_t(\delta, x) \leq \widetilde{\omega}_t(\delta) \leq 2a < +\infty.$$

6.2.2 Sensor Placement

We assume that we have a dense, noncooperative network of N sensors distributed uniformly over a hypercube partitioning of $G = [0, 1]^d$. The space $G = [0, 1]^d$ is uniformly partitioned into $L = l^d$ (where l is an integer) disjoint, hypercube supercells of side–length $(1/l)$. Each supercell is then further uniformly partitioned into $M = m^d$ (where m is an integer) hypercube subcells of side–length $(1/(lm))$, giving a total of LM subcells. In our distributed field coding and reconstruction scheme, described in Sect. 6.4, the field estimate for each snapshot is constant over each supercell and formed by averaging the measurements from a partial set of the sensors, determined by the subcells. This field reconstruction scheme requires knowledge of the sensor locations only up to subcell membership. Therefore, it has some natural robustness against sensor location uncertainty or error. The significance of the super and subcells will become clear in the sequel (Sects. 6.3 and 6.4).

We assume that the sensor deployment mechanism is able to uniformly distribute the sensors over the subcells. We define this uniform sensor deployment condition with:

Definition 3. (Uniform sensor deployment) *We say that a sensor deployment method is uniform if exactly* $n \triangleq (N/(LM))$ *sensors are located in each subcell.*

We also assume that each sensor is aware of which subcell it is in. See Fig. 6.2 for an illustration of the cell hierarchy and example sensor deployment for the $d = 2$ dimensional case.

6.2.3 Sensor Observation and Coding Models

Sensor Observation Noise

The sensor observations are corrupted by bounded, zero–mean additive noise which is independent across sensors, but can be arbitrarily correlated across

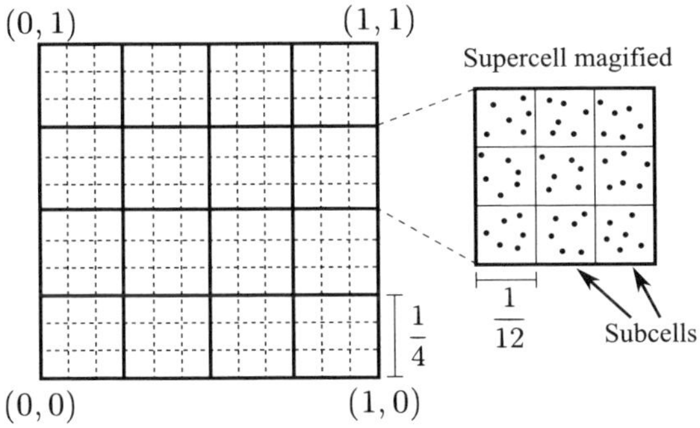

Fig. 6.2. *Example uniform sensor deployment and cell hierarchy* over $[0,1]^2$ ($d = 2$ dimensional case) with $N = 864$ sensors deployed over $L = 4^2$ supercells of side–length $(1/4)$ and $M = 3^2$ subcells per supercell of side–length $(1/(3 \cdot 4))$, resulting in 6 sensors per subcell.

field snapshots[4]. Let Z_{it} denote the noise affecting the observation of the tth snapshot by the ith sensor, and define the $\mathbf{Z} \triangleq \{Z_{it}\}_{i=1,t=1}^{N,T}$ (the collection of all of the noise random variables) and $\mathbf{Z}_i \triangleq \{Z_{it}\}_{t=1}^{T}$ (the collection of all of the noise random variables for a given sensor i). The noise $\mathbf{Z}$ has an unknown joint cumulative distribution function (cdf) $F_{\mathbf{Z}}(\mathbf{z})$ that can be arbitrary within the zero–mean, boundedness and independence constraints already stated. The maximum dynamic range of the noise $b \in [0, +\infty)$ is known. The noisy observation of field snapshot $t \in \{1, \ldots, T\}$ made by sensor $i \in \{1, \ldots, N\}$ is given by

$$Y_{it} = s_t(x_i) + Z_{it},$$

where x_i is the location of the ith sensor and $\mathbf{Z} \sim$ cdf $F_{\mathbf{Z}}(\mathbf{z})$. We use $\mathcal{F}$ to denote the set of all joint cdfs that are factorizable into N zero–mean joint cdfs on $\mathbb{R}^T$ with support within $[-b, +b]^T$, that is, $F_{\mathbf{Z}}(\mathbf{z}) = \prod_{i=1}^{N} F_{\mathbf{Z_i}}(\mathbf{z_i})$ where $F_{\mathbf{Z_i}}(\mathbf{z_i})$ is a zero–mean joint cdf (corresponding to the noise random variables for sensor i) with support within $[-b, +b]^T$. Note that $\mathcal{F}$ captures the feasible set of joint noise cdfs for the bounded–amplitude, zero–mean, and independence assumptions. Note that $|Y_{it}| \leq |s_t(x_i)| + |Z_{it}| \leq c \triangleq (a + b)$.

Randomized 1–bit SQ with Block Coding

Due to severe precision and reliability limitations, each sensor $i \in \{1, \ldots, N\}$, has access to only to a vector of unreliable binary quantized samples $\mathbf{B}_i \triangleq (B_{i1}, \ldots, B_{iT})$ for processing and coding and not direct access to the real–valued noisy observations $Y_{i1}, \ldots, Y_{iT}$. The quantized binary sample B_{it} is generated from the corresponding noisy observation Y_{it} through a randomized mapping $Q_{it} : [-c, c] \to \{0, 1\}$: for each $i \in \{1, \ldots, N\}$ and each

[4] The measurement snapshot timers of all the participating sensors are assumed to be synchronized.

$t \in \{1, \ldots, T\}$,

$$B_{it} = Q_{it}(Y_{it}),$$

where we assume that the mappings Q_{it} are independent across sensors i, but can be arbitrarily correlated across snapshots t. We denote the conditional marginal statistics of the quantized samples by $p_{B_{it}|Y_{it}}(y) \triangleq \mathbb{P}(B_{it} = 1|Y_{it} = y)$. We are specifically interested in cases where $p_{B_{it}|Y_{it}}(y)$ is an affine function of y since it allows estimates of the fields to be made from the B_{it}'s without knowledge of the noise distribution (see Appendix). Specifically we consider the conditional distribution

$$p_{B_{it}|Y_{it}}(y) = \left(\frac{y+c}{2c}\right).$$

This conditional distribution can be achieved by a quantization method which is based on comparing the noisy observation with a random uniformly distributed threshold given by

$$B_{it} = Q_{it}^{Th}(Y_{it}) \triangleq \mathbf{1}(Y_{it} > R_{it}), \tag{6.1}$$

where the R_{it}'s are Unif$[-c, c]$ random thresholds which are independent across sensors i, but arbitrarily correlated across snapshots t, and $\mathbf{1}(\cdot)$ denotes the indicator function:

$$\mathbf{1}(Y_{it} > R_{it}) = \begin{cases} 1 & \text{if } Y_{it} > R_{it}, \\ 0 & \text{otherwise.} \end{cases}$$

This uniform random–threshold 1–bit SQ model partially accounts for some practical scenarios that include (i) comparators with a floating threshold voltage, (ii) substantial variation of preset comparator thresholds accompanying the mass–manufacture of low–precision sensors, (iii) significant environmental fluctuations that affect the precision of the comparator hardware, or generally (iv) unreliable comparators with considerable sensing noise and jitter. An alternative justification is that the random thresholds are intentionally inserted as a random dither. Scenario (i) can be accommodated by independence across snapshots, scenario (ii) can be accommodated by complete correlation (fixed) across snapshots, and scenarios (iii) and (iv) can be accommodated by arbitrary correlation across snapshots.

Each sensor i utilizes a block encoder to "compress" its vector of T quantized samples $\mathbf{B}_i$ to a message $M_i \in \{1, 2, \ldots, 2^{rT}\}$ before transmitting to the fusion center. The block encoder and message for sensor i are given by

$$f_i : \{0,1\}^T \to \{1, 2, \ldots, 2^{rT}\}, \quad M_i = f_i(B_{i1}, \ldots, B_{iT}),$$

where r is the coding rate in bits per sensor per snapshot. For $r \geq 1$ compression is trivial since $\mathbf{B}_i$ can assume no more than 2^T distinct values which can be indexed using T bits.

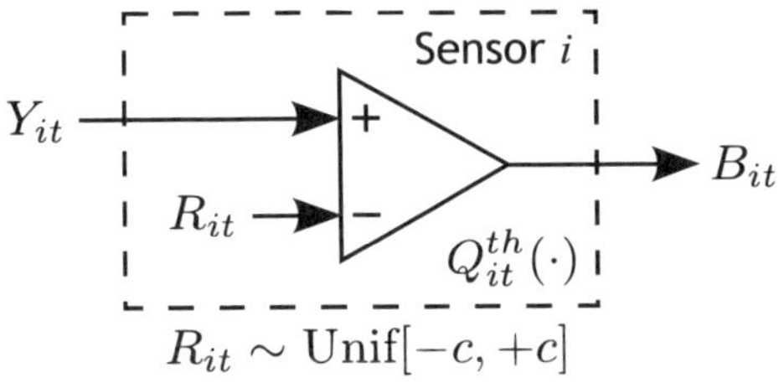

Fig. 6.3. *Quantizer Hardware Example:* The sensing model described by the $Q_{it}^{Th}(\cdot)$ function in (6.1) can be implemented by a comparator with a uniformly distributed threshold. These thresholds are independent across sensors, but arbitrarily correlated across snapshots, allowing many scenarios to be accommodated.

6.2.4 Transmission and Field Reconstruction

Our problem setup abstracts the underlying transmission network of sensors as a network of bit pipes that are capable of reliably delivering these N messages (the payloads) and the network addresses of the message origination nodes (the headers) to the fusion center. This enables the fusion center to correctly associate the spatial location information with the corresponding sensor field–measurement information for reliable field reconstruction. For our reconstruction scheme, the fusion center only needs to be able to differentiate which subcell each message originated from. This can be achieved by having each sensor append a $\log(LM)$ bits long label to its message. This results in a total sensor–location rate–overhead of $r_{ohd} = (N/T)\log(LM)$ bits per snapshot on the network information transport costs. This overhead will be negligible if $T \gg N\log(LM)$. If the underlying sequence of fields are spatially constant, then, the sensor location information is not needed at the fusion center (see Corollary 1 and Sect. 6.4).

The fusion center forms the estimates of the T fields based on the sensor messages using the reconstruction functions

$$g_t : G \times \{1, 2, \ldots, 2^{rT}\}^N \to [-a, a], \ \forall t \in \{1, \ldots, T\}.$$

The estimate of field t at point $x \in G$ is given by

$$\widehat{S}_t(x) = g_t(x, M_1, \ldots, M_N).$$

Definition 4. (Rate–r DFRS) *A rate–r Distributed Field Reconstruction Sensor–network (DFRS) based on randomized 1–bit SQ with block coding is defined by the set of rate–r encoder functions $\{f_i(\cdot)\}_{i=1}^N$ and the set of reconstruction functions $\{g_t(\cdot)\}_{t=1}^T$.*

Figure 6.1 depicts a rate–r DFRS using randomized 1–bit SQ with block coding.

Performance Criterion

Definition 5. (Pointwise MSE) *The pointwise MSE of the estimate of field t at location $x \in G$, for a given rate–r DFRS and a specific noise joint cdf $F_{\mathbf{Z}}(\mathbf{z}) \in \mathcal{F}$, is given by*

$$D_t(x; F_{\mathbf{Z}}) = \mathbb{E}[(\widehat{S}_t(x) - s_t(x))^2].$$

Since we are interested in schemes that will work for *any* noise cdf in $\mathcal{F}$, we consider the worst–case $D_t(x; F_{\mathbf{Z}})$ over all possible $F_{\mathbf{Z}} \in \mathcal{F}$. We also consider the maximization over all fields and all locations in G since we want to reconstruct every point of every field with high fidelity.

Definition 6. (Worst–case MSE) *The worst–case MSE D is given by*

$$D = \max_{t \in \{1,\ldots,T\}} \sup_{x \in G} \sup_{F_{\mathbf{Z}} \in \mathcal{F}} D_t(x; F_{\mathbf{Z}}).$$

Our objective is to understand the scaling behavior of MSE with N, T, and r. The next section summarizes our partial results in this direction.

6.3 Main Results

6.3.1 Achievable MSE Performance

Our first result gives an upper bound on the MSE achievable through a constructive DFRS based on randomized 1–bit SQ with block coding for rate $r = 1/M$, where M is the number of subcells per supercell. The actual scheme will be described in Sect. 6.4. The MSE analysis appears within the proof of the theorem detailed in Appendix. This achievable MSE upper bound can be made to decrease to zero as sensor–density goes to infinity (see (6.2)) without knowledge of the local or global smoothness properties of the sequence of fields. Furthermore, this scheme is universal in the sense that it does not assume knowledge of $F_{\mathbf{Z}}(\mathbf{z})$ beyond membership to $\mathcal{F}$.

Theorem 1. (Achievable MSE performance: Randomized 1–bit SQ and $r = 1/M$) *There exists a rate–$r = 1/M$ DFRS based on randomized 1–bit SQ with block coding (described in Sect. 6.4) such that for all $x \in G$, $t \in \{1,\ldots,T\}$, and $F_{\mathbf{Z}}(\mathbf{z}) \in \mathcal{F}$,*

$$\begin{aligned} D_t(x; F_{\mathbf{Z}}) &\le \omega_t^2\left(\frac{\sqrt{d}}{\sqrt[d]{L}}, x\right) + \left(\frac{LMc^2}{N}\right) \\ &\le \widetilde{\omega}_t^2\left(\frac{\sqrt{d}}{\sqrt[d]{L}}\right) + \left(\frac{LMc^2}{N}\right). \end{aligned}$$

Proof. See Appendix.

Note that Theorem 1 holds for arbitrary fields. The modulus of continuity terms in the local (first) and global (second) upper bounds of Theorem 1 are due to the bias of the field estimates and the $\left(\frac{LMc^2}{N}\right)$ term is due to the variance of the field estimates (see (6.4) in Sect. 6.4). From Theorem 1 and the properties of moduli of continuity (see Sect. 6.2.1), it follows that for the coding and reconstruction scheme of Sect. 6.4, as $N \longrightarrow \infty$, the estimate $\widehat{S}_t(x)$ uniformly converges, in a mean square sense, to $s_t(x)$ for all $x \in G$, provided that

$$\text{(i)} \left(\frac{N}{L}\right) \longrightarrow \infty, \text{ and (ii) } L \longrightarrow \infty. \tag{6.2}$$

It also follows that the worst–case MSE scaling behavior (see Definition 6) is bounded by

$$D \leq \max_{t \in \{1,\ldots,T\}} \left\{ \widetilde{\omega}_t^2 \left(\frac{\sqrt{d}}{\sqrt[d]{L}}\right) + \left(\frac{LMc^2}{N}\right) \right\} \tag{6.3}$$

and that $D \longrightarrow 0$ as N and L scale as in (6.2).

Implications: These results allow us to make the per sensor per snapshot bit rate r, worst–case MSE D, and sensor message ID overheads (given by $(N/T)\log(LM)$ bits) simultaneously smaller than any desired values $r^*, D^*, \epsilon > 0$, respectively. First, we can choose a sufficiently large number of subcells per supercell M^* such that the rate $r = 1/M^* < r^*$. Then we can choose a sufficiently large number of sensors N^* and number of supercells L^* such that the bound on D given by (6.3) is made less than D^*. Note that both N^* and M^* can be further increased while keeping the ratio M^*/N^* fixed without changing the bound on D. This corresponds to increasing the total number of sensors N, decreasing the per sensor rate $r = 1/M$, but keeping the total network per snapshot rate $Nr = N/M$ and distortion D fixed. Finally, we can look at a sufficiently large number of snapshots T^* such that network message overheads $(N^*/T^*)\log(L^*M^*) < \epsilon$.

In the constructive coding and field reconstruction scheme of Sect. 6.4, the field estimates are piecewise constant over the supercells. The estimate in each supercell is formed from only $n = (N/(LM))$ of the $Mn = (N/L)$ quantized observed values coming from the sensors located in that supercell. Since only $(1/M)$ of the total available quantized observed values for each snapshot are used, the transmission rate of $(1/M)$ is achievable by indexing only the necessary values (see Sect. 6.4 for details). As the number of supercells L increases, the piecewise constant estimate becomes finer and the bias is decreased. Also, as the number of sensors per supercell is increased, more observations are used thus decreasing the variance of the estimate.

Since the variance term $\frac{LMc^2}{N}$ in the upper bound of Theorem 1 can decrease no faster than $O(1/N)$, the decay of the global MSE upper bound, in the proposed constructive scheme, can be no faster than $O(1/N)$. However, the decay rate of $\frac{LMc^2}{N}$ is hindered by the fact that L simultaneously needs to approach infinity for the bias term $\widetilde{\omega}_t^2\left(\frac{\sqrt{d}}{\sqrt[d]{L}}\right)$ to decay to 0. When $\widetilde{\omega}_t(\cdot)$ is not identically zero, a bias–variance tradeoff exists and the appropriate relative growth rate for L with N that minimizes the decay rate of the global MSE upper bound of Theorem 1 is determined by the following condition

$$\widetilde{\omega}_t^2\left(\frac{\sqrt{d}}{\sqrt[d]{L}}\right) = \Theta\left(\frac{L}{N}\right).$$

For certain classes of fields for which the global modulus of continuity has a closed form, the optimum growth rate can be explicitly determined. For instance, if $d = 1$ and $\widetilde{\omega}_t(\delta) = \Delta \cdot \delta$ (Lipschitz–1 fields), $L_{opt}(N) = \Theta(N^{1/3})$ for which MSE $= O(N^{-2/3})$.

For the class of constant fields, we immediately have the following corollary.

Corollary 1. (Achievable MSE performance: Randomized 1–bit SQ, $r = 1/M$, and constant fields) *If for all $x \in G$ and all $t \in \{1, \ldots, T\}$, we have $s_t(x) = s_t$, or equivalently, for all $\delta \geq 0$ and all $t \in \{1, \ldots, T\}$, $\widetilde{\omega}_t(\delta) = 0$, then the result given by (6.3) reduces to*

$$D \leq \left(\frac{Mc^2}{N}\right),$$

where we can set $L = 1$ to minimize the bound.

Only $L = 1$ supercells are needed for an accurate piecewise constant reconstruction of a constant field. Furthermore, all snapshot–estimates given by the scheme from Sect. 6.4 are unbiased in this case. Also, the spatial locations of sensors are irrelevant: the MSE behavior is governed purely by the number of sensors N regardless of how they are distributed over the subcells. The N sensors must still be uniformly assigned to one of M groups (for the purpose of transmission coordination to achieve the compression factor of $1/M$), however these groups do not need to have any geographical significance.

6.3.2 Order–Optimal Minimax MSE for Constant Fields

The minimax reconstruction MSE over the class of constant fields is given by

$$\inf_{\{g_t\}_{t=1}^{t=T}} \sup_{F_{\mathbf{Z}} \in \mathcal{F}, s_t \in \mathcal{S}} D,$$

where the infimum is taken over all possible estimators and the supremum is taken over all noise distributions and fields from the class of constant fields which is denoted by $\mathcal{S}$. The achievable MSE result given by Corollary 1 establishes an upper bound on the minimax reconstruction MSE. Theorem 2 lower bounds the minimax reconstruction MSE for any rate r DFRS that produces unbiased estimates for the case of spatially constant fields.

Theorem 2. (Lower bound on MSE: Unbiased estimators for constant fields) *For a sequence of spatially constant fields and any DFRS which produces unbiased field estimates, there exists a joint cdf $F_{\mathbf{Z}} \in \mathcal{F}$ such that for noise distributed according to $F_{\mathbf{Z}}$ the MSE is lower bounded by*

$$\mathbb{E}[(\widehat{S}_t - s_t)^2] \geq \left(\frac{C_t}{N}\right), \quad \textit{for all } t \in \{1, \ldots, T\},$$

where C_t is finite, non–zero, and does not depend on N. Therefore,

$$\inf_{\{g_t\}_{t=1}^{t=T}} \sup_{F_{\mathbf{Z}} \in \mathcal{F}, s_t \in \mathcal{S}} D \geq \max_{t \in \{1,\ldots,T\}} \left(\frac{C_t}{N}\right).$$

Proof. Since $\{s_t\} \rightarrow \{Y_{it}\} \rightarrow \{B_{it}\} \rightarrow \{M_i\}$ forms a Markov chain, the estimates based on the sensor messages $\{M_1, \ldots, M_N\}$ cannot have a lower MSE than estimates based on the noisy observations $\{Y_{it}\}$. Let $F_{\mathbf{Z}} \in \mathcal{F}$ be any well–behaved, non–trivial, joint cdf such that the Z_{it} are iid and the conditional probabilities of Y_{it} given the fields satisfy the regularity conditions necessary for the Cramér-Rao bound [23] to be applied. By the Cramér-Rao bound, the MSE of each field estimate based on $\{Y_{it}\}$ is lower bounded by $\frac{C_t}{N}$ where C_t is finite, non–zero, and depends on $F_{\mathbf{Z}}$, but does not depend on N. Note that the bound also applies to general randomized 1–bit SQ functions $Q_{it}(\cdot)$ including those based on uniform random thresholds $Q_{it}^{Th}(\cdot)$ (see (6.1)).

Combining the results of Corollary 1 and Theorem 2 establishes that the order–optimal minimax MSE for spatially constant fields is $\Theta(1/N)$ and that the scheme of Sect. 6.4 achieves this order optimal performance.

6.4 Proposed Constructive Distributed Coding and Field Reconstruction Scheme

In this section we present the proposed DFRS scheme that was alluded to in Sect. 6.3. In this scheme, sensors create the quantized binary samples $\{B_{it}\}$ from their observations $\{Y_{it}\}$ through comparisons with the random thresholds $\{R_{it}\}$, as described in (6.1) of Sect. 6.2.3. The field estimates are piecewise constant over the supercells, where the estimate formed in each supercell is a function of only $(N/(LM))$ of the (N/L) quantized observed values coming from the sensors located in that supercell. This allows fractional transmission rates of $r = 1/M$ through a simple time–sharing based compression method. Note that there can be uncertainty in the sensor locations, within a degree given by the size of a subcell, at the fusion center, since it is only necessary for the fusion center to know which subcell each sensor is located in.

Each sensor i, instead of transmitting all of its T bits (the vector of its binary quantized observations $\mathbf{B}_i = (B_{i1}, \ldots, B_{iT})$), transmits only $rT = T/M$ of them and the remaining observations are dropped. The two–level hierarchy of supercells and subcells described in Sect. 6.2.2 is used in order to properly determine which bits sensors should drop or keep. Within each supercell, each sensor i from subcell $k \in \{1, \ldots, M\}$ communicates only every M^{th} bit (offset by k), that is $\{B_{i,k+Ml}\}_{l=0}^{l=(T/M)-1}$. These rT bits can be uniquely represented by the message $M_i \in \{1, \ldots, 2^{rT}\}$ and losslessly communicated to the fusion center. Thus for snapshot $t \in \{1, \ldots, T\}$, only the bits from senors in the $[((t-1) \bmod M) + 1]^{\text{th}}$ subcell of each supercell are communicated to the fusion center. We will denote the set of indices corresponding to the $n = (N/(LM))$ sensors belonging to the $[((t-1) \bmod M) + 1]^{\text{th}}$ subcell of supercell j by $I(j, t)$. This set of indices corresponds to the sensors that are located in supercell j and encode a bit for snapshot t.

For notational simplicity, the reconstruction $\widehat{S}_t(x) = g_t(x, M_1, \ldots, M_N)$ will be described directly in terms of the available binary quantized observations[5] B_{it}. The piecewise constant function $\widehat{S}_t(x)$ is described in two parts. First, the constant estimate of $s_t(x)$ over supercell j is given by

$$\widehat{S}_{tj} \triangleq 2c \left[\frac{1}{n} \sum_{i \in I(j,t)} B_{it} \right] - c,$$

which is the average (shifted and scaled into $[-c, +c]$) of the available quantized binary observations of snapshot t from sensors located in supercell j. The piecewise–constant estimate for $s_t(x)$ is then given by

$$\widehat{S}_t(x) = g_t(x, M_1, \ldots, M_N) \triangleq \sum_{j=1}^{L} \widehat{S}_{tj} \mathbf{1}(x \in \mathcal{X}_j), \tag{6.4}$$

where $\mathcal{X}_j \subseteq [0, 1]^d$ is the set of points within the jth hypercube supercell and $\mathbf{1}(x \in \mathcal{X}_j)$ is given by

$$\mathbf{1}(x \in \mathcal{X}_j) = \begin{cases} 1 & \text{if } x \in \mathcal{X}_j, \\ 0 & \text{otherwise.} \end{cases}$$

An alternative approach is for the fusion center to use a piecewise–linear (as used in [12]) or other higher–order interpolation algorithms such as those based on cubic B–splines. The resulting MSE will be of the same order. We use the former approach because its analysis is more compact. Appendix proves that the MSE of this constructive coding and reconstruction scheme is upper bounded by the result described in Theorem 1.

6.5 Related One-bit Estimation Problems

This section discusses the connections between the methods and results in [12], [13], and the present work. It is shown that the apparently different

---

[5] The set of binary quantized observations available at the fusion center is given by $\{B_{it}\}_{\{i \in \cup_{j=1}^{L} I(j,t)\}}$

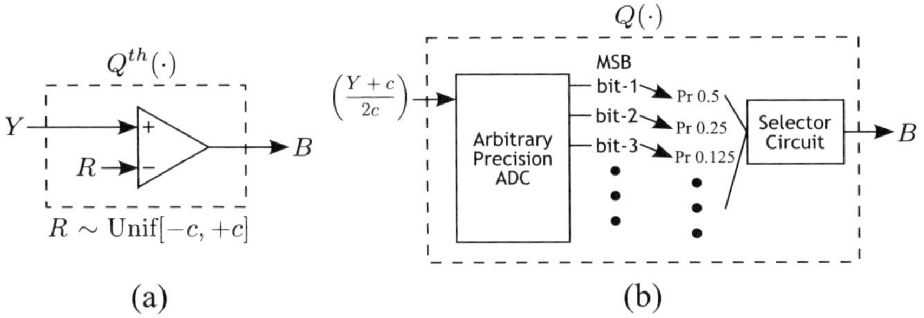

Fig. 6.4. *The* $Q_{it}^{Th}(\cdot)$ *function in (6.1) and the* $Q(\cdot)$ *function of [13]* suggest markedly different hardware implementations. The former naturally suggests (a), where the binary quantized value is produced by a simple comparison to a random threshold X. The latter suggests (b), where an arbitrarily–precise ADC circuitry probabilistically selects an arbitrary bit of the observed value. Interestingly, these two implementations produce statistically equivalent quantized outputs B given identical inputs Y.

randomized 1–bit field estimation schemes in these studies are in fact statistically and MSE performance equivalent. The general framework of the present work integrates the desirable field sensing and reconstruction properties and insights of the earlier studies and provides a unified view of the problem that simultaneously considers unreliable binary quantization, unknown arbitrary noise distributions, multiple snapshots of a temporally and spatially varying field, and communication rate issues. Since the work in both [12] and [13] deal with the reconstruction of only a single snapshot ($T = 1$), we will drop the snapshot indices t in our discussion to aid comparison.

6.5.1 One–Bit Randomized–Dithering

The problem setup of [12] considers the reconstruction of a single snapshot ($T = 1$) of a bounded, one–dimensional field ($d = 1$) from noiseless samples ($Z_i = 0$) that have been taken at deterministic sampling locations ($x_i = i/N$) and similarly randomly binary quantized by a threshold R_i that has a general distribution, which must satisfy certain technical conditions as described in [12, Sect. II.A]. These technical conditions include the uniform distribution (considered in this chapter) as a special case. An important conceptual difference is that in [12] the sensor quantization noise is viewed as a randomized dither signal which is intentionally added to the observations and that the dither cdf is known (it need not be uniform). The reconstruction explicitly exploits the knowledge of the dither statistics. Specifically, the noiseless observation Y_i, at sensor i, and the corresponding quantized binary sample B_i become

$$Y_i = s(x_i), \quad B_i = Q(Y_i) \triangleq \mathrm{sgn}(Y_i + X_i),$$

where X_i is iid dithering noise with a known distribution $p_X(\cdot)$ which satisfies certain technical assumptions as given in [12, Sect. II.A]. Note that taking the sign of the sum of the observation and random dither X_i is equivalent to comparing with the threshold $-X_i$. Thus the quantization function $Q(\cdot)$ of [12] is equivalent[6] to a comparator with a random threshold that is distributed according to $p_X(-x)$. The quantization function $Q_{it}^{Th}(\cdot)$ in (6.1) can be viewed as a special case of this where $p_X(-x)$ is the uniform distribution over $[-c, c]$. The constructive scheme of Sect. 6.4 and the analysis of this work shows that $Q_{it}^{Th}(\cdot)$ can in fact be used even on noisy field observations with an additive noise of *unknown* distribution.

6.5.2 Parameter Estimation with One–Bit Messages

The parameter estimation problem in [13] corresponds to the special case of a spatially constant field ($s(x_i) = s$ for all i where the index t is omitted since $T = 1$) which is addressed by Corollary 1. We summarize below the key features of the randomized binary quantizer proposed in [13] and show that the randomized 1–bit SQ function $Q(\cdot)$ of [13] is statistically and MSE performance–wise equivalent to the uniform random threshold quantizer $Q_{it}^{Th}(\cdot)$ in (6.1). However, the $Q(\cdot)$ function of [13] implicitly requires sensors of arbitrarily high precision, a property that is undesirable for sensor hardware implementations.

In [13], each sensor i first shifts and scales it observation Y_i into interval $[0, 1]$ creating the value $\widetilde{Y}_i \triangleq \left(\frac{Y_i+c}{2c}\right)$. Next, each sensor i generates an auxiliary random variable α_i, which is iid across sensors and is geometrically distributed over the set of all positive integers: $\mathbb{P}(\alpha_i = j) = 2^{-j}$ for all $j \in \{1, 2, 3, \ldots, \infty\}$. The final quantized binary sample B_i reported by sensor i is given by the α_i^{th} bit in the binary expansion of $\widetilde{Y}_i$:

$$B_i = Q(Y_i) \triangleq B(\widetilde{Y}_i, \alpha_i), \quad \text{where } \widetilde{Y}_i = \sum_{j=1}^{\infty} B(\widetilde{Y}_i, j) 2^{-j}. \tag{6.5}$$

Here, $B(\widetilde{Y}_i, j)$ denotes the jth bit of $\widetilde{Y}_i$. For example, if $\widetilde{Y}_i = 0.375$, then the first four bits of its binary expansion are given by $B(\widetilde{Y}_i, 1) = 0$, $B(\widetilde{Y}_i, 2) = 1$, $B(\widetilde{Y}_i, 3) = 1$, and $B(\widetilde{Y}_i, 4) = 0$. If $\alpha_i = 3$, then sensor i reports $B_i = 1$. This method for generating binary sensor messages requires sensors to have the operational ability to quantize an observed real number (the normalized values $\widetilde{Y}_i$) to an arbitrarily high bit–resolution. Note that the binary values B_i are iid across all sensors and that its expected value is given by

[6] The sign function maps to $\{-1, +1\}$ whereas a threshold comparator maps to $\{0,1\}$. However, the replacement of the -1 symbol with the 0 symbol is unimportant from an estimation viewpoint.

$$\begin{aligned}
\mathbb{E}[B_i] &= \mathbb{E}_{\widetilde{Y}_i}[\mathbb{E}_{B_i}[B_i|\widetilde{Y}_i]] \\
&= \mathbb{E}_{\widetilde{Y}_i}\left[\sum_{j=1}^{\infty} B(\widetilde{Y}_i, j)2^{-j}\right] \\
&= \mathbb{E}_{\widetilde{Y}_i}[\widetilde{Y}_i] = \mathbb{E}\left[\frac{Y_i + c}{2c}\right] \qquad (6.6) \\
&= \frac{\mathbb{E}[s + Z_i] + c}{2c} = \left(\frac{s+c}{2c}\right). \qquad (6.7)
\end{aligned}$$

In sharp contrast to the $Q(\cdot)$ function described above, which requires sensors to have the operational ability to resolve any arbitrary bit in the binary expansion of their normalized observations, $Q_{it}^{Th}(\cdot)$ requires only a noisy comparator. Despite the markedly different operational implementations of $Q(\cdot)$ and $Q_{it}^{Th}(\cdot)$ (see (6.5), (6.1), and Fig. 6.4 which depicts hardware implementations) they are in fact statistically identical: the binary quantized values B_i generated by the two schemes have the same $p_{B_{it}|Y_{it}}(\cdot)$ and $p_{B|s}(\cdot)$ functions where $p_{B_{it}|Y_{it}}(\cdot)$ is the conditional expectation of B_i given $Y_i = y_i$ and $p_{B|s}(\cdot)$ is the unconditional expectation of B_i parameterized by the underlying field value $s(x_i) = s$. These expectations have been evaluated in (6.6), (6.7), (6.8) and (6.9), and we see that for both functions

$$\mathbb{E}[B_i|Y_i = y_i] = p_{B_{it}|Y_{it}}(y_i) = \left(\frac{y_i + c}{2c}\right), \text{ and}$$

$$\mathbb{E}[B_i] = p_{B|s}(s(x_i)) = \left(\frac{s(x_i) + c}{2c}\right).$$

This statistical equivalence allows the two quantization functions $Q(\cdot)$ and $Q_{it}^{Th}(\cdot)$ to be interchanged without affecting the estimation performance.

6.6 Conclusions

The results of this work show that for the distributed field reconstruction problem, for every point of continuity of every field snapshot, it is possible to drive the MSE to zero with increasing sensor density while ensuring that the per–sensor bitrate and sensing–related network overhead rate simultaneously go to zero. This can be achieved with noisy threshold (one–bit) comparators with the minimal knowledge of signal and noise dynamic ranges provided that the noise samples are zero–mean, and independent across sensors and the underlying field, and the sensor placement and sampling schedule satisfy a certain uniformity property. Also, since knowledge of sensor locations need only to be within the precision of the size of the subcells, there is an inherent robustness to sensor location uncertainty. The rate of decay of MSE with increasing sensor density is related to the the local and global smoothness characteristics of the underlying fields and is order–optimal for the class of spatially constant fields. This work has further clarified the utility of randomization for signal acquisition in a distributed sensor network context and has attempted to systematically account for sensor placement and hardware issues in addition to the typical constraints encountered in related studies.

Acknowledgements

This material is based upon work supported by the US National Science Foundation (NSF) under award (CAREER) CCF0546598, ECS-0449194, . Any opinions, findings, and conclusions or recommendations expressed in this material are those of the authors and do not necessarily reflect the views of the NSF.

References

[1] J. Chou, D. Petrovic, and K. Ramchandran, "Tracking and exploiting correlations in dense sensor networks," in *Proc. Annual Asilomar Conference on Signals, Systems, and Computers*, Pacific Grove, CA, Nov. 2002.

[2] J. M. Kahn, R. H. Katz, and K. S. J. Pister, "Next century challenges: Mobile networking for "Smart Dust"," in *Proc. ACM International Conference on Mobile Computing and Networking (MOBICOM)*, Seattle, WA, Aug. 1999, pp. 271–278. [Online]. Available: {`citeseer.nj.nec.com/kahn99next.html`}

[3] T. M. Cover and J. A. Thomas, *Elements of Information Theory*, 1st ed. New York, NY: Wiley–Interscience, 1991.

[4] M. Gastpar and M. Vetterli, "Source–channel communication in sensor networks," *Lecture Notes in Computer Science*, vol. 2634, pp. 162–177, Apr. 2003.

[5] M. Gastpar, B. Rimoldi, and M. Vetterli, "To Code, or not to code: Lossy source–channel communication revisited," *IEEE Trans. Info. Theory*, vol. IT–49, pp. 1147–1158, May 2003.

[6] R. Nowak, U. Mitra, and R. Willet, "Estimating inhomogenous fields using wireless sensor networks," *IEEE J. Sel. Areas Commun.*, vol. 22, no. 6, pp. 999–1006, Aug. 2004.

[7] K. Liu, H. El–Gamal, and A. Sayeed, "On optimal parametric field estimation in sensor networks," in *Proc. IEEE/SP 13th Workshop on Statistical Signal Processing*, Jul. 2005, pp. 1170–1175.

[8] W. Bajwa, A. Sayeed, and R. Nowak, "Matched source–channel communication for field estimation in wireless sensor networks," in *Proc. Fourth Inttl. Symposium Information Processing in Sensor Networks*, Apr. 2005, pp. 332–339.

[9] N. Liu and S. Ulukus, "Optimal distortion–power tradeoffs in Gaussian sensor networks," in *Proc. IEEE International Symposium on Information Theory*, Seattle, WA, USA, Jul. 2006, pp. 1534–1538.
[10] Q. Zhao, A. Swami, and L. Tong, "The interplay between signal processing and networking in sensor networks," *IEEE Signal Processing Magazine*, vol. 23, no. 4, pp. 84–93, Jul. 2006.
[11] E. Masry and S. Cambanis, "Consistent estimation of continuous–time signals from nonlinear transformations of noisy samples," *IEEE Trans. Info. Theory*, vol. IT–27, pp. 84–96, Jan. 1981.
[12] E. Masry, "The reconstruction of analog signals from the sign of their noisy samples," *IEEE Trans. Info. Theory*, vol. IT–27, no. 6, pp. 735–745, Nov. 1981.
[13] Z. Q. Luo, "Universal decentralized estimation in a bandwidth constrained sensor network," *IEEE Trans. Info. Theory*, vol. IT–51, pp. 2210–2219, Jun. 2005.
[14] T. Berger, Z. Zhang, and H. Viswanathan, "The CEO problem [multiterminal source coding]," *IEEE Trans. Info. Theory*, vol. IT–42, pp. 887–902, May. 1996.
[15] H. Viswanathan and T. Berger, "The quadratic Gaussian CEO problem," *IEEE Trans. Info. Theory*, vol. IT–43, pp. 1549–1559, Sep. 1997.
[16] V. Prabhakaran, D. Tse, and K. Ramchandran, "Rate-region of the quadratic Gaussian CEO problem," in *Proc. IEEE International Symposium on Information Theory*, Chicago, IL, Jun. 2004, p. 119.
[17] A. Kashyap, L. A. Lastras-Montano, C. Xia, and Z. Liu, "Distributed source coding in dense sensor networks," in *Proc. Data Compression Conference*, Snowbird, UT, Mar. 2005, pp. 13–22.
[18] D. L. Neuhoff and S. S. Pradhan, "An upper bound to the rate of ideal distributed lossy source coding of densely sampled data," in *Proc. IEEE International Conference on Acoustics, Speech and Signal Processing*, Toulouse, France, May 2006, pp. 1137–1140.
[19] D. Marco, E. J. Duarte-Melo, M. Liu, and D. Neuhoff, "On the many–to–one transport capacity of a dense wireless sensor network and the compressibility of its data," in *Information Processing in Sensor Networks, Proceedings of the Second International Workshop, Palo Alto, CA, USA, April 22-23, 2003*, ser. Lecture Notes in Computer Science edited by L. J. Guibas and F. Zhao, Springer, New York, 2003, Apr., pp. 1–16.
[20] P. Ishwar, A. Kumar, and K. Ramchandran, "Distributed sampling for dense sensor networks: A "bit-conservation" principle," in *Information Processing in Sensor Networks, Proceedings of the Second International Workshop, Palo Alto, CA, USA, April 22-23, 2003*, ser. Lecture Notes in Computer Science edited by L. J. Guibas and F. Zhao, Springer, New York, 2003, pp. 17–31.
[21] A. Kumar, P. Ishwar, and K. Ramchandran, "On distributed sampling of smooth non-bandlimited fields," in *Proc. Third Inttl. Symposium In-*

formation Processing in Sensor Networks. New York, NY: ACM Press, 2004, pp. 89–98.

[22] C. D. Aliprantis and O. Burkinshaw, *Principles of Real Analysis.* San Diego, CA: Academic Press, 1990.

[23] S. M. Kay, *Fundamentals of Statistical Signal Processing, Volume I: Estimation Theory,* 1st ed. Upper Saddle River, NJ: Prentice–Hall, 1993, vol. 1.

[24] R. Durret, *Probability: Theory and Examples.* Thomson Learning College, 1990.

Appendix: Proof of Theorem 1

First, note that the expected value of the binary message B_{it} is given by

$$\begin{aligned}
\mathbb{E}[B_{it}] &= \mathbb{E}[\mathbf{1}(Y_{it} > R_{it})] \\
&= \mathbb{E}_{Y_{it}}[\mathbb{E}_{R_{it}}[\mathbf{1}(Y_{it} > R_{it})|Y_{it}]] \\
&= \mathbb{E}_{Y_{it}}[\mathbb{P}\left(R_{it} < Y_{it}|Y_{it}\right)] \\
&= \mathbb{E}_{Y_{it}}\left[\frac{Y_{it}+c}{2c}\right] \qquad (6.8) \\
&= \frac{\mathbb{E}[s_t(x_i) + Z_{it}] + c}{2c} \\
&= \left(\frac{s_t(x_i)+c}{2c}\right), \qquad (6.9)
\end{aligned}$$

which is the value of the field $s_t(\cdot)$ at location x_i shifted and normalized to the interval $[0,1]$. Note that this result holds for any $F_{\mathbf{Z}}(\mathbf{z}) \in \mathcal{F}$.

Using (6.9) we can bound the bias and the variance of the estimator $\widehat{S}_t(x)$. The bound on the MSE follows from the bounds on these values since, for any estimator of a non-random parameter, we have

$$\text{MSE}\left(\widehat{S}_t(x)\right) = \text{bias}^2\left(\widehat{S}_t(x)\right) + \text{var}\left(\widehat{S}_t(x)\right). \qquad (6.10)$$

Let $j \in \{1,\dots,L\}$ denote the index of the supercell that x falls in. We bound the magnitude of bias of the estimate $\widehat{S}_t(x)$ in the following way

$$\begin{aligned}
\left|\text{bias}\left(\widehat{S}_t(x)\right)\right| &= \left|\mathbb{E}\left[\widehat{S}_t(x) - s_t(x)\right]\right| \\
&= \left|\mathbb{E}\left[2c\left[\frac{1}{n}\sum_{i\in I(j,t)} B_{it}\right] - c - s_t(x)\right]\right| \\
&= \left|2c\left[\frac{1}{n}\sum_{i\in I(j,t)} \mathbb{E}\left[B_{it}\right]\right] - c - s_t(x)\right| \\
&\overset{(i)}{=} \left|2c\left[\frac{1}{n}\sum_{i\in I(j,t)} \left(\frac{s_t(x_i)+c}{2c}\right)\right] - c - s_t(x)\right| \\
&= \left|\frac{1}{n}\sum_{i\in I(j,t)} \left(s_t(x_i) - s_t(x)\right)\right| \\
&\le \frac{1}{n}\sum_{i\in I(j,t)} \left|s_t(x_i) - s_t(x)\right| \\
&\overset{(ii)}{\le} \frac{1}{n}\sum_{i\in I(j,t)} \omega_t\left(\|x - x_i\|, x\right) \\
&\overset{(iii)}{\le} \frac{1}{n}\sum_{i\in I(j,t)} \omega_t\left(\frac{\sqrt{d}}{\sqrt[d]{L}}, x\right) \\
&= \omega_t\left(\frac{\sqrt{d}}{\sqrt[d]{L}}, x\right) \overset{(iv)}{\le} \widetilde{\omega}_t\left(\frac{\sqrt{d}}{\sqrt[d]{L}}\right),
\end{aligned} \tag{6.11}$$

where (i) follows from (6.9), (ii) and (iv) follow from Definitions 1 and 2, and (iii) follows because the local modulus of continuity is a nondecreasing function of its first argument for each fixed value of its second argument and since any sensor in the supercell containing x is within distance $\frac{\sqrt{d}}{\sqrt[d]{L}}$ of x (the length of the diagonal of a supercell).

The variance of the estimate is bounded by

$$\begin{aligned}
\text{var}[\widehat{S}_t(x)] &= \text{var}\left[2c\left[\frac{1}{n}\sum_{i\in I(j,t)} B_{it}\right] - c\right] \\
&= \left(\frac{4c^2}{n^2}\right)\sum_{i\in I(j,t)} \text{var}[B_{it}] &(6.12) \\
&\le \left(\frac{4c^2}{n^2}\right)\left(\frac{n}{4}\right) = \left(\frac{LMc^2}{N}\right), &(6.13)
\end{aligned}$$

where we used standard properties of variance and the fact that $\{B_{it}\}$ are independent to obtain (6.12), and we used the fact the variance of a Bernoulli$\{0,1\}$ random variable is bounded by $(1/4)$ and that $n = (N/(LM))$ to obtain (6.13).

Combining these bounds for the bias and variance given in (6.11) and (6.13) of the estimator and using the identity in (6.10), we get the desired bound on the MSE for all $x \in G$, $t \in \{1, \dots, T\}$, and $F_{\mathbf{Z}}(\mathbf{z}) \in \mathcal{F}$.

7

On the Number of Bits to Encode the Outputs of Densely Deployed Sensors

David L. Neuhoff and S. Sandeep Pradhan

Electrical Engineering and Computer Science Department, University of Michigan, Ann Arbor, MI 48109, {neuhoff,pradhanv}@umich.edu *

7.1 Introduction

Suppose M sensors are densely deployed throughout some bounded geographical region in order to sample a stationary two-dimensional random field, such as temperature. Suppose also that each sensor encodes its measurements into bits in a lossy fashion for transmission to some collector or fusion center where the continuous-space field is reconstructed. We consider the following question. If the distortion in the reconstruction is required to be D or less, what happens to the total number of bits produced by the encoders as the sensors become more numerous and dense? Does the increasing number of sensors mean that the total number of bits increases without limit? Or does the increasing correlation between neighboring sensor values sufficiently mitigate the increasing number of sensors to permit the total number of bits to remain bounded as M increases?

A key constraint is that the encoders are *distributed*, i.e., each encodes its own data, without knowledge of the data from the other sensors, though with the knowledge that the decoder will have the encoded data from the other sensors, in addition to that from its own encoder. Without the distributed constraint, conventional rate-distortion theory (for *centralized*, as opposed to distributed coding) implies that the number of bits can remain bounded. This fact will be reviewed in the next section.

The goal of this chapter is to review recent work providing several answers to this question about the number of bits produced when encoding densely deployed sensors in a distributed fashion. The existence of different answers is due to the fact that different coding structures are assumed. We also provide pointers to related work in which the goal is to design systems whose distortion decreases to zero as sensor density increases.

* This work was supported by NSF Grants CCF-0329715 and CCF-04485115 (CAREER)

The first result, described in Section 7.3, is disappointing. It indicates that if each sensor uses an identical scalar quantizer, along with ideal distributed lossless source coding across all sensors, plus continuous-space field reconstruction after decoding, then the number of bits needed to attain distortion D grows without bound as sensor density increases. For the important case that the field is Gaussian, a bound is given on the rate at which the number of bits increases with sensor density.

The question then arises as to whether the catastrophically bad performance of scalar quantization and distributed lossless coding , when sensor density is large, is a characteristic of all distributed lossy coding systems, or just a characteristic of this type of system. And if the latter, what is it about the scalar quantization based system that makes it bad at large sensor densities?

Although this question has been raised in prior work, the fact that it has an immediate answer has been overlooked (by the present authors, among others). For one thing, the conclusion to be drawn about scalar quantization plus distributed lossless source coding is not that it is a bad approach, but rather that, with this approach, the density of samples should not be taken too large. For another, it has been overlooked that scalar quantization plus distributed lossless coding applied to samples with a given finite density can be considered to be a special case of lossy distributed coding applied to samples with high sensor density, albeit with subsampling embedded in the encoders. Therefore, the fact that a scalar quantizer based system operating at a finite sensor density can attain target distortion D while producing a finite number of bits implies that at all sufficiently high sensor densities, the best distributed lossy coding systems attaining distortion D produce that many bits or fewer. That is, unlike the scalar quantizer based system, the number of bits produced by the best distributed lossy coding systems remains bounded as sensor density increases. In a sense, the difficulty with the scalar quantization based system at high sampling densities is that it rules out subsampling, whereas ideal distributed lossy coding does not.

One characteristic of the aforementioned scalar quantization based distributed encoding system is that the encoders operating on each sensor produce the same numbers of bits per sensor value. This contrasts with the subsampling approach sketched in the previous paragraph in which only some of the sensors have encoders that produce bits. Thus, one may ask if it is the equal-bits characteristic of scalar quantization with distributed lossless coding that causes catastrophic behavior. However, this, too, is not the case, because without affecting performance one can use time-sharing to modify the system in such a way that all sensor encoders produce, on average, the same number of bits – simply time-share among various subsets of sensors in such a way that in the long run each sensor is encoded equally often. An explicit construction is described in Section 7.3 One concludes that it is not the equal-bits characteristic that causes catastrophic performance.

The next logical question is: How well can distributed lossy coding systems perform when operating on dense samples? Bounds to such performance are discussed in Section 7.4 for the case of a Gaussian field. Though the bounds of this section are primarily for stationary Gaussian fields, the fact that Gaussian is a "worst case" implies that they also provide upper bound for other stationary sources.

As a prologue to the main sections of this chapter, the next section introduces the random process and the distributed coding frameworks, including distortion and encoding rate performance measures, to be adopted in the chapter. There are a number of issues considered in this section.

- Although one is primarily interested in two-dimensional fields, the results we review are for one-dimensional fields, i.e., for conventional one-dimensional signals or waveforms. Nevertheless, it is believed that the behavior found for one-dimensional fields, such as whether the total number of encoded bits remains bounded or not, reflects the behavior of two-dimensional fields.
- Although we have so far envisioned sensors being deployed over a bounded region, in fact we will consider both bounded and unbounded deployment regions. One reason for the latter is mathematical tractability – nicer expressions can sometimes be obtained for the performance of a dense sensor network with an unbounded deployment region. Another reason derives from the fact that performance tends to improve as the deployment region becomes larger. Thus, it is natural to want to know performance in the limit of large deployment region. In a practical situation, the results of the unbounded case will be relevant when a bounded region is large relative to the maximum distance over which correlation remains significant.
- Although the encoders operate on samples of the field taken by the sensors, the task of the decoder is to provide a continuous-space waveform reconstruction. Therefore, there are two versions of the two performance measures (distortion and encoding rate): one for the discrete-space samples and the other for the continuous-space waveforms.
- Since the distributed coding methods considered rely on coding of temporal blocks, we introduce a temporal dimension to the field. That is, coding is considered in a spatio-temporal framework.
- The next section also reviews centralized theory. One reason is that we wish to compare results for distributed coding to those for centralized coding. Another is that the definitions and derivations of centralized theory are a model that can be paralleled, to a considerable degree, in distributed coding. For example, we review how, as sampling rate increases, the performance of ideal centralized coding of discrete-space samples followed by sample-and-hold waveform reconstruction approaches the performance of ideal centralized codes for the continuous-space source from which the samples are taken. Since this part of the section is a bit of a diversion from the main goals, the reader may wish to skim it on first reading and

return to it when a later section refers to it or when a later section introduces parallel definitions or discussions for the distributed coding scenario. Alternatively, the reader might appreciate the opportunity to review the classical derivation of centralized coding limits.

We conclude the introduction by mentioning that this chapter is intended to be tutorial, rather than to introduce new results. It focuses on key ideas, and provides proofs and derivations, or sketches thereof, when this is feasible, but leaves many details and proofs to references.

7.2 Random Field Model; Distributed and Centralized Encoding

7.2.1 Random Field Model

As mentioned in the introduction, we focus on one-dimensional fields, i.e. conventional one-dimensional signals or waveforms. As such, the model for a random field is simply a one-dimensional, spatially stationary, ergodic random process $X = \{X(s) : -\infty < s < \infty\}$, where $X(s)$ is a real-valued random variable representing the field outcome at location s. We assume its autocorrelation function $\rho_X(\tau)$ and power spectral density $S_X(\Omega)$ are known. We make no assumption about it being bandlimited or not. We assume for convenience that it has mean zero and variance 1. We also assume it is mean-square continuous; equivalently $\rho_X(\tau) \to 1$ as $\tau \to 0$.

Given a positive integer N, we assume that sensors are uniformly spaced $1/N$ apart and that the nth sensor takes the sample $Y_n = X(\frac{n}{N})$ of X. Thus N is viewed as a sampling rate, and $Y = \{Y_n : -\infty < n < \infty\}$ is a stationary discrete-space process with mean 0, variance 1, autocorrelation function $\rho_Y(n) = \rho_X(n/N)$, and power spectral density

$$\Phi_N(\omega) = N \sum_{k=-\infty}^{\infty} S_X((\omega - 2\pi k)N) \ .$$

Moreover, for some positive integer M, we assume we have access only to the sensor values $Y_1, \ldots, Y_M$, i.e. to the sensor values in the spatial interval $(0, \frac{M}{N}]$.

We will consider what happens as the sampling rate N tends to infinity, and either $M = N$ or $M \gg N$. These model the situations that sensors are deployed in bounded and unbounded regions, respectively. Note that due to the continuity of the autocorrelation function at $\tau = 0$, the correlation between adjacent samples, i.e. $\rho_Y(1) = \rho_X(1/N)$, approaches one as N increases. In addition, $\frac{1}{N}\Phi_N\left(\frac{\Omega}{N}\right) \to S_X(\Omega)$ pointwise as $N \to \infty$. For tractability, we often assume that X, and consequently Y, is Gaussian.

7.2.2 Distributed Encoding

As illustrated in Figure 7.1a, each of the M sensors has an encoder that performs lossy encoding of the sample taken by that sensor. Viewed as a group, these encoders perform *distributed lossy encoding* of the samples. On the one hand, this is a limitation because the data at neighboring samples is correlated, yet each encoder sees only the samples at its own location. On the other hand, there are distributed encoding techniques that nevertheless can exploit the correlation, to varying degrees [1, 2, 3, 4]. Such methods generally operate on temporal blocks of samples. Accordingly, as in [5, 6, 7, 8, 9], we hypothesize a temporal sequence of independent random processes. That is, for each $t \in \mathbb{Z}_+ \triangleq \{1, 2, \ldots\}$, $X^{(t)} = \{X^{(t)}(s) : s \in (0, \frac{M}{N}]\}$ is a random process identical to X, and the random processes corresponding to different times t are independent[2]. We think of $X^{(t)}$ as the snapshot of the field at time t. The encoder for sensor n operates on the IID discrete-time sequence $\{Y_n^{(t)} : t \in \mathbb{Z}_+\}$, where $Y_n^{(t)} = X^{(t)}(\frac{n}{N})$. Note: we will generally omit the superscript (t) when all variables under discussion correspond to the same value of t.

In the generic distributed coding system shown in Figure 7.1b, there is a temporal blocklength L, and for each $n \in \{1, \ldots, M\}$, the nth sensor encodes the first temporal block $\underline{Y}_n = ({Y_n}^{(1)}, \ldots, {Y_n}^{(L)})$ into a binary sequence $\underline{Z}_n = (Z_n^{(1)}, \ldots, Z_n^{(b_n)})$ of some length b_n. A single decoder receives all M binary sequences $\underline{Z}_1, \ldots, \underline{Z}_M$ and outputs a reproduction $\widehat{Y}_n^{(t)}$ of $Y_n^{(t)}$ for each $t \in \{1, \ldots, L\}$ and $n \in \{1, \ldots, M\}$. Next, for each $t \in \{1, \ldots, L\}$, the decoder output $\widehat{Y}^{(t)}$ is used to reconstruct a waveform $\widehat{X}^{(t)}$ as a reproduction of $X^{(t)}$.

To keep things simple and because it does not entail much loss in performance, we will generally presume sample-and-hold waveform reconstruction, which produces

$$\widehat{X}^{(t)}(s) = \widehat{Y}_n^{(t)}, \text{ when } \frac{n-1}{N} < s \leq \frac{n}{N} .$$

Notice that $\widehat{X}^{(t)}$ is defined to equal the reproduction of the *next* sample rather than the *previous* sample, as is more common, because this will simplify notation a bit.

Though we focused on encoding and decoding the first temporal block, $(X^{(1)}, \ldots, X^{(L)})$, clearly subsequent temporal block can be encoded and decoded in the same fashion.

---

[2] In a real situation, such as measuring a temperature field, the $X^{(t)}$'s would not ordinarily be independent. However, we make the assumption for tractability. It is not anticipated that the main conclusions of this chapter would change significantly if this assumption were omitted.

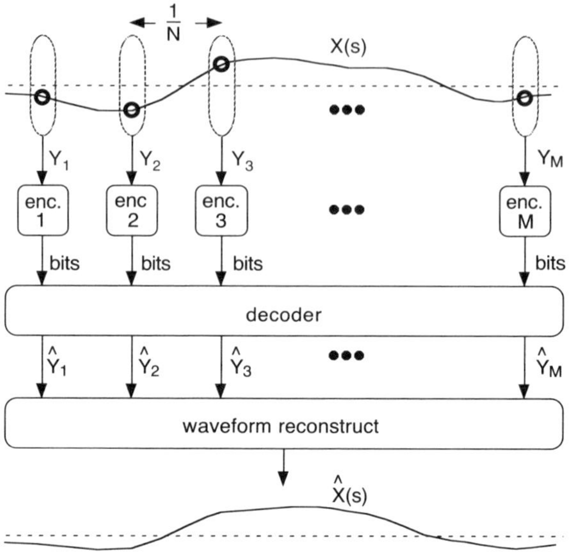

(a) one sample at a time

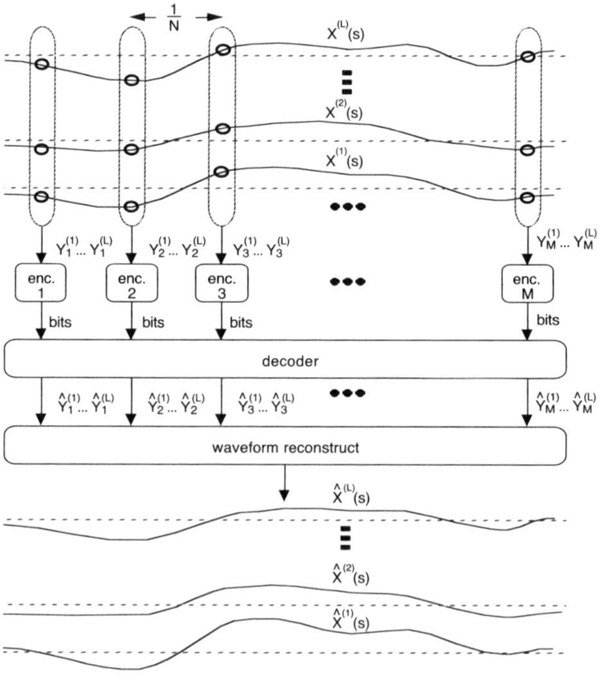

(b) L temporal samples at a time

Fig. 7.1. Distributed encoding of sensor values

7.2.3 Performance Measures: Rate and Distortion

For such a system we consider two performance measures: *encoding rate* and *distortion*. As mentioned in the introduction, there are actually two versions of each — one for the sampled process and another for the continuous-space waveform.

The first and most important version of encoding rate or simply *rate*[3], is the *rate per unit distance*, denoted by script $\mathcal{R}$. It is the sum of the average numbers of bits produced by each of the M encoders per snapshot divided by the length M/N of the spatial interval. The *average* is over time t and the randomness in X. It will also be called the *per snapshot encoding rate*. The second version of rate, called the *per sample encoding rate* and denoted by the conventional R, is the average number of bits per sample produced by one encoder in one time instant, where the average is over time t, sensor n, and the randomness in X. The relationship between the two rates is

$$\mathcal{R} = NR \ .$$

We use the terms *encoding rate* or *rate* without qualifier, when the intended version is clear from context.

To quantify distortion, we use *mean squared error* (MSE). Again, there are two versions. The *sample distortion/MSE* is

$$D = \frac{1}{LM} \sum_{t=1}^{L} \sum_{n=1}^{M} E(Y_n^{(t)} - \widehat{Y}_n^{(t)})^2 \ ,$$

where L and M are the temporal and spatial blocklengths of the system. The *waveform distortion/MSE* is

$$\mathcal{D} = \frac{1}{L} \sum_{t=1}^{L} \int_0^{M/N} E(X^{(t)}(s) - \widehat{X}^{(t)}(s))^2 \, ds \ .$$

If sample-and-hold reconstruction is used, then it is easy to argue that when N is large

$$D \approx \mathcal{D} \ .$$

In particular, it can be shown that $\mathcal{D}$ - $D \to 0$ as $N \to \infty$. Moreover, for any $D_o > 0$, the convergence is uniform on the class of all systems with $D \leq D_o$.

While we focus mainly on sample-and-hold reconstruction, it is easy to specify, at least in principle, optimal reconstruction: it is the minimum MSE estimator for $X^{(t)}(s)$ based on $Y^{(t)}$, i.e. for $t \in \mathbb{Z}_+$ and $s \in \left(0, \frac{M}{N}\right]$,

$$\widehat{X}^{(t)}(s) = E\left[X^{(t)}(s) \,\middle|\, \widehat{Y}_n^{(t')} : n = 1, \ldots, N, t' = 1, \ldots, L\right] \ .$$

[3] Though we may sometimes use the shorter term *rate* we will mostly use the longer term *encoding rate* to distinguish it from both *sampling rate* and the *rate* at which *encoding rate* might increase as *sampling rate* increases.

Another noteworthy reconstruction is the linear minimum MSE estimator of $X^{(t)}(s)$ based on $\widehat{Y}^{(t)}$.

Note that we will consistently use script font to denote waveform, i.e. continuous-space, quantities, and conventional symbols for discrete-space, i.e. per sample quantities. We will often subscript rate and distortion with the sampling rate N, when they depend substantially upon it.

7.2.4 Benchmarks from Centralized Coding

In order to provide benchmarks for the performance of the distributed coding systems considered in subsequent sections and exemplars for the definitions and derivations for distributed coding systems, in this subsection we review certain aspects of centralized coding based on sampling[4]. In the centralized case, one encoder sees and encodes the outputs of all sensors, and as before, one decoder produces reproductions of all sensor outputs. The optimal performances of such systems are determined by information theoretic *rate-distortion functions.*

Given sampling rate N and another positive integer M, the lossy source coding theorem for discrete-time processes (c.f. [11, 12]) shows that the least per sample encoding rate obtainable by any centralized coding system applied to the discrete-time random process $\{Y_n^{(t)} : n = 1, \ldots, M, t = 1, 2, \ldots\}$ attaining sample distortion D or less is given by the *Mth-order rate-distortion function* of Y

$$R_N(M, D) \triangleq \inf_{q:\frac{1}{M}\sum_{n=1}^{M} E_q(Y_n - \widetilde{Y}_n)^2 \le D} \frac{1}{M} I_q(Y_1, \ldots, Y_M; \widetilde{Y}_1, \ldots, \widetilde{Y}_M) \, , \quad (7.1)$$

where q is a conditional distribution, called a *forward test channel*, on hypothetical random variables $\widetilde{Y}_1, \ldots, \widetilde{Y}_M$ given actual random variables $Y_1, \ldots, Y_M$; $I_q(Y_1, \ldots, Y_M; \widetilde{Y}_1, \ldots, \widetilde{Y}_M)$ is the mutual information between $(Y_1, \ldots, Y_M)$ and $(\widetilde{Y}_1, \ldots, \widetilde{Y}_M)$ assuming test channel q; the subscript N indicates the sampling rate with which Y is generated from X; and superscript (t)'s have been omitted from the Y's because they would be the same for all and because the specific value of t is immaterial. For any N, M, the performance represented by (7.1) can be attained to arbitrary tolerance with an encoder[5] that encodes a spatio-temporal block $\{Y_n^{(t)} : n = 1, \ldots, M, t = 1, \ldots, L\}$, for any sufficiently large L.

---

[4] As indicated in the introduction, one might skim this subsection and return to it when it is referenced or when parallel discussions occur for distributed coding.

[5] In conventional centralized lossy coding, where spatial, but not temporal blocks, from the source are available, unless M is large, $R_N(M, D)$ is an unattainable lower bound to the performance of block codes operating (only) on M spatial samples $Y_1, \ldots, Y_M$. However, in the situation considered here, the presence of temporal samples lends $R_N(M, D)$ its direct operational significance. Specifically, one can show that performance cannot be better than $R_N(M, D)$ using conven-

Since $R_N(M, D)$ is subadditive[6] over M, taking the limit of (7.1) as $M \to \infty$ yields the least per sample encoding rate of any centralized coding system with sample distortion D or less applied to the discrete-space process Y produced by sampling X at rate N:

$$R_N(D) \triangleq \lim_{M \to \infty} R_N(M, D) , \tag{7.2}$$

which is the conventional *rate-distortion function* of Y. This performance is attained, approximately, by centralized coders that encode[7] $M \times L$ spatio-temporal blocks of samples, for all sufficiently large values of M and L.

The systems whose performances are characterized by (7.1) and (7.2) can also be considered as codes for the continuous-space process X. In this case, their performances are measured by per snapshot encoding rate and waveform MSE. For the performance represented by (7.1), the per snapshot encoding rate is

$$\mathcal{R} = N R_N(M, D) , \tag{7.3}$$

and if N is large and sample-and-hold reconstruction is used, the waveform MSE is

$$\mathcal{D} \approx D . \tag{7.4}$$

For the limiting performance represented by (7.2), the per snapshot encoding rate is

$$\mathcal{R} = N R_N(D) , \tag{7.5}$$

and again (7.4) applies.

Now consider what happens as the sampling rate N increases. To our knowledge, it is not known if $N R_N(D)$ is decreasing or subadditive in N. However, by an argument sketched in the Appendix, one can show the weaker property that for any D, $0 < D < 1$, there exists N_o such that for all positive integers k and $N \geq N_o$

$$k N R_{kN}(D) \leq N R_N(D) + \epsilon_N(D) , \tag{7.6}$$

tional approaches to deriving converses (c.f. [11, 12]), and one can show that such performance is attainable in the same way that in conventional rate-distortion theory one shows that performance at least as good as the Mth-order rate-distortion function is attainable with a code that simultaneously encodes a large number of spatially successive blocks of length M (c.f. [11, 12]). In the present situation, temporal samples substitute for spatially successive blocks.

6 To say that $R_N(M, D)$ is subadditive over M is to say that

$$R_N(M_1 + M_2, D) \leq \frac{M_1}{M_1 + M_2} R_N(M_1, D) + \frac{M_2}{M_1 + M_2} R_N(M_2, D) \text{ for all } M_1, M_2 .$$

It implies that $R_N(M, D)$ has a limit with respect to M and $\inf_M R_N(M, D) = \lim_{M \to \infty} R_N(M, D)$, (c.f. [11], p. 112).

7 In conventional centralized lossy coding, this performance is attained by coding a sufficiently large spatial block, i.e., the presence of a temporal sequence is not required to attain the limiting performance in (7.2) (c.f. [12]).

where $\epsilon_N(D) \rightarrow 0$ as $N \rightarrow \infty$. This is sufficient to establish that $\liminf_{N\to\infty} NR_N(D) = \inf_N NR_N(D)$. Further, we make the reasonable assumption[8] that

$$\lim_{N\to\infty} NR_N(D) = \inf_N NR_N(D) \; . \tag{7.7}$$

We conclude that the least per snapshot encoding rate of any centralized coding system operating on regularly spaced samples of the continuous-space process X and attaining waveform distortion $\mathcal{D}$ or less is

$$\mathcal{R}(\mathcal{D}) \triangleq \lim_{N\to\infty} NR_N(\mathcal{D}) = \lim_{N\to\infty} \lim_{M\to\infty} NR_N(M,\mathcal{D}) \; , \tag{7.8}$$

which is called the rate-distortion function of the continuous-time process X. Because this rate is given as a limit over N, we conclude that it can be attained to arbitrary tolerance by encoding samples taken at any sufficiently high rate. Moreover, because its derivation presumed sample-and-hold reconstruction, it can be attained to arbitrary tolerance with this simple reconstruction method.

As a sidelight, we note that lossy source coding of a continuous-time source X can also be accomplished without sampling as the first step. For example, one might apply a linear filter to X before sampling and encoding; or one might project X onto a set of basis functions $\phi_1(s), \phi_2(s), \ldots$, and then encode the resulting projections. While it seems generally to be taken for granted that $\mathcal{R}(\mathcal{D})$, as defined above, represents the best possible performance of any lossy source coding system (whether based on sampling first or not) for a continuous-time source X, the theory of such is not entirely well developed. When X is Gaussian, it is known that $\mathcal{R}(\mathcal{D})$ defined above does indeed represent the best possible performance. Moreover, although it appears that this also holds for nonpathological continuous-time sources, it is not clear what characterizes the sources for which it does and does not hold. Since in this chapter, we are focused exclusively on codes that sample first, the question of what constitutes the best possible performance of any type of lossy source code is not central. However, if a source is such that $\mathcal{R}(\mathcal{D})$ represents the best possible performance, then we can also be sure that such performance can be attained with sample-and-hold reconstruction. That is, at high sampling rates, sample-and-hold reconstruction engenders no loss.

Gaussian Sources

For a stationary Gaussian source, the above rate-distortion function can be reduced to the following parametric form [13] (c.f. [11, 12]):

[8] Though we do not offer a proof, we believe that the mean-square continuity of X should be sufficient to insure $\lim_{N\to\infty} NR_N(D) = \inf_N NR_N(D)$. An alternative to accepting this conjecture is to replace $\lim_{N\to\infty} NR_N(D)$ with $\liminf_{N\to\infty} NR_N(D)$ throughout the remainder of the chapter.

$$\mathcal{R}(\theta) = \frac{1}{2\pi}\int_{-\infty}^{\infty} \max\left\{\frac{1}{2}\log_2 \frac{S(\Omega)}{\theta}, 0\right\} d\Omega \tag{7.9}$$

$$\mathcal{D}(\theta) = \frac{1}{2\pi}\int_{-\infty}^{\infty} \min\left\{S(\Omega), \theta\right\} d\Omega \ , \tag{7.10}$$

where $S(\Omega)$ is the power spectral density of X, and θ is a free parameter, $0 \leq \theta \leq \max_\Omega S(\Omega)$. We now sketch the derivation of this result, since a similar derivation will be needed for the distributed case.

We begin by considering the definition of $R_N(M,D)$ given in (7.1). One can straightforwardly show that if T is an $M \times M$ orthogonal matrix, i.e. a transform, then $R_N(M,D)$ for $\underline{U} = T\underline{Y}$ is the same as $R_N(M,D)$ for $\underline{Y}$, where $\underline{U} = (U_1,\ldots,U_M)'$ and $\underline{Y} = (Y_1,\ldots,Y_M)'$, and $'$ denotes vector transpose. This is because for any test channel q on $\underline{Y}$ with output $\widetilde{\underline{Y}}$, the test channel $\widetilde{q}$ on $\underline{U}$ defined as having output $\widetilde{\underline{U}} = T\widetilde{\underline{Y}}$ has the properties that $I_{\widetilde{q}}(\underline{U};\widetilde{\underline{U}}) = I_q(\underline{Y};\widetilde{\underline{Y}})$ and $E_{\widetilde{q}}\|\underline{U} - \widetilde{\underline{U}}\|^2 = E_q\|\underline{Y} - \widetilde{\underline{Y}}\|^2$, where $\|\cdot\|$ denotes the Euclidean norm. As an orthogonal matrix T, let us choose a Karhunen-Loeve transform, which is defined as an $M \times M$ matrix whose rows are an orthonormal set of eigenvectors of the $M \times M$ covariance matrix $K_{N,M}$ of $\underline{Y}$. As is well-known and can be easily shown, this causes the components of $\underline{U} = T\underline{Y}$ to be uncorrelated. Indeed, since they inherit the Gaussianity of $\underline{Y}$, they are also independent. Next, since $\underline{U}$ has independent components, one can straightforwardly show that the inf in the definition of $R_N(M,D)$ for $\underline{U}$ can be attained by a test channel $\widetilde{q}$ that is memoryless in the sense that for each n, given U_n, random variable $\widetilde{U}_n$ is conditionally independent of all $U_{n'}$'s and $\widetilde{U}_{n'}$'s, $n' \neq n$. In this case, $I_{\widetilde{q}}(\underline{U};\widetilde{\underline{U}}) = \sum_{n=1}^M I_{\widetilde{q}_n}(U_n;\widetilde{U}_n)$, where $\widetilde{q}_n$ is the nth marginal of q. It follows that one may decouple the inf's over the M marginal test channels and write

$$\begin{aligned} R_N(M,D) &= \inf_{d_1\geq 0,\ldots,d_M\geq 0:\frac{1}{M}\sum_{n=1}^M d_n \leq D} \frac{1}{M}\sum_{n=1}^M \inf_{q_n:E_{q_n}(U_n-\widetilde{U}_n)^2\leq d_n} I_{q_n}(U_n;\widetilde{U}_n) \\ &= \inf_{d_1\geq 0,\ldots,d_M\geq 0:\frac{1}{M}\sum_{n=1}^M d_n \leq D} \frac{1}{M}\sum_{n=1}^M \frac{1}{2}\log_2 \frac{\lambda_{N,M,n}}{d_n} \ , \end{aligned} \tag{7.11}$$

where the second equality follows from the facts that the rate-distortion function of a single Gaussian random variable with variance σ^2 is $\frac{1}{2}\log_2\frac{\sigma^2}{D}$, and that the variance of U_n is the eigenvalue $\lambda_{N,M,n}$ of $K_{N,M}$ corresponding to the eigenvector that is the nth row of T. Note the test channels q_n that attain the inf's in the summand are Gaussian, as is consequently the test channel q that attains the inf in (7.1).

We now find a parametric expression for $R_N(M,D)$. Given a parameter $\phi \geq 0$, one can use the Karush-Kuhn-Tucker conditions [14] to show that the following choices of $d_1,\ldots,d_M$ achieve the inf in (7.11) for some D:

$$d_n = \begin{cases} \lambda_{N,M,n}, & \lambda_{N,M,n} \leq \phi \\ \phi, & \lambda_{N,M,n} > \phi \end{cases} .$$

This leads to the following parametric expressions for per sample encoding rate and distortion that attain points on $R_N(M, D)$:

$$r_{N,M}(\phi) = \frac{1}{M}\sum_{i=1}^{M} \max\left\{\frac{1}{2}\log_2 \frac{\lambda_{N,M,i}}{\phi}, 0\right\} \tag{7.12}$$

$$d_{N,M}(\phi) = \frac{1}{M}\sum_{i=1}^{M} \min\left\{\lambda_{M,N,i}, \phi\right\} . \tag{7.13}$$

With sampling rate N fixed, we now take the limit of these parametric expressions as $M \to \infty$. The key is the Grenander-Szego asymptotic eigenvalue distribution theorem [15] (see also [16]), which shows that for any continuous function g

$$\lim_{M\to\infty} \frac{1}{M}\sum_{n=1}^{M} g(\lambda_{N,M,n}) \;=\; \frac{1}{2\pi}\int_{-\pi}^{\pi} g(\Phi_N(\omega))\, d\omega \;,$$

where $\Phi_N(\omega)$ is the power spectral density of the discrete-space process Y obtained by sampling X at rate N. Applying this theorem to (7.12) and (7.13) yields per sample encoding rate and distortion

$$r_N(\phi) \;\triangleq\; \lim_{M\to\infty} r_{N,M}(\phi) \;=\; \frac{1}{2\pi}\int_{-\pi}^{\pi} \max\left\{\frac{1}{2}\log_2 \frac{\Phi_N(\omega)}{\phi}, 0\right\} d\omega$$

$$d_N(\phi) \;\triangleq\; \lim_{M\to\infty} d_{N,M}(\phi) \;=\; \frac{1}{2\pi}\int_{-\pi}^{\pi} \min\left\{\Phi_N(\omega), \phi\right\} d\omega \;.$$

Finally, we consider per snapshot encoding rate and waveform distortion, and we let the sampling rate N approach infinity. With the goal of having $Nr_N(\phi)$ and $d_N(\phi)$ approach constants, we fix a new parameter $\theta \geq 0$, and let $\phi = N\theta$. Changing variables in the above integrals gives

$$Nr_N(N\theta) \;=\; \frac{1}{2\pi}\int_{-N\pi}^{N\pi} \max\left\{\frac{1}{2}\log_2 \frac{\Phi_N(\Omega/N)}{N\theta}, 0\right\} d\Omega \tag{7.14}$$

$$d_N(N\theta) \;=\; \frac{1}{2\pi}\int_{-N\pi}^{N\pi} \min\left\{\frac{\Phi_N(\Omega/N)}{N}, \theta\right\} d\Omega \;. \tag{7.15}$$

Using the fact that $\Phi_N(\Omega/N)/N \to S(\Omega)$ pointwise and an integral convergence theorem [17], one obtains (7.9) and (7.10) in the limit of large N.

We comment that sample-and-hold reconstruction is presumed in arguing that if the per sample distortion is given by the formula (7.15) for $d_N(N\theta)$, then as $N \to \infty$, the waveform distortion converges to $\mathcal{D}(\theta)$ given by (7.10). Since one can envision better waveform reconstruction methods, one might wonder if better performance is attainable than that expressed by (7.9) and (7.10). However, for the Gaussian case being considered, it can be shown that such performance expressions are the best attainable by any coding technique. (c.f. [11, 12]). Hence, in this case there is definitely no loss in the presumption of sample-and-hold reconstruction.

Bounded Deployment Region

Now consider the situation in which the field X is restricted to the spatial interval $(0, 1]$ and, consequently, $M = N$. In this case, the least per sample encoding rate obtainable by any centralized coding system applied to the discrete-time process $\{Y_n^{(t)} : n = 1, \ldots, N,\ t = 1, 2, \ldots\}$ attaining sample distortion D or less is given by

$$R_N(N, D) \ . \tag{7.16}$$

Considered as a code for the continuous-space process X, a system attaining (7.16) has per snapshot encoding rate

$$\mathcal{R} \ = \ N R_N(N, D) \ , \tag{7.17}$$

and when N is large, its waveform distortion is

$$\mathcal{D} \ \approx \ D \ . \tag{7.18}$$

As with $NR_N(D)$, it is not known if $NR_N(N, D)$ is decreasing or subadditive in N. However, similar to (7.6), one can show straightforwardly the weaker property[9] that for any D, $0 < D < 1$, there exists N_o such that for all positive integers k and $N \geq N_o$,

$$kNR_{kN}(kN, D) \leq NR_N(N, D) + \epsilon_N(D) \ , \tag{7.19}$$

where $\epsilon_N(D) \ \rightarrow \ 0$ as $N \ \rightarrow \ \infty$. This is sufficient to establish that $\liminf_N NR_N(N, D) = \inf_N NR_N(N, D)$. Further, we make the reasonable assumption[10] that $\lim_N NR_N(N, D) = \inf_N NR_N(N, D)$. We conclude that the least per snapshot encoding rate of any centralized coding system applied to the continuous-space process $\{X^{(t)}(s) : s \in (0, 1], t = 1, 2, \ldots\}$ attaining waveform distortion $\mathcal{D}$ or less is

$$\overline{\mathcal{R}}(\mathcal{D}) \ \triangleq \ \lim_{N \rightarrow \infty} NR_N(N, \mathcal{D}) \ . \tag{7.20}$$

While we have presumed sample-and-hold reconstruction, it is our belief that this represents the best possible performance of any code with any reconstruction method. Note that we will consistently use $\overline{\mathcal{R}}$ to indicate a rate for a bounded deployment situation.

Unfortunately, there is no name for the function $\overline{\mathcal{R}}(\mathcal{D})$ in the literature, nor is there an expression, parametric or otherwise, for evaluating it in the Gaussian case, probably due to the fact that there is no analog of the Grenander-Szego theorem when the sampling rate N changes as M increases. Instead,

---

[9] The derivation is essentially the same as that of (7.6) sketched in the Appendix.

[10] As before, an alternative to accepting this conjecture is to replace $\lim_{N \rightarrow \infty} NR_N(N, D)$ with $\liminf_{N \rightarrow \infty} NR_N(N, D)$ throughout the remainder of the chapter.

in the Gaussian case, all one can do is use (7.12) and (7.13) to estimate $NR_N(N,\mathcal{D})$ for large values of N. Clearly,

$$\mathcal{R}(\mathcal{D}) \leq \overline{\mathcal{R}}(\mathcal{D}) < \infty ,$$

where the latter follows from the fact that $N_o R_{N_o}(N_o,\mathcal{D}) < \infty$ for some N_o.

Given a physical situation in which a large number of sensors are densely deployed, the question arises as to whether $\overline{\mathcal{R}}(\mathcal{D})$ or $\mathcal{R}(\mathcal{D})$ represents the appropriate performance benchmark for centralized coding. We offer the following guideline. If X is such that its variables separated by unit distance or more are nearly independent, then $\mathcal{R}(\mathcal{D}) \approx \overline{\mathcal{R}}(\mathcal{D})$, so either is appropriate, but the former is more tractable, at least in the Gaussian case; otherwise to be conservative, $\overline{\mathcal{R}}(\mathcal{D})$ is appropriate.

7.3 Identical Scalar Quantization with Slepian-Wolf Coding

7.3.1 The System

The first distributed coding scheme to consider is scalar quantization followed by Slepian-Wolf distributed lossless coding [1], as introduced and analyzed in [5, 6, 7] for a bounded deployment region, which as indicated previously we model by choosing $M = N$. Specifically, a scalar quantization is characterized by a finite or countably infinite set of levels $C \subset \mathbb{R}$ containing no points of accumulation and a *quantization rule* $Q : \mathbb{R} \to C$ such that for each n, t, the sample $Y_n^{(t)}$ is quantized into

$$\widetilde{Y}_n^{(t)} = Q(Y_n^{(t)}) .$$

Then for some positive integer L, some $r_N > 0$, and each $n \in \{1,\ldots,N\}$, the nth sensor uses Slepian-Wolf encoding to encode the temporal block $(\widetilde{Y}_n^{(1)},\ldots,\widetilde{Y}_n^{(L)})$ into a binary sequence $\underline{Z}_n = (Z_n^{(1)},\ldots,Z_n^{(b_N)})$ of length $b_N = r_N L$ (each sensor is encoded at the same rate). A single decoder receives the N binary sequences $\underline{Z}_1,\ldots,\underline{Z}_N$ and outputs a reproduction $\widehat{Y}_n^{(t)}$ of $\widetilde{Y}_n^{(t)}$ for each $t \in \{1,\ldots,L\}$ and $n \in \{1,\ldots,N\}$. After the decoder outputs $\widehat{Y}_n^{(t)}, n = 1,\ldots,N, t = 1,\ldots,L$, some type of waveform reconstruction is used to reconstruct sample function approximations $\widehat{X}^{(t)}$ to $X^{(t)}$, $t = 1,\ldots,L$.

7.3.2 Performance

Let $H(\widetilde{Y}_1,\ldots,\widetilde{Y}_N)$ denote the entropy of $(\widetilde{Y}_1,\ldots,\widetilde{Y}_N)$. According to the theory of [1], if $r_N > \frac{1}{N} H(\widetilde{Y}_1,\ldots,\widetilde{Y}_N)$, then for any $\epsilon > 0$ and all sufficiently large L, there exists encoding and decoding rules such that

$$\Pr\left(\widehat{Y}_n^{(t)} = \widetilde{Y}_n^{(t)}, n = 1, \ldots, N,\ t = 1, \ldots, L\right) \geq 1 - \epsilon . \tag{7.21}$$

Accordingly, we consider the per snapshot encoding rate of this coding scheme to be

$$\mathcal{R}_N = H(\widetilde{Y}_1, \ldots, \widetilde{Y}_N) ,$$

where the subscript N emphasizes the dependence on sampling rate and the number of samples.

Let D_N and $\mathcal{D}_N$ denote, respectively, the per sample distortion and the waveform distortion assuming sample-and-hold reconstruction, where N is once again a reminder of the sampling rate. Then using stationarity, (7.21) and the fact mentioned in Section 7.2.3 that $\mathcal{D}$ converges uniformly to D, it is easy to see that one can choose r_N's so that $r_N > H(\widetilde{Y}_1, \ldots, \widetilde{Y}_N)$ for each N, $r_N - H(\widetilde{Y}_1, \ldots, \widetilde{Y}_N) \to 0$ as $N \to \infty$, and

$$\lim_{N\to\infty} \mathcal{D}_N = \lim_{N\to\infty} D_N = E(Y_1 - Q(Y_1))^2 = E(X(0) - Q(X(0)))^2 . \tag{7.22}$$

That is, the limiting distortion equals the distortion of the scalar quantizer applied directly to the continuous-space process X. Later we comment on the potential to benefit from more sophisticated waveform reconstruction methods.

We see from the above that the distortion of this system is determined by the scalar quantizer applied to a single sample. Thus, if we wish to fix the distortion, we need only fix the scalar quantizer. We now consider what happens to the encoding rate $\mathcal{R}_N = H(\widetilde{Y}_1, \ldots, \widetilde{Y}_N)$ for a fixed scalar quantizer. It is helpful to think of this as the product of two terms:

$$\mathcal{R}_N = H(\widetilde{Y}_1, \ldots, \widetilde{Y}_N) = N \times \frac{1}{N} H(\widetilde{Y}_1, \ldots, \widetilde{Y}_N) ,$$

with the first term being the sampling rate in samples/unit distance and the second being the per sample encoding rate in bits/sample. As easily shown in [7], the per sample encoding rate decreases to zero as N increases, due to the increasing correlation between neighboring samples. The question is: does it decrease rapidly enough that $N \times \frac{1}{N} H(\widetilde{Y}_1, \ldots, \widetilde{Y}_N)$ remains bounded. In [5, 6, 7], using an idea attributed to Hajek[11], it is shown that

$$H(\widetilde{Y}_1, \ldots, \widetilde{Y}_N) \to \infty . \tag{7.23}$$

The only condition (aside from purely technical ones) is that the quantizer have a threshold that is crossed by X with positive probability, which for example will happen in the commonly occurring case that X is ergodic and $E(X - Q(X))^2 < \mathrm{var}(X)$ or, equivalently, that X is ergodic and there are at least two cells with positive probability. Note that this happens whenever the quantizer has a nontrivial effect.

[11] Professor Bruce Hajek, University of Illinois.

The proof involves a few technicalities. However, to indicate the main idea behind this disappointing result, let us sketch a proof under the stronger assumption that the quantizer has a threshold u that X crosses *with probability one* in the interval $(0, 1]$, as opposed to just *with positive probability.* Then the location at which X first crosses the threshold u within the interval $(0, 1]$ is a continuous random variable T and, consequently, has infinite entropy. As illustrated in Figure 7.2, this crossing location T can be approximated by a random variable $\widetilde{T}_N$ that can be determined from $(\widetilde{Y}_1, \ldots, \widetilde{Y}_N)$ in such a way that $\left|T - \widetilde{T}_N\right| \le \frac{1}{2N}$. In particular, let $\widetilde{T}_N$ be the average of the first pair of adjacent sampling locations, $\frac{n}{N}$ and $\frac{n+1}{N}$, such that either $\widetilde{Y}_n < u < \widetilde{Y}_{n+1}$ or $\widetilde{Y}_n > u > \widetilde{Y}_{n+1}$. As discussed below, the fact that the infinite entropy variable T can be approximated with increasing accuracy by the $\widetilde{T}_N$'s, implies that $H(\widetilde{T}_N) \to \infty$. Since $\widetilde{T}_N$ is a function of $(\widetilde{Y}_1, \ldots, \widetilde{Y}_N)$, it must happen that $H(\widetilde{Y}_1, \ldots, \widetilde{Y}_N) \to \infty$, as well.

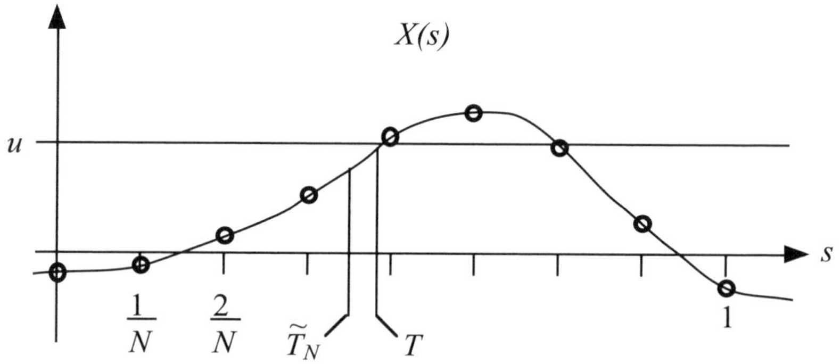

Fig. 7.2. The first crossing time T of the threshold u, and the estimate $\widetilde{T}_N$ derived from the quantized samples.

There are a number of ways to argue that $H(\widetilde{T}_N) \to \infty$. For example, using elementary information theory arguments

$$H(\widetilde{T}_N) \ \ge \ I(T; \widetilde{T}_N) \ = \ h(T) - h(T|\widetilde{T}_N) \ ,$$

where $h(T)$ and $h(T|\widetilde{T}_N)$ denote differential entropy and conditional differential entropy, respectively. Using Fano's inequality for MSE (c.f. [18], p. 255)

$$h(T|\widetilde{T}_N) \ \le \ \frac{1}{2}\log_2 2\pi e E(T - \widetilde{T}_N)^2 \ \to -\infty \quad \text{as } N \to \infty \ .$$

Substituting the above into the previous equation shows that $H(\widetilde{T}_N) \to \infty$.

In addition to being disappointing, (7.23) is surprising in view of the fact that scalar quantization with entropy coding is normally considered to have performance close to that of optimal centralized coding [19], which in turn requires a finite encoding rate. In particular, the per sample encoding rate of scalar quantization with entropy coding[12] is $R_N(D)+\epsilon_N(D)$ for some function $\epsilon_N(D)$ that approaches 0.255 as $D \to 0$, and zero as $D \to 1$. Now consider the effect of multiplying by N and letting N become large. According to (7.8) the first term $NR_N(D)$ converges to $\mathcal{R}(D)$, the rate-distortion function of the continuous-space process X, which is ordinarily finite. Evidently, it is the second term $N\epsilon_N(D)$ that tends to infinity.

It is somewhat ironic that the problem with simple scalar quantization is that it provides too much information, i.e. more than is needed simply to get good MSE. Instead it provides enough to capture level crossing locations as well.

As mentioned in the introduction, this disappointing result does not actually say that scalar quantization plus distributed lossless coding is a bad approach. Rather it says that this approach should not be used with too high a sampling rate. It also happens that the encoding rate required to attain distortion D increases to infinity as the sampling rate decreases toward the minimal value, denoted N_o, such that reconstruction from unquantized samples is possible with distortion D. Therefore, there must be a finite, nonzero sampling rate that minimizes the encoding rate required for such a system to attain distortion D.

7.3.3 The Rate at which Entropy Increases

Given that the entropy $H(\widetilde{Y}_1, \ldots, \widetilde{Y}_N)$ increases without bound, it is natural to ask about the rate at which it increases. This question has been explored in a couple of ways, each giving partial answers. In [6, 7] an asymptotic expression was found for $H(\widetilde{Y}_2|\widetilde{Y}_1)$ for a special, but important case. This induces an upper bound to the rate of increase since

$$H(\widetilde{Y}_1, \ldots, \widetilde{Y}_N) \ \leq \ H(\widetilde{Y}_1) + (N-1)H(\widetilde{Y}_2|\widetilde{Y}_1) \ .$$

Specifically, the asymptotic result is for the case that X is stationary and Gaussian with zero mean, variance one, and autocorrelation function $\rho_X(\tau)$, and the scalar quantizer has uniformly spaced thresholds, separated by some Δ, with one threshold located at $-\theta\Delta$, where $0 \leq \theta < 1$ is the quantizer offset. Uniform threshold quantizers are a natural choice to consider because when distortion must be small, they have essentially minimal output entropy [19].

For this situation, a rather complex analysis showed

$$\lim_{N\to\infty} \frac{H(\widetilde{Y}_2|\widetilde{Y}_1)}{-\sqrt{1-\rho_X(1/N)}\log_2\sqrt{1-\rho_X(1/N)}} \;=\; \frac{\sqrt{2}}{\pi}\sum_{k=-\infty}^{\infty} e^{-\frac{(k-\theta)^2\Delta^2}{2}} \;\triangleq\; S_\Delta \ . \tag{7.24}$$

[12] This is entropy coding that exploits the correlation between successive quantized samples.

It was also shown that $S_\Delta \Delta \to \frac{2}{\sqrt{\pi}}$ uniformly in θ as $\Delta \to 0$.

Although the above leads only to an upper bound to $H(\widetilde{Y}_1, \ldots, \widetilde{Y}_N)$, it is quite possible that for most autocorrelation functions it reflects the actual rate at which entropy-rate increases.

To give examples of the rate of increase of $H(\widetilde{Y}_2|\widetilde{Y}_1)$ and $H(\widetilde{Y}_1, \ldots, \widetilde{Y}_N)$, consider the case that $\rho_X(\tau)$ has well defined first and second derivatives from the right at 0, denoted $\rho'_X(0)$ and $\rho''_X(0)$, respectively. Then when τ is small

$$\rho_X(\tau) \;\approx\; 1 - \tau \left|\rho'_X(0)\right| - \frac{\tau^2}{2}\left|\rho''_X(0)\right| \;.$$

On the one hand, if $\rho'_X(0) = 0$, as for example when $\rho_X(\tau) = e^{-\tau^2}$ or $\sin(\tau)/\tau$, then for large N

$$H(\widetilde{Y}_1, \ldots, \widetilde{Y}_N) \;\le\; H(\widetilde{Y}_1) + NH(\widetilde{Y}_2|\widetilde{Y}_1) \;\approx\; S_\Delta \frac{1}{\sqrt{2}} \sqrt{\left|\rho''_X(0)\right|}\, \log_2 N \;. \quad (7.25)$$

On the other hand, if $\rho'_X(0) \neq 0$, as for example when $\rho_X(\tau) = e^{-|\tau|}$, then for large N

$$H(\widetilde{Y}_1, \ldots, \widetilde{Y}_N) \;\le\; H(\widetilde{Y}_1) + NH(\widetilde{Y}_2|\widetilde{Y}_1) \;\approx\; S_\Delta \frac{1}{2} \sqrt{\left|\rho'_X(0)\right|}\, \sqrt{N} \log_2 N \;. \quad (7.26)$$

We observe that when X is bandlimited, e.g., $S_X(\Omega) = 1$, $|\Omega| \le \pi$, and $\rho_X(\tau) = \sin(\pi\tau)/\pi\tau$, or when the power spectral density has a sufficiently light tail, e.g., $S_X(\Omega) = e^{-\Omega^2}$ and $\rho_X(\tau) = e^{-\tau^2}$, the entropy $NH(\widetilde{Y}_2|\widetilde{Y}_1)$ increases as $\log N$. However, when $\rho_X(\tau) = e^{-|\tau|}$ and the spectrum has the heavier tail $2/(\Omega^2+1)$, then $NH(\widetilde{Y}_2|\widetilde{Y}_1)$ increases at the faster rate $\sqrt{N}\log N$.

For a bandlimited process, the $\log N$ rate of increase found for $NH(\widetilde{Y}_2|\widetilde{Y}_1)$ is consistent with what Shamai [20] found for $NH_\infty(\widetilde{Y})$ for a binary quantizer with threshold at the origin, where $H_\infty(\widetilde{Y}) \triangleq \lim_{M\to\infty} \frac{1}{M} H(\widetilde{Y}_1, \ldots, \widetilde{Y}_M)$. Since

$$NH_\infty(\widetilde{Y}) \le H(\widetilde{Y}_1, \ldots, \widetilde{Y}_N) \le H(\widetilde{Y}_1) + (N-1)H(\widetilde{Y}_2|\widetilde{Y}_1) \;,$$

one may conclude that for a bandlimited Gaussian process and a binary quantizer with threshold at the origin, $H(\widetilde{Y}_1, \ldots, \widetilde{Y}_N)$ increases as $\log N$. That is, in this case we know the precise rate of increase.

Order $\log N$ is also characteristic of the per snapshot encoding rates attained by several coding schemes in the literature. For example, encoding rates on this order were attained by Cvetkovic and Vetterli [21, 22] with a simple scheme for losslessly encoding the output of a sampler and scalar quantzier applied to a bounded, bandlimited deterministic signal. For the same class of signals, Cvetkovic and Daubechies [23] and Ishwar et al. [24] attained order $\log N$ encoding rates with schemes involving dithered scalar quantizers. While these are not results about the entropy of the quantized samples of a random process, they nevertheless show comparable behavior. Finally, we mention that Kumar et al. [25] described a similar method for bounded, non-bandlimited deterministic signals with a encoding rate depending on the tail of the signal spectrum. For example, if the tail decays exponentially, the encoding rate grows as $(\log N)^2$, which like (7.26) indicates a growth rate faster than $\log N$ for a source that is not bandlimited.

7.3.4 Potential Improvements

We now explore various modifications to the scalar quantizer based system to see if they might permit rate to have finite limit as N goes to infinity.

Temporal block quantization

Supposed that instead of fixing a scalar quantizer at each sensor, one fixes a block quantizer that quantizes over a temporal block of some fixed length $L > 1$. Then one can make the same sort of argument that with ideal distributed lossless coding the per snapshot rate will again grow without bound as the sampling rate N increases. In this case the per snapshot rate is

$$\mathcal{R}_N = \frac{1}{L} H(\widetilde{Y}_n^{(t)} : t = 1, \dots, L, n = 1, \dots, N) ,$$

and one can again show that the location T at which $X(s)$ first crosses a threshold u can be approximated arbitrarily well by a random variable T_N that is determined from the samples [26]. It follows as before that $\mathcal{R}_N \to \infty$, while the distortion $\mathcal{D}_N$ goes to a nonzero value (under normal circumstances). Therefore, for any positive integer L, using block quantization with block-length L suffers the same disappointing behavior as scalar quantization.

Improved reconstruction methods

As discussed in [7], the significance of the fact that $H(\widetilde{Y}_1, \dots, \widetilde{Y}_N)$ grows without bound rests considerably on the fact stated in (7.22) that the distortion $\mathcal{D}_N$ produced by sample-and-hold reconstruction approaches a nonzero limit as $N \to \infty$. If some other waveform reconstruction could drive $\mathcal{D}_N$ to zero as $N \to \infty$, then $H(\widetilde{Y}_1, \dots, \widetilde{Y}_N)$ increasing to infinity would not be so significant or disappointing, because it might be that it increases at the rate at which $\mathcal{R}(\mathcal{D}_N)$ or $\overline{\mathcal{R}}(\mathcal{D}_N)$ increase to infinity. We now discuss the potential for other reconstruction methods to drive $\mathcal{D}_N$ to zero, based on the discussion in [7].

If linear reconstruction methods are considered, then

$$\mathcal{D}_N \geq \mathcal{D}_{\mathrm{W}} ,$$

where $\mathcal{D}_{\mathrm{W}}$ denotes the minimum MSE when linearly estimating $X(s)$ from $\{Q(X(s')) : s' \in \mathbb{R}\}$. Indeed, since $\mathcal{D}_{\mathrm{W}}$ is the MSE of a Wiener filter operating on $Q(X(t))$, it is ordinarily nonzero. There is a well known formula for $\mathcal{D}_{\mathrm{W}}$ in terms of the power spectral density of $Q(X(s))$ and the cross power spectral density of $X(s)$ and $Q(X(s))$. However, it is difficult to use it to specify a condition on X and Q to determine when, if ever, $\mathcal{D}_{\mathrm{W}} = 0$. Nevertheless, we believe that $\mathcal{D}_{\mathrm{W}} > 0$ in all but exceptional circumstances. Therefore, generally speaking, linear reconstruction cannot drive $\mathcal{D}_N$ to zero.

If nonlinear reconstruction methods are considered, then as above one can write

$$\mathcal{D}_N \geq \mathcal{D}_{\mathrm{opt}} ,$$

where $\mathcal{D}_{\mathrm{opt}}$ is the MSE of the optimum estimator for $X(s)$ from $\{Q(X(s')) : s' \in \mathbb{R}\}$, i.e. for $\widehat{X}^{(t)}(s) = E\left[X^{(t)}(s) \,\middle|\, \widehat{Y}_n^{(t)} : n = 1, \ldots, N\right]$. Unfortunately, we are aware of no generally applicable method for distinguishing when $\mathcal{D}_{\mathrm{opt}}$ is zero from when it is not.

For bounded, bandlimited deterministic signals there are results describing nonlinear reconstruction methods that cause the MSE of quantized samples to approach zero as sampling rate increases [27, 28, 29]. Since these results are based on the fact that such signals cross quantizer thresholds a minimum number of times in a finite interval, they do not translate to statements that bandlimited random processes can also be so reconstructed, because such processes do not have the requisite threshold crossing property with probability one. Thus, one would not suspect that it is possible to drive $\mathcal{D}_N$ to zero for most bandlimited processes. Indeed, Bar-David [30] demonstrates the existence of a bandlimited Gaussian random process for which $\mathcal{D}_N$ cannot go to zero.

Considering, now, nonbandlimited random processes, there are elementary examples for which the reconstruction MSE can be made to go to zero. For example, if X_s switches exclusively between values of plus and minus one, as in a random telegraph process, and if the quantizer has a threshold at zero, then X can be easily reconstructed from quantized samples with MSE approaching zero as sampling rate increases. Indeed, linear sample-and-hold reconstruction will suffice. On the other hand, if one adds random amplitudes to this example, then MSE cannot go to zero. As another example, Slepian [31] has shown that $\mathcal{D}_N$ cannot go to zero for a Gauss-Markov (i.e. Ornstein-Uhlenbeck) process, and Marco [32] has recently generalized this result to any Markov process that is continuous in probability and has absolutely continuous second-order distribution. Therefore, we believe that $\mathcal{D}_N$ can go to zero only in very special cases. Indeed, we conjecture that it cannot happen, for example, if X is Gaussian with power spectral density that satisfies the Paley-Wiener condition for a process to be nondeterministic, i.e., $\int_{-\infty}^{\infty} \frac{|\log S_X(\Omega)|}{1+\Omega^2}\, d\Omega < \infty$.

In summary, we assert that the prospects for the situation to be improved by better reconstruction methods are dim.

Adapting the quantizer to the sampling rate

Another potential improvement is to adapt the quantizer to the sampling rate N. The idea is that as sampling rate increases, one might be able to use increasingly coarse quantizers, and use the increasing correlation to maintain distortion D while decreasing entropy, perhaps to the degree that $\lim_{N\to\infty} H(\widetilde{Y}_1,\ldots,\widetilde{Y}_N)$ becomes finite. To make this concrete, note that what can be deduced from (7.22) and (7.23) is that when $0 < D < \mathrm{var}(X)$,

$$\inf_{Q:E(X-Q(X))^2\leq D}\ \lim_{N\to\infty} H(\widetilde{Y}_1,\ldots,\widetilde{Y}_N) \;=\; \infty\ .$$

On the other hand, it was claimed in [6] (the proof was omitted) that under some additional conditions, such as that the process is ergodic with piecewise continuous sample functions,

$$\lim_{N\to\infty}\ \inf_{Q:\mathcal{D}_\mathrm{W}\leq D} H(\widetilde{Y}_1,\ldots,\widetilde{Y}_N) \;=\; \infty\ ,$$

where, as before, $\mathcal{D}_\mathrm{W}$ is the distortion of minimum MSE linear estimation of $X(s)$ from $\{X(s') : s' \in \mathbb{R}\}$, which is a lower bound to $\mathcal{D}_N$ for all N and all linear reconstruction methods. This shows that even when the scalar quantizer is adapted to the sampling rate, the problem still exists. Although this was derived for linear reconstruction, as before, we do not expect nonlinear estimation to change the conclusion.

Unbounded deployment: $M \gg N$

A fourth potential improvement is to allow the Slepian-Wolf distributed lossless coding of the scalar quantizer output to code across $M \gg N$ sensor values, instead of requiring $M = N$, as so far assumed. In this case, sensors extend over a region whose size grows with N, and in the limiting (most advantageous) situation the per snapshot encoding rate can be taken to be

$$\mathcal{R}_N \;=\; NH_\infty(Y) \;\triangleq\; N\times\lim_{M\to\infty}\frac{1}{M}H(\widetilde{Y}_1,\ldots,\widetilde{Y}_M)\ .$$

Although the analysis is not entirely complete [26], it appears that under fairly general conditions, with this definition of encoding rate, we again have $\mathcal{R}_N \to \infty$. Moreover, the discussion about $\mathcal{D}_N$ not generally converging to zero applies here, because it was not tailored to reconstructions being based only on $\widetilde{Y}_n, n = 1,\ldots,N$. Therefore, it does not appear that allowing $M \gg N$ will remedy the situation.

Space varying quantization

A final improvement to consider is that instead of using the same scalar quantizer at all sensor locations, one might use different quantizers. For example, the quantizers at neighboring sensors, whose measured sensor values are highly correlated and hence nearly the same, might use quantizers that are the same except for a small shift δ of their levels and thresholds relative to one another. Since adding δ to the levels and thresholds of a scalar quantizer is equivalent to subtracting δ from Y before quantizing and adding δ after quantizing, quantizer shifting is essentially the same as what is often called *dithering*, though the latter term is usually applied when the shifts are random or time/space varying. In any event, shifting a quantizer Q by δ creates a quantizer with quantization rule

$$Q^{(\delta)}(y) \;=\; Q(y-\delta)+\delta \; .$$

Now suppose, for example, that given a nominal uniform quantizer Q with step size Δ, and given some $K < N$, to the nth sample Y_n we apply the quantizer

$$Q_n(y) \;=\; Q^{(\delta_n)}(y) \; ,$$

where

$$\delta_n \;=\; \frac{\Delta}{K} \times (n \mod K) \; .$$

If N and K are large, yet $K << N$, then K samples in the vicinity of X_s are all nearly equal to X_s, and after quantizing them with the Q_n's described above, one can obtain an estimate of X_s that is accurate to within, approximately, $\pm\frac{\Delta}{2K}$. From this, we see that if we let K increase appropriately with N, then the distortions D_N and $\mathcal{D}_N$ can be made to go to zero as N increases. Although it is easy to argue, again, that $H(\widetilde{Y}_1,\ldots,\widetilde{Y}_N) \to \infty$, this is not disappointing since distortion is going to zero. Indeed this is the nature of the results in [21, 23, 24, 25], which describe coding techniques based on dithered quantization for bounded deterministic signals with distortion $\mathcal{D}_N$ going to zero and encoding rates increasing at rates similar to the rate at which $\mathcal{R}(\mathcal{D}_N)$ increases as $N \to \infty$ for certain random process sources. This is the work, mentioned in the introduction, whose goal is to have distortion decreasing to zero as sensor density increases.

However, our interest in this chapter is in fixing a target distortion $\mathcal{D}$, rather than having distortion go to zero as N increases. In this case, one might consider a dithered quantization scheme like that just described, but with Δ_N increasing in order to keep distortion constant. Increasing Δ_N with N will reduce $H(\widetilde{Y}_1,\ldots,\widetilde{Y}_N)$ relative to the fixed Δ case. However, it is not known if it will reduce it sufficiently so that it remains bounded. This is an important open question.

7.4 Bounds to the Performance of Ideal Distributed Lossy Coding

7.4.1 Distributed Rate-Distortion Functions

We begin by introducing rate-distortion functions that describe the best possible (i.e., ideal) performance of distributed lossy codes. Since, unlike centralized coding, there is no information theoretic characterization of the performance of the best distributed codes, these are operational definitions, rather than information theoretic definitions like (7.1), (7.2), (7.8), (7.20).

Accordingly, let $R_N^d(M,D)$ denote the least per sample encoding rate of any distributed lossy source code with sample MSE D or less that operates on the discrete-time process $\{Y_n^{(t)} : t \in \mathbb{Z}_+, n = 1, \ldots, M\}$, where the superscript d stands for "distributed", and $Y_n^{(t)}$ is obtained by sampling $X^{(t)}(s)$ at rate N. The performance represented by $R_N^d(M,D)$ can be attained to arbitrary tolerance with a distributed encoder operating on $\{Y_n^{(t)} : n = 1, \ldots, M, t = 1, \ldots, L\}$ for all sufficiently large L. That is, there is a separate encoder at each of the M sensor locations operating on a temporal block of length L. The decoder receives the outputs from all encoders and produces reproductions $\{\widehat{Y}_n^{(t)} : t \in \mathbb{Z}_+, n = 1, \ldots, M\}$ of $\{Y_n^{(t)} : t \in \mathbb{Z}_+, n = 1, \ldots, M\}$.

Since $R_N^d(M,D)$ is subadditive over M, the limit as $M \to \infty$ yields the least per sample encoding rate of any distributed lossy source coding system with sample distortion D or less operating on the samples taken at rate N:

$$R_N^d(D) \;=\; \lim_{M\to\infty} R_N^d(M,D) \;. \tag{7.27}$$

The performance represented by $R_N^d(D)$ can be attained to arbitrary tolerance with a distributed encoder operating on $\{Y_n^{(t)} : n = 1, \ldots, M, t = 1, \ldots, L\}$ for all sufficiently large M and L.

Let us now consider the coding of the continuous-space process X based on distributed coding of its samples taken at rate N. For the performance represented by $R_N^d(M,D)$, one attains per snapshot encoding rate

$$\mathcal{R} \;=\; N R_N^d(M,D) \;, \tag{7.28}$$

and if N is large, sample-and-hold reconstruction results in waveform MSE

$$\mathcal{D} \;\approx\; D \;. \tag{7.29}$$

For the performance represented by (7.27), the per snapshot encoding rate is

$$\mathcal{R} \;=\; N R_N^d(D) \;, \tag{7.30}$$

and again (7.29) applies.

Now consider what happens as N increases. As with centralized coding, it is not known if $NR_N^d(D)$ is decreasing or subadditive. However, as with (7.6) and (7.19), it can be shown[13] that for any D, $0 < D < 1$, there exists N_o such that for all positive integers k and $N \geq N_o$

$$kNR_{kN}^d(D) \leq NR_N^d(D) + \epsilon_N(D) \;,$$

[13] As sketched in the Appendix, the proof is similar, but involves codes rather than test channels.

where $\epsilon_N(D) \to 0$ as $N \to \infty$. This is sufficient to imply $\liminf_{N\to\infty} NR^d_N(D) = \inf_N NR^d_N(D)$. Further, we make the assumption that $\lim_{N\to\infty} NR^d_N(D) = \inf_N NR^d_N(D)$. We conclude that the least per snapshot encoding rate of any distributed coding system operating on samples of the continuous-space process X and attaining waveform distortion $\mathcal{D}$ or less is

$$\mathcal{R}^d(\mathcal{D}) \triangleq \lim_{N\to\infty} NR^d_N(\mathcal{D}) \ , \tag{7.31}$$

where $\mathcal{R}^d(\mathcal{D})$ is the distributed rate-distortion function of the continuous-space process X. While we have presumed sample-and-hold reconstruction, it is again our belief that this represents the best possible performance of any code with any reconstruction method.

Bounded deployment

As with centralized coding, one can consider the performance of distributed codes restricted to the spatial interval (0,1] and choose $M = N$. In particular, the least per sample encoding rate obtainable by any distributed coding system applied to the discrete-time process $\{Y^{(t)}_n : n = 1, \ldots, N, t = 1, 2, \ldots\}$ attaining sample distortion D or less is given by

$$R^d_N(N, D) \ . \tag{7.32}$$

Considered as a code for the continuous-space process X, a system attaining (7.32) has per snapshot encoding rate

$$\mathcal{R} \ = \ NR^d_N(N, D) \ , \tag{7.33}$$

and when N is large, its waveform distortion is

$$\mathcal{D} \ \approx \ D \ . \tag{7.34}$$

Once again, we make the reasonable assumption[14] that $\lim_N NR^d_N(N, D) = \inf_N NR^d_N(N, D)$. We conclude that the least per snapshot encoding rate of any distributed coding system operating on samples of the continuous-space process $\{X^{(t)}(s) : s \in (0, 1], t = 1, 2, \ldots\}$ attaining waveform distortion $\mathcal{D}$ or less is

$$\overline{\mathcal{R}}^d(\mathcal{D}) \ \triangleq \ \lim_{N\to\infty} NR^d_N(N, \mathcal{D}) \ . \tag{7.35}$$

While we have presumed sample-and-hold reconstruction, it is again our belief that this represents the best possible performance of any code with any reconstruction method.

[14] As before, an alternative to accepting this conjecture is to replace $\lim_{N\to\infty} NR^d_N(N, D)$ with $\liminf_{N\to\infty} NR^d_N(N, D)$.

7.4.2 The Finiteness of $\mathcal{R}^d(\mathcal{D})$ and $\overline{\mathcal{R}}^d(\mathcal{D})$

Let us fix a target per snapshot MSE $\mathcal{D} > 0$. In this subsection, we sketch the simple time-sharing argument, mentioned in the introduction, that demonstrates that $\mathcal{R}^d(\mathcal{D})$ and $\overline{\mathcal{R}}^d(\mathcal{D})$ are finite. It also shows that when the sampling rate N is large, there exists a distributed coding scheme attaining distortion $\mathcal{D}$ or less that encodes each sample at the same finite rate. This is, essentially, a simplified version of a construction given in [9].

It suffices to consider the bounded deployment case, as it is more restrictive. We begin by fixing a sampling rate N_o such that sample-and-hold reconstruction attains a per-snapshot MSE $\mathcal{D}_o < \mathcal{D}$, when applied to N_o unquantized samples taken at rate N_o. Since $\mathcal{D}_o < \mathcal{D}$, one can design a scalar quantizer that introduces so little distortion that when applied to the samples taken at rate N_o, the overall distortion after sample-and-hold reconstruction is again less than $\mathcal{D}$. Let the outputs of the scalar quantizers be separately losslessly encoded at some rate $R_o < \infty$ bits/sample. The lossless coding, which can be of the fixed- or variable-length variety, can be viewed as a kind of distributed coding. This leads to a per snapshot encoding rate of $N_o R_o < \infty$ bits per unit distance.

Now suppose that for some $N \gg N_o$, we desire a distributed scheme that encodes samples taken at rate N with identical finite encoding rates and with resulting per snapshot MSE of $\mathcal{D}$ or less. To simplify the discussion assume $N = KN_o$ with K an integer. (An argument with a little more complexity is needed when N is not a multiple of N_o.) Consider the following time-sharing argument. For the sample function $X^{(t)}$ at time t, apply the scalar quantization and the lossless encoding to the N_o samples at locations

$$\tau,\ K+\tau,\ 2K+\tau,\ \ldots,\ (N_o-1)K+\tau\ ,$$

where $\tau = t \bmod K$. Since these samples are spaced $\frac{1}{N_o}$ apart, the sample-and-hold reconstruction will have distortion $\mathcal{D}$ or less. Moreover, each sample is encoded once every K time units at rate R_o. for an average rate of $R_o/K = N_o R_o$ bits/sample, which is the same for every sensor. This completes the construction of a distributed coding system that for a large sampling rate N quantizes and encodes each sensor output with the same finite rate, and attains per snapshot MSE $\mathcal{D}$ or less. It follows that $\overline{\mathcal{R}}(D) < \infty$, and, of course, $\mathcal{R}(D) \le \overline{\mathcal{R}}(D) < \infty$, as well.

Remark

As discussed in the first bulleted item of Section 7.3.4, for any $L > 0$, temporal quantization with blocklength L has rate increasing to infinity, rather than staying bounded as with ideal distributed lossy coding. We conclude, therefore, that to approach the finite rates $\overline{\mathcal{R}}^d(\mathcal{D})$ or $\mathcal{R}^d(\mathcal{D})$, the temporal blocklength L must grow without bound as the sampling rate increases.

7.4.3 Berger-Tung Upper Bounds

This subsection describes the information theoretic upper bound to $R_N^d(M,D)$, known as the Berger-Tung bound. It originated in [2, 33] and was extended in [34, 35]. For our purposes, we state it without derivation as

$$\begin{aligned} R_N^d(M,D) &\le R_N^{d,BT}(M,D) \\ &\triangleq \inf_{\substack{q,g_1,\ldots,g_M: \\ \frac{1}{M}\sum_{n=1}^M E_q\left(Y_n - g_n(\widetilde{Y}_1,\ldots,\widetilde{Y}_M)\right)^2 \le D}} \frac{1}{M} I(Y_1,\ldots,Y_M;\widetilde{Y}_1,\ldots,\widetilde{Y}_M) \end{aligned} \quad (7.36)$$

This is just like the information theoretic definition of $R_N(M,D)$ for centralized coding, except: (1) the test channel is constrained to be memoryless in the sense that for each n, given Y_n, random variable $\widetilde{Y}_n$ is conditionally independent of all $Y_{n'}$'s and $\widetilde{Y}_{n'}$'s, $n' \neq n$, and (2) for each n, the MSE is computed between Y_n and a function $g_n(\widetilde{Y}_1,\ldots,\widetilde{Y}_M)$. This bound has been shown to be tight in the case of two Gaussian random variables [36]. However, it is not believed to be tight for more than two Gaussian random variables.

We also note that any choice of $q, g_1, \ldots, g_M$ satisfying the constraint in the definition of $R_N^{d,BT}(M,D)$ yields an upper bound to $R_N^d(M,D)$, which we will also call a Berger-Tung bound.

Since $R^{d,BT}(M,D)$ is subadditive in M, its limit exists (and equals its inf) and so

$$R_N^d(D) \;\le\; R_N^{d,BT}(D) \;\triangleq\; \lim_{M\to\infty} R_N^{d,BT}(M,D) \;.$$

Consider now the coding of the continuous-space process X based on distributed coding of its samples taken at rate N. For the performance represented by $R_N^d(M,D)$, one attains per snapshot encoding rate

$$\mathcal{R} \;=\; NR_N^d(M,D) \;\le\; NR_N^{d,BT}(M,D) \;, \quad (7.37)$$

and if N is large, sample-and-hold reconstruction yields waveform MSE is

$$\mathcal{D} \;\approx\; D \;. \quad (7.38)$$

For the performance represented by (7.27), the per snapshot encoding rate is

$$\mathcal{R} \;=\; NR_N^d(D) \;\le\; NR_N^{d,BT}(D) \;, \quad (7.39)$$

and again (7.38) applies.

Now consider the dependence on N. As before, it is not known if $NR_N^{d,BT}(D)$ is decreasing or subadditive. However, with a proof like that of (7.6), it can be shown that for any D, $0 < D < 1$, there exists N_o such that for all positive integers k and $N \ge N_o$

$$kNR_{kN}^{d,BT}(D) \le NR_N^{d,BT}(D) + \epsilon_N(D) \;,$$

where $\epsilon_N(D) \rightarrow 0$ as $N \rightarrow \infty$. This implies $\liminf_{N\rightarrow\infty} NR_N^{d,BT}(D) = \inf_N NR_N^{d,BT}(D)$. Making the reasonable assumption that $\lim_{N\rightarrow\infty} NR_N^{d,BT}(D) = \inf_N NR_N^{d,BT}(D)$, we conclude that the least per snapshot encoding rate of any distributed coding system applied to the continuous-space process X that attains waveform distortion $\mathcal{D}$ or less is

$$\mathcal{R}^d(\mathcal{D}) = \lim_{N\rightarrow\infty} NR_N^d(\mathcal{D}) \leq \mathcal{R}^{d,BT}(\mathcal{D}) \triangleq \lim_{N\rightarrow\infty} NR_N^{d,BT}(\mathcal{D}) \ . \quad (7.40)$$

We note that the Berger-Tung bound is clearly applicable to the bounded deployment case.

Gaussian Sources

In the Gaussian case, the Berger-Tung bound can be used as follows to obtain a parametric upper bound to $R_N^d(M,D)$. Let us fix a parameter $\phi \geq 0$. Since a Gaussian test channel is optimal when evaluating the centralized rate-distortion function, it is natural to choose q to be Gaussian. It must also be memoryless. To make things tractable, let us also choose q to have identical components. In this case, the effect of q can be modeled as producing

$$\widetilde{Y}_n = Y_n + Z_n \ , \ n = 1,\ldots,M \ , \quad (7.41)$$

where $Z_1,\ldots,Z_M$ are IID Gaussian $\mathcal{N}(0,\phi)$. Finally, let us choose

$$g_n(\widetilde{y}_1,\ldots,\widetilde{y}_M) = E\left[Y_n \middle| \widetilde{Y}_1 = \widetilde{y}_1,\ldots,\widetilde{Y}_M = \widetilde{y}_M\right] . \quad (7.42)$$

For these choices it can be shown that

$$r_{N,M}^{d,BT}(\phi) \triangleq \frac{1}{M} I_q(Y_1,\ldots,Y_M;\widetilde{Y}_1,\ldots,\widetilde{Y}_M) = \frac{1}{M}\sum_{i=1}^{M} \frac{1}{2}\log_2\left(\frac{\lambda_{N,M,i}}{\phi}+1\right) , \quad (7.43)$$

and (cf [37], p. 157)

$$d_{N,M}^{d,BT}(\phi) \triangleq \frac{1}{M}\sum_{n=1}^{M} E_q\left(Y_n - g_n(\widetilde{Y}_1,\ldots,\widetilde{Y}_M)\right)^2 = \frac{1}{M}\sum_{i=1}^{M}\frac{\lambda_{N,M,i}}{\lambda_{N,M,i}+1} , \quad (7.44)$$

where as in (7.11)-(7.13), $\lambda_{N,M,i}$ is one of the M eigenvalues of the covariance matrix $K_{N,M}$ of $Y_1,\ldots,Y_M$. It follows that for any $\phi \geq 0$, the performance point $(d_{N,M}^{d,BT}(\phi), r_{N,M}^{d,BT}(\phi))$ lies above the $R_N^d(\cdot)$ curve in the sense that

$$R_N^d(M, d_{N,M}^{d,BT}(\phi)) \leq r_{N,M}^{d,BT}(\phi) \ .$$

In other words, $(d_{N,M}^{d,BT}(\phi), r_{N,M}^{d,BT}(\phi))$ is a sample distortion and per sample encoding rate pair that is achievable with distributed coding over M samples taken at rate N. When N is large, it is also an achievable waveform distortion and per snapshot encoding rate for the continuous-space process X.

Note that this discussion has not claimed to have found the test channel that minimizes the Berger-Tung bound. However, it has described a reasonable upper bound to $R_N^d(M, D)$.

Note also that if X is not Gaussian, one can upper bound $R_N^d(M, D)$ by the same quantity for a Gaussian process with the same power spectral density. This follows by the same sort of argument used to show this property in the centralized coding case.

7.4.4 Bounded Deployment Region

For a Gaussian field in a bounded deployment region, Kashyap et al. [8] used the Berger-Tung bound in the following way. Given N and $M = N$, they chose a test channel as in (7.41), parameterized by ϕ, and g_n's as in (7.42), and used a bounding approach from [38] to show that for any D

$$\frac{1}{N} I_q(Y_1, \ldots, Y_N; \widetilde{Y}_1, \ldots, \widetilde{Y}_N) \;\le\; R_N(N, D) + \frac{1}{2} \log_2 \left(1 + \frac{D}{\phi}\right) .$$

Next, choosing $\phi = \phi_N^*(D)$ such that $d_{N,N}^{d,BT}(\phi_N^*(D)) = D$ yields

$$R_N^{d,BT}(N, D) \;\le\; R_N(N, D) + \frac{1}{2} \log_2 \left(1 + \frac{D}{\phi_N^*(D)}\right) . \tag{7.45}$$

They then showed that $\phi_N^*(D) \ge \theta N$ for all sufficiently large N, where θ is the largest number such that (1) $\rho_X(\tau)$ is monotonic on $[-\theta, \theta]$, and (2) $1 - \frac{\rho_X^2(\theta)}{4} \le D$. Using this and $\ln(1 + x) \le x$ in (7.45), they found that for all sufficiently large N

$$N R_N^{d,BT}(N, D) \;\le\; N R_N(N, D) + \frac{1}{\theta 2 \ln 2} . \tag{7.46}$$

They also showed that $N R_N(N, D)$ remains bounded as a function of N. From this, one can conclude that $N R_N^{d,BT}(N, D)$, and consequently $N R_N^d(N, D)$, remain bounded as N increases. It follows that

$$\overline{\mathcal{R}}^d(\mathcal{D}) < \infty ,$$

which was one of their main goals.

In [9], the argument is strengthened by removing the assumption that the $\rho_X(\tau)$ is monotonic near $\tau = 0$.

While the bound in (7.46) and the analogous one in [9] are useful for making the argument that $\overline{\mathcal{R}}^d(\mathcal{D}) < \infty$, they are actually rather loose. As a result, [8] plots $R_N^{d,BT}(N, D)$ instead, as given for example by (7.43) and (7.44); [9] takes a similar approach.

7.4.5 Unbounded Deployment Region, Gaussian Case

In [10] the authors of the present chapter relaxed the bounded deployment assumption and used the Berger-Tung bound and the Grenander-Szego Theorem to find a tractable parametric upper bound to $\mathcal{R}^d(\mathcal{D})$ for the Gaussian case. We now review this result.

Applying the Grenander-Szego Theorem to (7.43) and (7.44), one finds in the limit as $M \to \infty$, the following per sample encoding rate and sample distortion are achievable with distributed coding of samples taken at rate N when coding over a block of M sensors and L time instants, when M and L are sufficiently large:

$$r_N^{d,BT}(\phi) \triangleq \frac{1}{2\pi}\int_{-\pi}^{\pi}\frac{1}{2}\log_2\left(\frac{\Phi_N(\omega)}{\phi}+1\right)d\omega$$

$$d_N^{d,BT}(\phi) \triangleq \frac{1}{2\pi}\int_{-\pi}^{\pi}\frac{\Phi_N(\omega)}{\Phi_N(\omega)+1}\,d\omega\ ,$$

where as before $\Phi_N(\omega)$ is the power spectral density of the discrete-space process Y.

When a system with the above performance is used to encode the continuous-space process X, the per snapshot encoding rate is

$$\mathcal{R} = N r_N^{d,BT}(\phi) = N\frac{1}{2\pi}\int_{-\pi}^{\pi}\frac{1}{2}\log_2\left(\frac{\Phi_N(\omega)}{\phi}+1\right)d\omega\ ,$$

and when N is large, the waveform distortion with sample-and-hold reconstruction is

$$\mathcal{D} \approx d_N^{d,BT}(\phi) = \frac{1}{2\pi}\int_{-\pi}^{\pi}\frac{\Phi_N(\omega)}{\Phi_N(\omega)+1}\,d\omega\ .$$

Finally, let the sampling rate N approach infinity. As in (7.14) and (7.15), with the goal of having $Nr_N(\phi)$ and $d_N(\phi)$ approach constants, fix a new parameter $\theta \geq 0$ and let $\phi = N\theta$. Changing variables, passing to the limit, and using the facts that $\mathcal{D} \to D$ and $\frac{1}{N}\Phi_N(\Omega) \to S(\Omega)$, one finds that the following per snapshot encoding rate and waveform distortion are achievable with distributed coding of samples of X at high sampling rates:

$$\mathcal{R}^{d,BT}(\theta) \triangleq \lim_{N\to\infty} N r_N^{d,BT}(\phi) = \frac{1}{2\pi}\int_{-\infty}^{\infty}\frac{1}{2}\log_2\left(\frac{S(\Omega)}{\theta}+1\right)d\Omega \quad (7.47)$$

$$\mathcal{D}^{d,BT}(\theta) \triangleq \lim_{N\to\infty} d_N^{d,BT}(\phi) = \frac{1}{2\pi}\int_{-\infty}^{\infty}\frac{S(\Omega)}{\frac{S(\Omega)}{\theta}+1}\,d\Omega\ . \quad (7.48)$$

Tightness

There were three places in the derivation of the parametric formulas (7.47), (7.48) at which upper bounds were used that might cause the formulas to be loose rather than tight. (1) the use of the Berger-Tung bound, (2) the choice of q as Gaussian with identical components, and (3) the use of sample-and-hold reconstruction. Nevertheless, it is possible that the bound is tight, i.e., that it gives the actual form of $\mathcal{R}^d(\mathcal{D})$. Though, as mentioned earlier, the Berger-Tung bound is not believed to be tight for $N \geq 3$, it is quite conceivable that it is asymptotically tight for large N and $M \gg N$. Moreover, the choices of a Gaussian test channel with identical components and sample-and-hold reconstruction might also entail no loss. One specific case in which the bound is known to be tight is discussed below.

Rate and distortion profiles

It is interesting to compare the parametric expressions (7.47), (7.48) for the upper bound to the distributed rate-distortion function to the parametric expressions (7.9), (7.10) for the centralized rate-distortion function. It is customary to interpret the integrands of (7.9) and (7.10) at frequency Ω as indicating the rate and distortion, respectively, at which the component of X at frequency Ω are encoded. (They are actually *densities* of such, in that they are quantities that must be integrated to yield rate and distortion.) We call them *rate* and *distortion profiles*, respectively.

In considering such profiles one first notices from (7.10) that with no distributed coding constraint, optimal centralized coding makes the distortion profile as flat as possible. Then at frequencies Ω such that $S(\Omega)$ becomes too small to support a flat distortion profile, one sees in (7.9) that optimal coding assigns zero rate. From the derivation of the centralized rate-distortion function, one sees that such optimal shaping of the rate and distortion profiles is made possible by the transform, which exposes frequency-like components. In a practical system, such bandlimiting and profile shaping might also be accomplished by bandpass filtering X before sampling and coding. In contrast, with distributed coding, each encoder has access to the samples from just one sensor. As a result, no spatial transform or filtering of any type is possible. Instead, the distortion created at one sensor will be, essentially, uncorrelated with that at another sensor. For this reason, one can think of distributed coding as adding spatially white noise. Now, as is well known, if one estimates a random process X from $Y = X + V$, where V is additive white noise with power spectral density ϕ, then the optimal linear filter, i.e., the Wiener filter, has mean squared error

$$\frac{1}{2\pi} \int_{-\infty}^{\infty} \frac{S(\Omega)}{\frac{S(\Omega)}{\phi} + 1} \, d\Omega \ . \tag{7.49}$$

which is precisely the form of (7.48). Our interpretation, then, is that when sampling rate is high, the effect of distributed coding is like adding spatially white noise. And unlike the centralized case, explicit shaping of the rate and distortion profiles is not possible.

Nevertheless, there is a special case where the distributed and centralized rate-distortion functions are easily seen to be identical. This is the case that there is a positive constant W such that $S(\Omega) = \frac{2\pi}{W}$ on a set of frequencies of measure W (e.g. on the union of a finite set of finite intervals of total length W) and 0 elsewhere. In this case, to attain $\mathcal{R}(\mathcal{D})$, no transform or filtering is needed since the source is already bandlimited and since the spectrum is constant (where not zero), as are the ideal rate and distortion profiles. Indeed one may straightforwardly verify that (7.47) and (7.48) lead to the rate-distortion function given by (7.9) and (7.10), namely,

$$\mathcal{R}(\mathcal{D}) \;=\; \mathcal{R}^{d,BT}(\mathcal{D}) \;=\; \frac{W}{4\pi} \log_2 \frac{1}{\mathcal{D}} \;.$$

Since the upper bound $\mathcal{R}^{d,BT}(\mathcal{D})$ equals the centralized rate-distortion function, in this case it must also equal the ideal distributed rate-distortion function $\mathcal{R}^{d}(\mathcal{D})$; so the upper bound is tight.

In the special case that $S(\Omega) = \frac{2\pi}{W}$, $|\Omega| < \frac{W}{2}$, and zero elsewhere, it is easy to see why the distributed rate-distortion function should equal the centralized rate-distortion function. In this case, sampling at the Nyquist rate $\frac{W}{2\pi}$ produces independent and identical samples, which can be independently coded without loss, given the availability of temporal samples at each sensor location.

Equal rates

Notice that when deriving bounds to $\mathcal{R}^{d}(\mathcal{D})$ for a Gaussian source, no effort was made to require each sensor to encode with the same rate. Nevertheless, one can see from the derivations that the performance specified by the bounds is attainable with an equal rate strategy. On the other hand, as discussed above, in the special case that $S(\Omega) = \frac{2\pi}{W}$, $|\Omega| < \frac{W}{2}$, optimal performance can also be attained by sampling at a high rate, discarding all samples but a subset taken at the Nyquist rate, and independently coding the Nyquist samples. This strategy can be viewed as coding the original high rate samples with unequal rates. Thus, for this example, optimal performance can also be attained using very unequal sensor encoding rates.

7.4.6 Gaussian Examples

We consider two stationary Gaussian processes: (1) Markov (Ornstein-Uhlenbeck) with autocorrelation function $\rho_X(\tau) = e^{-|\tau|}$ and power spectral density

$$S(\Omega) \;=\; \frac{2}{1+\Omega^2} \;,$$

and (2) bandlimited with the flat power spectral density

$$S(\Omega) = \begin{cases} 1\,, & |\Omega| \le \pi \\ 0\,, & \text{else} \end{cases}\,,$$

with autocorrelation function $\rho_X(\tau) = \text{sinc}(\tau) = \frac{\sin(\pi\tau)}{\pi\tau}$.

Figure 7.3 plots their rate and distortion profiles at $\mathcal{D} = 1/3$ for distributed and centralized coding, where the former are based on the upper bound given by (7.47), (7.48). Also shown are the distributed and centralized rate-distortion functions $\mathcal{R}^{d,BT}(\mathcal{D})$ and $\mathcal{R}(\mathcal{D})$.

Table 7.1 gives parametric and closed form expressions for $\mathcal{R}^{d,BT}(\mathcal{D})$ and $\mathcal{R}(\mathcal{D})$. For the Markov source, we give an approximate closed form expression for $\mathcal{R}(D)$ that applies for small $\mathcal{D}$ ([12], p. 145). To facilitate concrete comparisons, these expressions are evaluated at $\mathcal{D} = 0.1$. One can see that for the Markov source and small distortion, the upper bound to the distributed rate-distortion function is approximately 25% larger than the centralized rate-distortion function (compare $\frac{1}{\mathcal{D}}$ to $\frac{8}{\pi^2\mathcal{D}} = \frac{0.81}{\mathcal{D}}$). For the bandlimited flat spectrum, the distributed and centralized rate-distortion functions are the same, as expected.

Also shown in Table 7.1 shown are numerical results for bounded deployment taken from [8, 9]. Specifically, the last row for each source gives $\mathcal{R}_N^{d,BT}(N,\mathcal{D})$ taken from Figure 2 of [9] at $D = 0.1$ and $N = 100$. It also shows the computed value of $\mathcal{R}_N(N,\mathcal{D})$ taken from Figure 2 of [8] for the same D and N.

It is interesting to observe that the effect of bounded deployment on the upper bound to distributed coding is similar to its effect on centralized coding. Specifically, for the Markov source, $\frac{\mathcal{R}_{100}^{d,BT}(100,0.1)}{\mathcal{R}^{d,BT}(0.1)} = 2.5$, while $\frac{\mathcal{R}_{100}(100,0.1)}{\mathcal{R}(0.1)} = 3.8$. For the bandlimited source, these ratios are even closer. It is also interesting that while we knew a priori that distributed coding is as good as centralized for the case of unbounded deployment, we see from the table that it is nearly as good for the bounded deployment case, as well. Indeed, the numerical results do not rule out the possibility that for this source distributed and centralized are equally good for bounded deployment.

7.5 Conclusions

This chapter has reviewed several recent results regarding the performance of distributed lossy source codes operating on a dense sensor network in the limit as density increases without bound.

The first result, in Section 7.3, showed that if identical scalar quantizers are used by each sensor, then even if followed by correlation exploiting Slepian-Wolf distributed lossless coding, the encoding rate in bits per snapshot required to attain a fixed target $\mathcal{D}$ grows to infinity. Bounds to the rate at which the encoding rate grows were discussed for the important special case of a Gaussian source and a uniform scalar quantizer.

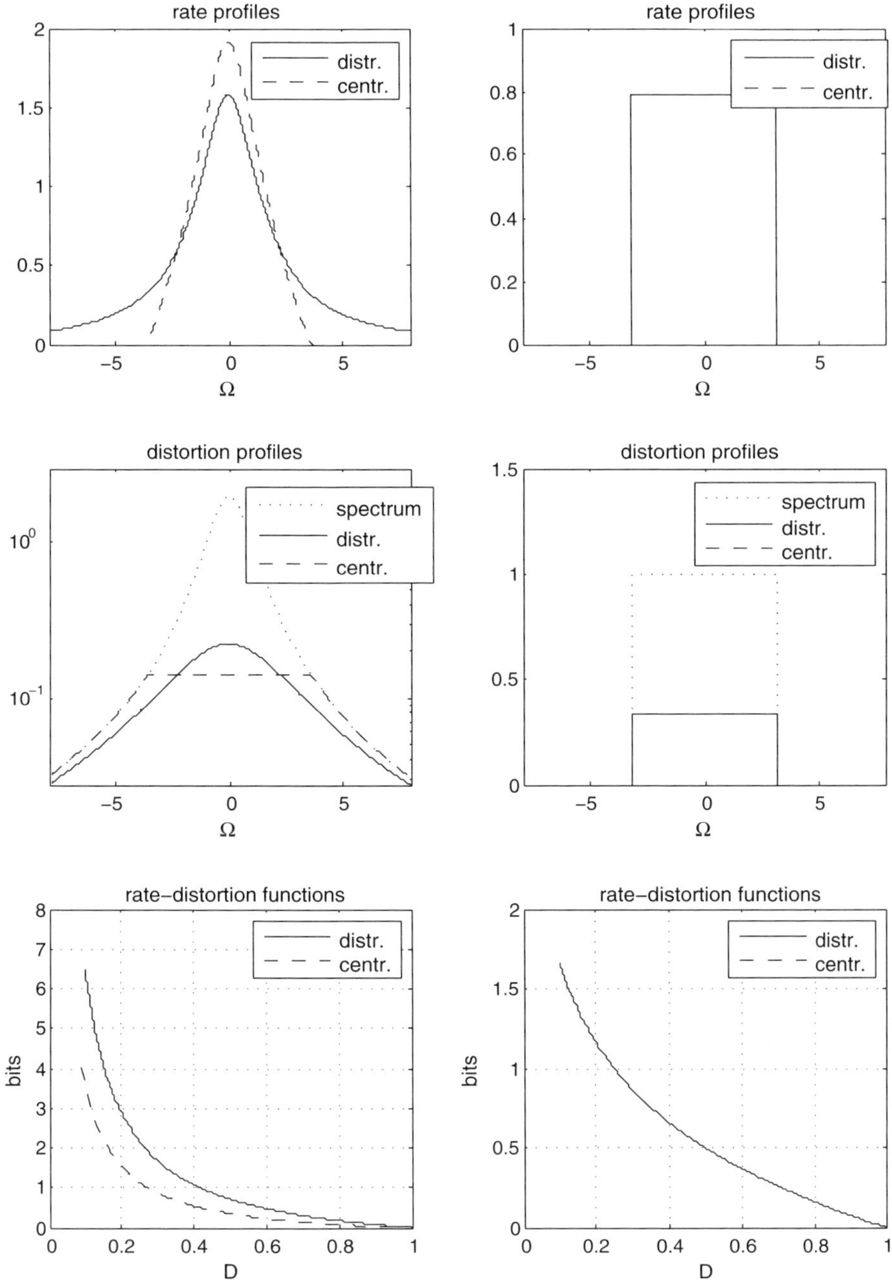

Fig. 7.3. Rate profiles (top), distortion profiles (middle), and rate-distortion functions (bottom) for the Markov source (left) and the bandlimited source (right). The rate and distortion profiles are plotted for $\mathcal{D} = 1/3$. For the Markov source the corresponding encoding rates are 0.74 for centralized and 1.44 for distributed. For the bandlimited source both encoding rates are 0.79 .

Table 7.1. Comparison of distributed and centralized coding of Gaussian sources at high sampling rates.

Distributed (upper bound to optimal)	Centralized (optimal)
Markov, $S(\Omega) = \frac{2}{1+\Omega^2}$	
$\theta \geq 0$	$0 \leq \theta \leq 2$
$\mathcal{D}^{d,BT}(\theta) = \sqrt{\frac{\theta}{\theta+2}}$	$\mathcal{D}(\theta) = 1+\frac{1}{\pi}\sqrt{2\theta - \theta^2} - \frac{2}{\pi}\tan^{-1}\sqrt{\frac{2-\theta}{\theta}}$
$\mathcal{R}^{d,BT}(\theta) = \frac{1}{2\ln 2}\left(\sqrt{\frac{\theta+2}{\theta}} - 1\right)$	$\mathcal{R}(\theta) = \frac{1}{\pi \ln 2}\left(\sqrt{\frac{2-\theta}{\theta}} - \tan^{-1}\sqrt{\frac{2-\theta}{\theta}}\right)$
$\mathcal{R}^{d,BT}(\mathcal{D}) = \frac{1}{2\ln 2}\left(\frac{1}{\mathcal{D}} - 1\right)$	$\mathcal{R}(\mathcal{D}) \approx \frac{1}{2\ln 2}\left(\frac{8}{\pi^2 \mathcal{D}} - 1\right)$, for small D
$\mathcal{R}^{d,BT}(0.1) = 6.49$	$\mathcal{R}(0.1) = 3.56$
$\mathcal{R}^{d,BT}_{100}(100, 0.1) = 16.5$	$\mathcal{R}_{100}(100, 0.1) = 13.4$
Bandlimited Flat Spectrum, $S(\Omega) = 1, \lvert\Omega\rvert \leq \pi$	
$\theta \geq 0$	$0 \leq \theta \leq 1$
$\mathcal{D}^{d,BT}(\theta) = \frac{\theta}{\theta+1}$	$\mathcal{D}(\theta) = \theta$
$\mathcal{R}^{d,BT}(\theta) = \frac{1}{2}\log_2 \frac{\theta+1}{\theta}$	$\mathcal{R}(\theta) = \frac{1}{2}\log_2 \frac{1}{\theta}$
$\mathcal{R}^{d,BT}(\mathcal{D}) = \frac{1}{2}\log_2 \frac{1}{\mathcal{D}}$	$\mathcal{R}(\mathcal{D}) = \frac{1}{2}\log_2 \frac{1}{\mathcal{D}}$
$\mathcal{R}^{d,BT}(0.1) = 1.70$	$\mathcal{R}(0.1) = 1.70$
$\mathcal{R}^{d,BT}_{100} = (100, 0.1) = 3.4$	$\mathcal{R}_{100}(100, 0.1) = 3.3$

It is important to note that the result of this section does not say that scalar quantization plus Slepian-Wolf coding is a bad approach. It merely says that when this approach is used, sampling rate should not be taken too large.

It was also remarked that changing from scalar to L-dimensional temporal block quantization will not remedy the situation if L is fixed.

Section 7.4 considered bounds to the performance of ideal distributed lossy coding, which in effect uses temporal block quantization with blocklength growing as sensor density increases. First it is argued that with ideal coding, the per snapshot rate can remain bounded as sensor density increases without bound. Moreover, such performance can be attained with simple schemes that encode each sensor output at the same rate. Next, Berger-Tung type bounds to the performance of ideal distributed lossy coding were discussed. These were for bounded, as well as unbounded, deployment. In the latter case, nice parametric formulas are obtained that can be compared to the parametric formulas for conventional centralized coding. They also show that in the limit of high sampling rate distributed coding need not lose much relative to optimal centralized coding. Indeed, when the source has a flat bandlimited spectrum, there is no loss whatsoever.

References

[1] D. Slepian and J. Wolf, "Noiseless coding of correlated information sources," *IEEE Trans. Inform. Theory.*, vol. 19, pp. 471-480, Jul. 1973.

[2] T. Berger, "Multiterminal source coding," *Information Theory Approach to Communications, CISM Courses and Lecture Notes No. 229*, G. Longo, Ed., New York: Springer-Verlag, 1977.

[3] S. S. Pradhan, J. Kusuma, K. Ramchandran, "Distributed compression in a dense microsensor network," *IEEE Sig. Proc. Mag.*, pp. 51-60, Mar. 2002.

[4] S. S. Pradhan and K. Ramchandran, "Distributed source coding using syndromes (DISCUS): Design and construction," *IEEE Trans. Inform. Theory*, vol. 49, pp. 626-643, Mar. 2003.

[5] D. Marco, E. J. Duarte-Melo, M. Liu, and D. L. Neuhoff, "On the many-to-one transport capacity of a dense wireless sensor network and the compressibility of its data," *IPSN*, Palo Alto, pp. 1-16, Apr. 2003.

[6] D. Marco and D.L. Neuhoff, "Entropy of quantized data at high sampling rates," *IEEE Int. Symp. Inform. Thy.*, Adelaide, Aus., pp. 342-346, Sept. 2005.

[7] D. Marco and D.L. Neuhoff, "Entropy of quantized data at high sampling rates," submitted to *IEEE Trans. Inform. Thy.*, Sept. 2006.

[8] A. Kashyap, L. A. Lastras-Montano, C. Xia, and Z. Liu, "Distributed source coding in dense sensor networks," *Data Compression Conference (DCC)*, pp. 13-21, Snowbird, UT, Mar. 2005.

[9] A. Kashyap, L. A. Lastras-Montano, C. Xia, and Z. Liu, "Distributed source coding in dense sensor networks," http://www.ifp.uiuc.edu/~kashyap/research/tosn_submission.pdf, 2006.

[10] D. L. Neuhoff and S. S. Pradhan, "An Upper Bound to the Rate of Ideal Distributed Lossy Source Coding of Densely Sampled Data," ICASSP, Toulouse, France, May 2006.

[11] R. G. Gallager, *Information Theory and Reliable Communication.* New York: Wiley, 1968.

[12] T. Berger, *Rate distortion theory: A mathematical basis for data compression.* Englewood Cliffs: Prentice Hall, 1971.

[13] A. N. Kolmogorov, "On the Shannon theory of information transmission in the case of continuous signals," *IRE Trans. Inform. Thy.*, vol. 2, pp. 102-108, 1956.

[14] D. P. Bertsekas, *Nonlinear Programming.* Belmont, MA: Athena Sc., 2003.

[15] U. Grenander and G. Szego, *Toeplitz Forms and Their Applications.* Berkeley, CA: University of California Press, 1958.

[16] R. M. Gray, "Toeplitz and Circulant Matrices: A Review," *Foundation and Trends in Comm. and Inform. Thy.*, vol. 2, no. 3, pp. 155-239, 2006.

[17] H. L. Royden, *Real Analysis, 2nd Ed.*. New York: Macmillan, 1968.

[18] T. M. Cover and J. A. Thomas, *Elements of Information Theory, 2nd Ed.*. Hoboken, NJ: John Wiley and Sons, 2006.

[19] H. Gish and J. N. Pierce, "Asymptotically efficient quantization," *IEEE Trans. Inform. Thy.*, vol. 14, pp. 676-683, Sept. 1968.

[20] S. Shamai, "Information rates by oversampling the sign of a bandlimited process," *IEEE Trans. Inform. Thy.*, vol. 40, pp. 1230-1236, July 1994.

[21] Z. Cvetkovic and M. Vetterli, "Error-rate characteristics of oversampled analog-to-digital conversion," *IEEE Trans. Inform. Thy.*, vol. 44, pp. 1961-1964, Sept. 1998.

[22] Z. Cvetkovic and M. Vetterli, "On simple oversampled A/D conversion in $L^2(\mathbb{R})$," *IEEE Trans. Inform. Thy.*, vol. 47, pp. 146-154, Jan. 2001.

[23] Z. Cvetkovic and I. Daubechies, "Single-bit oversampled A/D conversion with exponential accuracy in the bit-rate," *DCC*, Snowbird, UT, pp. 343-352, Mar. 2000.

[24] P. Ishwar, A. Kumar, and K. Ramchandran, "Distributed sampling for dense sensor networks: A 'bit-conservation principle'," *IPSN*, Palo Alto, CA, pp. 17-31, April 2003.

[25] A. Kumar, P. Ishwar, and K. Ramchandran, "On distributed sampling of smooth non-bandlimited fields," *IPSN*, Berkeley, CA, pp. 89-98, Apr. 2004.

[26] D. Marco and D. L. Neuhoff, paper in preparation.

[27] N.T. Thao and M. Vetterli, "Reduction of the MSE in R-times oversampled A/D conversion $O(1/R)$ to $O(1/R^2)$," *IEEE Trans. Sig. Proc.*, vol. 42, pp. 200-203, Jan. 1994.

[28] N.T. Thao and M. Vetterli, "Deterministic analysis of oversampled A/D conversion and decoding improvement based on consistent estimates," *IEEE Trans. Sig. Proc.*, vol. 42, pp. 519-531, Mar. 1994.

[29] Z. Cvetkovic and M. Vetterli, "Error analysis in oversampled A/D conversion and quantization of Weyl-Heisenberg frame expansions," *ICASSP*, vol. 3, pp. 1435-1438, May 1996.
[30] I. Bar-David, "Sample functions of a Gaussian random process cannot be reconstructed from their zero crossings," *IEEE Trans. Inform. Thy.*, vol. 21, pp. 86-87, Jan. 1975.
[31] D. Slepian, "Estimation of the Gauss-Markov Process from observation of its sign," *Stoch. Proc. Appl.*, vol. 14, pp. 249-265, 1983.
[32] D. Marco, "Markov random processes are not recoverable after quantization and mostly not recoverable from samples," *IEEE Int. Symp. Inform. Thy.*, Nice, France, pp. 2886-2890 June 2007.
[33] S.-Y. Tung, "Multiterminal source coding," Ph.D. Dissertation, Cornell University, Ithaca, NY, 1978.
[34] R. Zamir and T. Berger, "Multiterminal source coding with high resolution," *IEEE Trans. Inform. Thy.*, vol. 45, pp. 106-1127, Jan. 1999.
[35] P. Viswanath, "Sum rate of multiterminal Gaussian source coding," *DIMACS Series in Discrete Math. and Theoretical Computer Science*, 2002.
[36] A. .B. Wagner, S. Tavildar and P. Viswanath, "Rate region for the quadratic Gaussian two-encoder source-coding problem," *IEEE Int. Symp. Inform. Thy.*, pp. 1404-1408, Seattle, July 2006.
[37] H. L. van Trees, *Detection, Estimation, and Modulation Theory, Part I*, New York: Wiley, 1968.
[38] R. Zamir, "Rate loss in the Wyner-Ziv problem," *IEEE Trans. Inform. Thy.*, vol. 42, pp. 2073-2084, Nov. 1996.

Appendix

Sketch of proof of (7.6)

Recall (7.6), namely, that for any $D > 0$, there exists $N_o > 0$ such that for all positive integers k and $N \geq N_o$, $kNR_{kN}(D) \leq NR_N(D) + \epsilon_N(D)$, where $\epsilon_N(D) \to 0$ as $N \to \infty$.

The idea is that given N and k, for some $D'_N < D$ to be chosen later, we take M large and find an M-dimensional test channel q that approximately attains $NR_N(D'_N)$ for sampling rate N. Then from this test channel, we create a kM-dimensional test channel q' for sampling rate kN by applying q to the M variables $Y_1, Y_{k+1}, Y_{2k+1}, \ldots, Y_{(M-1)k+1}$ consisting of every kth sample, and letting each of the $kM - M$ other test channel outputs equal the most recent output in the aforementioned collection. The distortion induced by q' will be D'_N plus an interpolation error that becomes arbitrarily small when N is sufficiently large. Therefore, we can choose N_o so large that for all $N \geq N_o$, it is possible to choose $D'_N < D$ so that the test channel q' attains distortion at most D for all k. Moreover, we can have $D'_N \to D$ as $N \to \infty$. It then follows that for $N \geq N_o$, $kNR_{kN}(D) \leq I_{q'}(Y_1,\ldots,Y_{kM};\widetilde{Y}_1,\ldots,\widetilde{Y}_{kM}) = I_q(Y_1,\ldots,Y_M;\widetilde{Y}_1,\ldots,\widetilde{Y}_M) \approx NR_N(D'_N) = NR_N(D)+\epsilon_N(D)$, where $\epsilon_N(D) \triangleq NR_N(D'_N) - NR_N(D) \to 0$ as $N \to \infty$.

Sketch of proof of (7.31)

Recall (7.31), namely, that for any $D > 0$, there exists $N_o > 0$ such that for all positive integers k and $N \geq N_o$, $kNR^d_{kN}(D) \leq NR^d_N(D) + \epsilon_N(D)$, where $\epsilon_N(D) \to 0$ as $N \to \infty$.

The idea is that given N and k, for some $D'_N < D$ to be chosen later, we take M large and find a code with spatial blocklength M and some sufficiently large temporal blocklength L that approximately attains $NR^d_N(D'_N)$ for sampling rate N. We then use this code to create a new code with spatial blocklength kM and temporal blocklength L that can be used to encode the kM samples taken at rate kN by applying the original code to every kth sample. At the decoder, the reproduction produced at sampling rate N is upsampled to sampling rate kN using sample-and-hold. The per snapshot rate of the new system for sampling rate kN is the same as for the original code for sampling rate N, i.e., it is approximately $NR_N(D'_N)$. The distortion of the new code will be D'_N plus an interpolation error that becomes arbitrarily small as N becomes large. Therefore, we can choose N_o so large that for all $N \geq N_o$, it is possible to choose $D'_N < D$ so that the new code attains distortion at most D for all k. It then follows that $kNR^d_{kN}(D) \leq$ rate of new code $=$ rate of original code $\approx NR^d_N(D'_N) = NR^d_N(D) + \epsilon_N(D)$ for all positive integers k, where $\epsilon_N(D) = NR^d_N(D'_N) - NR^d_N(D) \to 0$ as $N \to \infty$.

8

Separation Theorems And Partial Orderings For Sensor Network Problems

Michael C. Gastpar[1]

University of California, Berkeley gastpar@berkeley.edu

8.1 Introduction

In this chapter, we discuss information-theoretic techniques to understand sensor network performance. From an information-theoretic perspective, sensor network problems are typically *joint source-channel coding* problems: The goal is to recover an approximate version of the underlying source information (by contrast to, for example, the standard channel coding problems where the goal is to communicate bits at the smallest possible error probability). Hence, the overall encoding process maps a sequence of source observations into a suitable sequence of channel inputs in such a way that the decoder, upon observing a noisy version of that sequence, can get an estimate of the source observations at the highest possible fidelity. Successful code constructions must exploit the structure of the underlying source (and the mechanism by which the source is observed) and the communication channel. Designing codes that simultaneously achieve both should be expected to be a rather difficult task, and it is therefore somewhat surprising that Shannon [27] found a very elegant solution for the case of point-to-point communication (as long as both the source and the channel are stationary and ergodic, and cost and fidelity are assessed by per-letter criteria). The solution consists in a *separation* of the overall task into two separate tasks. Specifically, an optimal communication strategy can be designed in two parts, a source code, exploiting the structure of the source and the observation process, followed by a channel code, exploiting the structure of the communication channel. The two stages are connected by a *universal* interface — *bits* — that *does not* depend on the precise structure. For the purpose of this chapter, we will refer to such an architecture as *separation-based.*

Some of the most relevant communication systems of today operate in a distributed fashion. This means that there is not just a single encoder that makes all the observations and has direct access to the full communication infrastructure. Rather, there are multiple *distributed* terminals, each having partial observations. There are certain such scenarios where again, a separation theorem can be established, and an optimal overall code can be implemented separately by a source coding stage followed by a channel coding stage. We present a non-exhaustive list of such cases in Section 8.2. However, more generally, there are counter-examples where such a design strategy fails rather dramatically. We again present a non-exhaustive list that illustrates some of the ways in which such failure occurs in Section 8.3. Then, in Section 8.4, we discuss a few examples where an *approximate* separation theorem holds. That is, separate source and channel code designs are not strictly optimal, but they are close (at least in a certain sense).

It should be noted that we selected those examples for which separation theorems can be established in a simple and elegant fashion from *single-letter* mutual information expressions. There are rather straightforward extensions of many of the examples to slightly more intricate models, but for which one has to argue via infinite-letter information expressions.

In the final section (Section 8.5), we slightly change the focus of the discussion by asking the question of how to order such joint source-channel communication problems according to their difficulty. In cases where a separation theorem applies, this ordering follows from capacity and rate-distortion functions. However, for other cases, no global ordering appears to exist. Instead, we present some very tentative approaches of *partial orderings.*

8.2 Exact Separation Theorems

We first discuss Shannon's (stationary, ergodic) source/channel separation theorem. Then, we present a non-exhaustive list of a few examples of network situations for which an appropriate separation theorem holds.

8.2.1 Point-to-point communication problems

In this section, we provide a brief review of the information-theoretic results for the point-to-point source-channel communication system, illustrated in Figure 8.1. We also use this opportunity to introduce the basic notation for the remainder of this chapter.

Fig. 8.1. The general point-to-point source-channel communication problem.

We first need the concept of an information source, which will be a sequence S^n drawn randomly from a distribution $p(s^n)$. For most parts of the chapter, this distribution will be taken to be an n-fold product of a fixed underlying scalar distribution,

$$p(s^n) = \prod_{i=1}^{n} p_S(s_n), \tag{8.1}$$

in which case we will refer to the source as being *stationary and memoryless.* The second integral part of a source is the *distortion measure* $d(S^n, \hat{S}^n)$ with respect to which the source sequence S^n must be reconstructed at the receiver. Often, we will consider *single-letter* distortion measures, which means that

$$d(S^n, \hat{S}^n) = \frac{1}{n}\sum_{i=1}^{n} d(S_i, \hat{S}_i). \tag{8.2}$$

The (stationary and memoryless) channel is characterized by a conditional distribution $P_{Y|X}(y|x)$. The second integral part of a channel is the *cost function,* which assigns, to every channel input symbol x, a cost $\rho(x)$. The cost of a channel input sequence x^n is then simply given by

$$\rho(X^n) = \frac{1}{n}\sum_{i=1}^{n} \rho(X_i). \tag{8.3}$$

For the purpose of this brief review, we suppose that the encoder F maps a sequence of n source symbols onto a sequence of n channel input symbols. We also suppose that the decoder is synchronized with the encoder, and maps a sequence of n channel output symbols onto a sequence of n source reconstruction symbols. The goal of the code (F, G) is to produce a minimum distortion,

$$D = Ed\left(S^n, \hat{S}^n\right), \tag{8.4}$$

using, simultaneously, a minimum cost on the channel,

$$P = E\rho(X^n). \tag{8.5}$$

The key problem of source-channel communication is to determine the optimal cost-distortion pairs (P, D). We consider this problem in the information-theoretic sense, i.e., we are interested in the optimum *irrespective of the coding complexity and delay.*

To bound this trade-off, a very simple argument can be made using the *data processing inequality* (see e.g. [5]), as follows: If n source symbols are encoded into n channel inputs,[1] then we must have

$$I(S^n; \hat{S}^n) \leq I(X^n; Y^n). \tag{8.6}$$

At this point, we will use the standard concepts of the *rate-distortion function,* usually denoted by $R(D)$ (see e.g. [5, p.306ff.]), and of the *capacity-cost*

[1] Some communication scenarios may allow the code designer to select the ratio of source symbols to channel inputs. The arguments discussed here can be easily extended to this case, see e.g. [7].

function, usually denoted by $C(P)$ (see e.g. [6, p.108]). Specifically, the rate-distortion theorem [5, Thm.10.2.1] implies that since our code attains a resulting distortion of D, the left hand side of the previous equation is lower bounded by $nR(D)$. By analogy, the capacity-cost theorem [6, Thm.2.1.11] implies that since our code uses a cost of P, the right hand side of the previous equation is upper bounded by $nC(P)$. Therefore, we find that *any* overall strategy that attains distortion D and incurs cost P *must* satisfy

$$R(D) \leq C(P). \tag{8.7}$$

The remaining question is whether there is indeed a communication strategy that attains this upper bound. Shannon showed that one such strategy is given by first compressing the source down to $R(D)$ bits per symbol. By the operational meaning of the rate-distortion function, this is possible and incurs a distortion arbitrarily close to D (as $n \to \infty$). The resulting bits are then reliably communicated across the channel. By the operational meaning of the capacity-cost function, it is possible to send $C(P)$ bits per channel use reliably (as $n \to \infty$). In conclusion, the following theorem was obtained by Shannon [27, Thm.21]:

Theorem 1 (separation theorem). *A (stationary, ergodic) source can be communicated at distortion D across a (stationary, ergodic) channel incurring cost P if and only if*

$$R(D) \leq C(P). \tag{8.8}$$

Simple extensions of this separation theorem apply to scenarios with side information (in the style of Gel'fand and Pinsker, as well as of Wyner and Ziv), see e.g. [7, 19]. Moreover, an alternative perspective on the separation theorem has been presented recently [7, 10, 11].

8.2.2 Parallel Channels With A Common Decoder

A particularly simple extension of the point-to-point separation theorem concerns parallel channels with separate encoders, but a common decoder. Specifically, we consider the scenario in Figure 8.2. Note that if the encoder is *also* common, then the scenario is a point-to-point communication problem. Moreover, if *only* the encoder is common, but the decoders separate, then we have merely *multiple* separate point-to-point communication problems.

In order to state a separation theorem, we first define the source coding problem as the problem of encoding all M sources without loss, using a rate of R_m bit per symbol at encoder m, for $m = 1, 2, \ldots, M$. The region of rate vectors $(R_1, R_2, \ldots, R_M)$ that permit to attain this goal will be denoted by $\mathcal{R}$, and has been found by Slepian and Wolf [28] (see also [5, ch.15]). We also define the *capacity region* $\mathcal{C}$ of the communication network as the set of rate vectors $(R_1, R_2, \ldots, R_M)$ that can be simultaneously attained in a

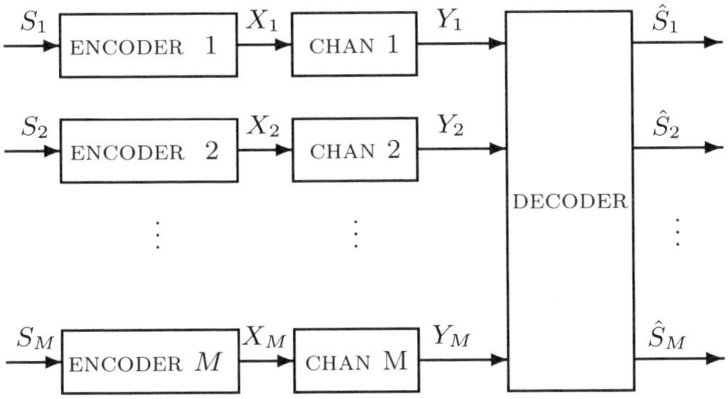

Fig. 8.2. Parallel channels.

reliable fashion, where R_m is the rate at which encoder m communicates to the receiver. Note that for the case of parallel channels, this region is of trivial rectangular shape. Then, the following general theorem holds (a proof can be found e.g. in [3]):

Theorem 2. *The lossless communication of arbitrarily dependent sources across parallel channels to a single receiver is feasible if and only if*

$$\mathcal{R} \cap \mathcal{C} \neq \emptyset. \tag{8.9}$$

It is important to note that this is the natural extension of Theorem 1 in the sense that it also requires a source coding rates to be *no larger* than the highest rates across the channel. That is, any such joint source-channel communication problem can be understood via *bits*, making them a universal currency for this special case.

The proof of the particular separation theorem given in Equation (8.9) is entirely straightforward. Achievability follows directly from the result of Slepian and Wolf [28], and the converse follows from a simple application of Fano's inequality.

8.2.3 Multiple Access with independent sources and separate criteria

Perhaps a slightly more interesting general separation theorem for network scenarios concerns the multiple-access channel with independent sources, as long as these sources need to be recovered with respect to *separate* criteria. More precisely, consider the scenario in Figure 8.3: There are M encoders, each with an independent stream of source information, denoted by S_m, for

$m = 1, 2, \ldots, M$. The decoder needs to recover all of these streams, either exactly or with respect to *separate* distortion criteria of the form

$$E[d_m(S_m, \hat{S}_m)] \leq D_m. \tag{8.10}$$

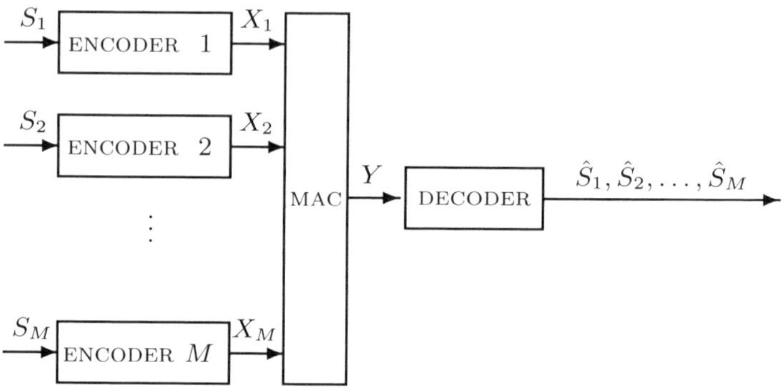

Fig. 8.3. Multiple access.

More precisely, for given costs $P_1, P_2, \ldots, P_M$ at the respective inputs of the multi-access channel, the capacity region (for the transmission of independent messages)

$$\mathcal{C}(P_1, P_2, \ldots, P_M) \tag{8.11}$$

has been found [1, 17], see e.g. [5, Thm. 15.3.1]. Moreover, since the sources are assumed to be independent, the region of rates required to attain distortions $D_1, D_2, \ldots, D_M$ is given by

$$\mathcal{R}(D_1, D_2, \ldots, D_M) = \{(R_1, R_2, \ldots, R_M) : R_m \geq R_{S_m}(D_m)\}, \tag{8.12}$$

where $R_{S_m}(D_m)$ denotes the rate-distortion function of source m. This follows directly from the standard rate-distortion theorem, see e.g. [5, Thm.10.2.1].

For this scenario, the following general separation theorem applies [7]:

Theorem 3 (independent sources on the MAC). *The sources $S_1, S_2, \ldots, S_M$ can be communicated across the MAC, resulting in distortions $D_1, D_2, \ldots, D_M$ and incurring costs $P_1, P_2, \ldots, P_M$ if and only if*

$$\mathcal{R}(D_1, D_2, \ldots, D_M) \cap \mathcal{C}(P_1, P_2, \ldots, P_M) \neq \emptyset. \tag{8.13}$$

Proof Outline. The achievability follows immediately from the operational meaning of the rate-distortion function and of the capacity region. For the converse, suppose that our code achieves $(P_1, P_2, \ldots, P_M)$ and $(D_1, D_2, \ldots, D_M)$. Then,

$$\begin{aligned} nR_{S_1}(D_1) &\overset{(a)}{\leq} I(S_1^n; \hat{S}_1^n) \\ &\overset{(b)}{\leq} I(S_1^n; Y^n) \\ &\overset{(c)}{\leq} I(S_1^n; Y^n | X_2^n, X_3^n, \ldots, X_M^n) \\ &\overset{(d)}{\leq} I(X_1^n; Y^n | X_2^n, X_3^n, \ldots, X_M^n), \end{aligned}$$

where (a) follows from the definition of the rate-distortion function, (b) is the data processing inequality, (c) holds because S_1 and $(X_2, X_3, \ldots, X_M)$ are independent and (d) is again the data processing inequality. By analogy, one can find the following bounds:

$$n \sum_{m \in \mathcal{T}} R_{S_m}(D_m) \leq I(\{X_m^n\}_{m \in \mathcal{T}}; Y^n | \{X_m^n\}_{m \notin \mathcal{T}}),$$

for all subsets $\mathcal{T}$ of the set $\{1, 2, \ldots, M\}$. To complete the proof, one has to argue that the information expressions on the right hand side in the above inequality must describe a point inside the capacity region. This can be done along the lines of the proof given in [5, p.539].

This theorem can also be extended to the case of the multiple access channel with causal feedback. For the scope of this note, we only discuss the special case of the Gaussian MAC below in Section 8.2.5, a case for which a simple argument can be given.

8.2.4 Modulo-Additive Multiple-Access Channels

Our next example is a particularly simple class of multiple access channels where all input alphabets and the output alphabet are discrete, finite, and of the same size. Moreover, there is no cost constraint in this problem, i.e., the function $\rho(x) = 0$ for all x (see Equation (8.3)). Without loss of generality, we take this alphabet to be the set $\{1, 2, \ldots, L\}$, and the output of the multiple access channel is characterized by

$$Y = X_1 \oplus X_2 \oplus \cdots \oplus X_M \oplus Z, \qquad (8.14)$$

where $\oplus$ denotes addition modulo L, and Z is a random variable arbitrarily distributed over the same alphabet $\{1, 2, \ldots, L\}$.

Theorem 4. *Arbitrarily dependent sources $S_1, S_2, \ldots, S_M$, can be losslessly transmitted across a modulo-additive MAC with arbitrary causal feedback signals if and only if*

$$\mathcal{R} \cap \mathcal{C} \neq \emptyset. \qquad (8.15)$$

This theorem can be proved from a simple maximum entropy argument (applied to the output of the multiple-access channel) and has appeared (modulo the causal feedback signals) e.g. in [25].

8.2.5 Gaussian Multiple-Access Channels

We close our multiple access examples with the special case of the Gaussian MAC (see e.g. [5, p.544]), for which a few special cases may be of interest. The output of the Gaussian MAC is characterized by

$$Y = Z + \sum_{m=1}^{M} X_m, \tag{8.16}$$

where Z is white Gaussian noise of variance σ_Z^2.

Under Received-Signal Constraints

The first special case may be viewed as the Gaussian equivalent of the modulo-additive MAC. Rather than imposing a cost constraint on the *transmitted signals* (as in Equation (8.3)), we now impose the constraint that the transmitted signals must be chosen in such a way as to ensure

$$E[Y^2] \leq Q + \sigma_Z^2, \tag{8.17}$$

where Q is a constant, i.e., the coding must satisfy a constraint on the *received* signal. This problem is discussed in detail in [8]. For example, consider the source scenario as in Section 8.2.2, denoting the source coding rate region by $\mathcal{R}$. Then, the following theorem can be given [8]:

Theorem 5. *For the Gaussian MAC under a received-power constraint, the lossless transmission of arbitrarily dependent sources, even if arbitrary causal feedback signals are available, is feasible if and only if*

$$\mathcal{R} \cap \mathcal{C} \neq \emptyset. \tag{8.18}$$

As shown in [8], the capacity region in this case is merely a simplex,

$$\mathcal{C} = \left\{ (R_1, R_2, \ldots, R_M) : \sum_{m=1}^{M} R_m \leq \frac{1}{2} \log_2 \left(1 + \frac{Q}{\sigma_Z^2} \right) \right\}. \tag{8.19}$$

Under Fast Fading

As another case of potential interest, consider the following slight variation: The channel output signal is now given by

$$Y = Z + \sum_{m=1}^{M} A_m X_m, \tag{8.20}$$

where A_m are independent circularly symmetric complex random variables, freshly chosen for each channel use. The decoder gets access to the augmented channel output signal $(Y, A_1, A_2, \ldots, A_m)$. We consider standard transmit power constraints (see e.g. [5, Eqn. (15.134)]). For a more detailed discussion of this fading model, see e.g. [29]. The following separation theorem can be established:

Theorem 6. *For the Gaussian MAC under fast fading, the lossless transmission of arbitrarily dependent sources, even if arbitrary causal feedback signals are available, is feasible if and only if*

$$\mathcal{R} \cap \mathcal{C} \neq \emptyset. \tag{8.21}$$

This theorem can be proved by noting that due to the fading, the optimal input distribution is to select independent signals. But this is also feasible in a distributed fashion.

With Noiseless Causal Feedback

Let us now consider the standard Gaussian MAC (see e.g. [5, p.544]) with cost constraints imposed in the shape of the regular power constraints on the transmitted signals (i.e., the function $\rho(x) = x^2$, see Equation (8.3) or [5, Eqn. (15.134)]). Moreover, we also allow noiseless causal feedback, that is, each encoder, when generating the channel input at any time instant n, knows the precise values of all past channel outputs up to (and including) time $n-1$. Consider two arbitrary independent sources (not necessarily with the same statistics), S_1 and S_2, and suppose that they need to be reconstructed with respect to *separate* distortion criteria according to $Ed_m(S_m, \hat{S}_m) \leq D_m$ for $m = 1, 2$. For this special case, we can establish the equivalent of Theorem 3, as follows:

Theorem 7 (independent sources, Gaussian MAC with noiseless feedback). *The independent sources S_1 and S_2 can be communicated across the Gaussian MAC, resulting in distortions D_1 and D_2 and using powers P_1 and P_2 if and only if*

$$\mathcal{R}(D_1, D_2) \cap \mathcal{C}(P_1, P_2) \neq \emptyset. \tag{8.22}$$

Using the same arguments as in the derivation of Theorem 3, we first find the bound

$$nR_1(D_1) \leq I(S_1^n; Y^n | S_2^n). \tag{8.23}$$

Note that the argument is yet in terms of the source sequences, rather than the channel input sequences, a trick that appears (e.g.) in [18]. But then, we can further upper bound

$$\begin{aligned} I(S_1^n; Y^n | S_2^n) &\overset{(a)}{=} \sum_{i=1}^{n} h(Y_i | Y_1^{i-1}, S_2^n) - \sum_{i=1}^{n} h(Y_i | Y_1^{i-1}, S_1^n, S_2^n, X_{1,i}, X_{2,i}) \\ &\overset{(b)}{=} \sum_{i=1}^{n} h(Y_i | Y_1^{i-1}, S_2^n, X_{2,i}) - \sum_{i=1}^{n} h(Y_i | X_{1,i}, X_{2,i}) \\ &\overset{(c)}{\leq} \sum_{i=1}^{n} I(X_{1,i}; Y_i | X_{2,i}), \end{aligned} \tag{8.24}$$

where (a) follows because $X_{1,i}$ and $X_{2,i}$ are uniquely determined by Y_1^{i-1}, S_1^n, S_2^n; for (b), the first term follows again because $X_{2,i}$ is uniquely determined by Y_1^{i-1}, S_2^n and the second term from the assumption that the channel is memoryless; and (c) follows by omitting conditioning in the first term. Analogous arguments can be used for the second source, and for the sum rate term. Then, we note that the mutual information term (8.24) and its corresponding terms in the remaining two bounds precisely describe the noiseless casual feedback capacity region of the 2-user Gaussian MAC (as given in [23]), completing the argument.

8.3 Counter-Examples

As mentioned earlier, there is no general source/channel separation theorem for networks, and as a matter of fact, the separate design of the source code and the channel code generally leads to severe performance penalties. In this section, we present a selection of the key counter-examples. The first is a classical example, and the remaining three each illustrate a different aspect of the breakdown of the separation theorem.

8.3.1 Multiple Access with correlated sources

A classical example illustrating the fact that source/channel separation does not hold for networks was given in [4]. Here, the multiple-access channel is the binary adder multiple access channel, taking two binary $\{0,1\}$ inputs and providing as its output their sum $\{0,1,2\}$. The capacity region $\mathcal{C}$ of this channel has the pentagonal shape given in Figure 8.4, see [5], Fig.15.13, for more details. Now suppose that the two transmitting terminals each observe a binary sequence, call them S_1^n and S_2^n. The two sequences are correlated with each other such that for each time instant, the events $(S_1, S_2) = (0,0), (0,1)$, and $(1,1)$ are all equally likely, and $(1,0)$ does not occur. Clearly, at least $H(S_1, S_2) = \log_2 3 \approx 1.58$ bits per source sample are required. The full

Slepian-Wolf rate region $\mathcal{R}$ is also given in Figure 8.4. The two regions do not intersect, and hence, one is tempted to guess that these two sources *cannot* be transmitted across this MAC. However, this conclusion is wrong: While there is no separation-based architecture that achieves this, there is a simple "analog" strategy: pure uncoded transmission will always permit to recover both source sequences without error, due to the fact that the dependence structure of the sources is perfectly matched to the channel. This illustrates that no separation theorem applies to general networks. Considerable generalizations of this have appeared in [2].

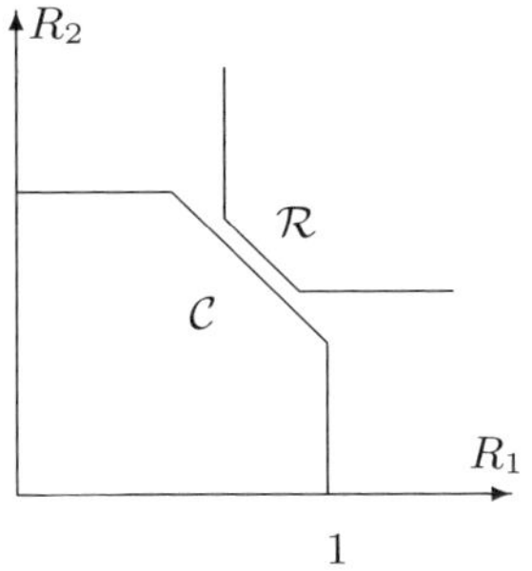

Fig. 8.4. Capacity region $\mathcal{C}$ and rate-distortion region $\mathcal{R}$ do not intersect in this example.

8.3.2 Computation over Multiple Access Channels

A second counter-example appears when we drop the assumption of *separate reconstruction criteria,* such as in the problem of computation. Specifically, let us consider two binary memoryless sources S_1 and S_2. S_1 is simply a Bernoulli-1/2 process, and S_2 can be defined as $S_2 = S_1 + E$, where E is Bernoulli-p and independent of S_1. The two sources must be encoded separately for transmission over a multiple-access channel that takes binary inputs and outputs their modulo-2 sum. The goal of the final receiver is not to obtain the actual values of S_1 and S_2, but merely their modulo-2 sum, i.e., the memoryless sequence E.

For this simple scenario, it is immediately clear what to do: simply transmit the source outputs without any further coding. The channel output itself will already provide the desired modulo-2 sum, and any coding is entirely unnecessary.

The interesting question, of course, is whether this same performance can be attained by clever separate source and channel codes. A first candidate code may be the Slepian-Wolf construction (see e.g. [5, Thm.15.4.1]) to encode the two sources, requiring a sum rate of

$$R_1 + R_2 = H(S_1, S_2) = 1 + H_b(p). \tag{8.25}$$

The multiple-access channel at hand can carry at most a sum rate of 1 bit per channel use (since its output is binary), and hence, for any non-trivial value of p, this code does not work.

Now, the problem with this code is that it provides the decoder with too much information, namely enough to recover both S_1 and S_2. The question is whether there is a more efficient code that indeed only retains the modulo-2 sum. Such a code was found by Körner and Marton [16].

To permit the receiver to recover the modulo-2 sum, the following rate conditions need to be satisfied:

$$R_1 \geq H(S_1|S_2) = h_b(p)$$
$$R_2 \geq H(S_2|S_1) = h_b(p)$$

This strategy works only as long as $p \leq 0.11$ (or equivalently, $p \geq 0.89$).

The arguments in [16] imply that this is the best performance we can ever hope to attain via the separate design of a source and a channel code. In other words, for $0.11 < p < 0.89$, uncoded transmission is strictly superior to any separate source and channel code.

Interesting follow-up questions include whether anything can be done if the multiple-access channel is noisy. Clearly, uncoded transmission no longer works: It will not permit to reliably recover the modulo-2 sum of the sources. However, more elaborate code constructions have been found in [20, 21].

8.3.3 Multiple Access with the CEO source

Let us now consider a network topology that may model a simple sensor network: One single underlying Gaussian source of interest is observed by many sensors, subject to independent Gaussian observation noise. The sensors communicate over a standard Gaussian multiple-access channel to a fusion center. The *sensing task* is to estimate the underlying source with respect to mean-squared error. This example, and extensions thereof, have been discussed in [12, 13, 14]. We start the discussion with a simple riddle, illustrated in Figure 8.5: For one and the same sensing task, namely a sequence of unit-variance Gaussians X^n, observed three times in independent unit-variance Gaussian noise, we have to choose between two different suggested communication networks, one of total capacity 1/2 bits per network use, the other of total capacity 3/4 bits per network use. The network may be used on average *once* per source sample, and *arbitrary coding* is permitted in either case. That is, each sensor can accumulate as many observation as it needs and jointly map them into a suitable sequence for transmission over the multiple-access channel. Which communication infrastructure is preferable?

By now, it is perhaps not surprising anymore that the capacity does *not* characterize the relevant performance. Indeed, one finds that for the communication network on the left in Figure 8.5, the smallest attainable end-to-end mean-squared error is given by

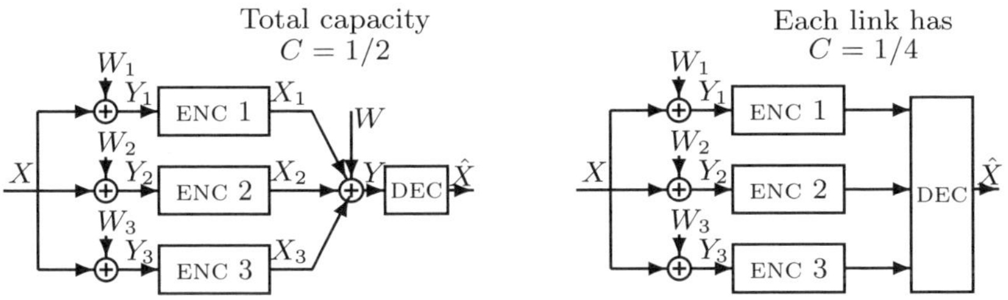

Fig. 8.5. A riddle. Although not explicitly drawn in the figure, each encoder has access to an arbitrarily long sequence of observations, and is allowed to arbitrarily map this into a suitable channel input sequence.

$$D = \frac{1}{2}, \tag{8.26}$$

which can be found in [9], whereas for the communication network on the right in Figure 8.5, it is

$$D \geq \frac{1}{1.75} \approx 0.57, \tag{8.27}$$

which follows from the rate-distortion function for the CEO problem, see e.g. [22].

More generally, going beyond $M = 3$ sensors, let us now consider the sensor network topology in Figure 8.6, where we assume that the source sequence S^n is memoryless and unit-variance Gaussian, and so are the observation noise sequences $W_1^n, W_2^n, \ldots, W_M^n$ as well as the communication noise sequence W^n. Moreover, whatever code is used, it must satisfy the power constraint

$$\sum_{m=1}^{M} P_m \leq P_{tot}, \tag{8.28}$$

where P_m is the average power consumed by encoder m. Finally, the goal of the decoder is to produce a source reconstruction sequence $\hat{S}^n$ at the smallest possible mean-square error $D = (1/n)E[\|S^n - \hat{S}^n\|^2]$. Then, we have recently proved the following theorem:

Proposition 1. *For the simple linear Gaussian sensor network, the minimum distortion (using arbitrary encoding and decoding) is*

$$D = \frac{1}{M+1}\left(1 + \frac{M}{\frac{M+1}{2}P_{tot} + 1}\right). \tag{8.29}$$

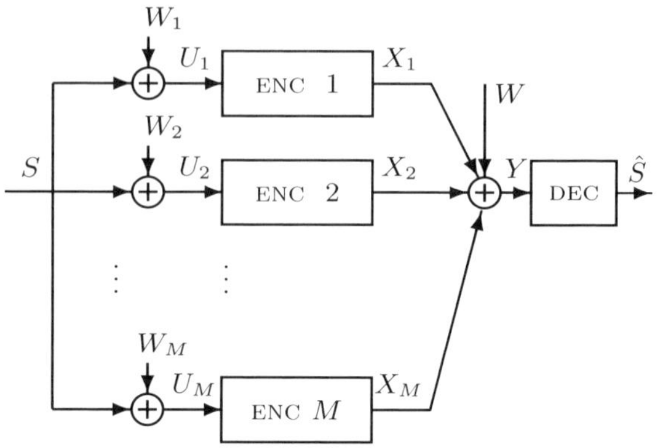

Fig. 8.6. The simple linear Gaussian sensor network.

A more general version of this result was obtained in [9].

By contrast, if source code design is separated from channel code design, we obtain the following:

Proposition 2 (no separation theorem). *For the simple linear Gaussian sensor network, the smallest distortion for a "separate" code that first compresses the observations into bit streams and then delivers these bit streams reliably satisfies*

$$D_{separation} \geq \frac{1}{\log_2(1 + MP_{tot}) + 1}. \tag{8.30}$$

The most interesting aspect of this insight is the comparison between Equation (8.29) and Equation (8.30): In the former, the distortion scales like $1/M$, whereas in the latter, it scales like $1/\log M$. Hence, for this simple example, the separation-based architecture not only performs suboptimally, it entails an *unbounded* penalty, exponential in the number of nodes M.

8.3.4 Single-source Broadcast

The next example we consider is initially obvious and unsurprising. Nevertheless, at a deeper level, it is one of the most difficult to understand. This concerns broadcasting a single source to many receivers to within a fidelity criterion. The case of two receivers is illustrated in Figure 8.7. A memoryless Gaussian source sequence S^n of mean zero and variance σ_S^2 is to be transmitted across a Gaussian two-user broadcast channel (see [5, p.570]) with power constraint P and with noise variances $\sigma_1^2 < \sigma_2^2$. For ease of reference, denote the capacities of the two underlying point-to-point channels by $C_1 = 1/2 \log_2(1 + P/\sigma_1^2)$ and $C_2 = 1/2 \log_2(1 + P/\sigma_2^2)$. To illustrate the point, suppose that we want the decoder with the stronger noise to decode the source at its smallest possible distortion, i.e., $D_2 = D(C_2) = \sigma_S^2\sigma_2^2/(P + \sigma_2^2)$, where $D(\cdot)$ denotes the distortion-rate function of the Gaussian source (see e.g. [5, Eqn. (10.37)]). What is the smallest achievable distortion for user 1, denoted by D_1?

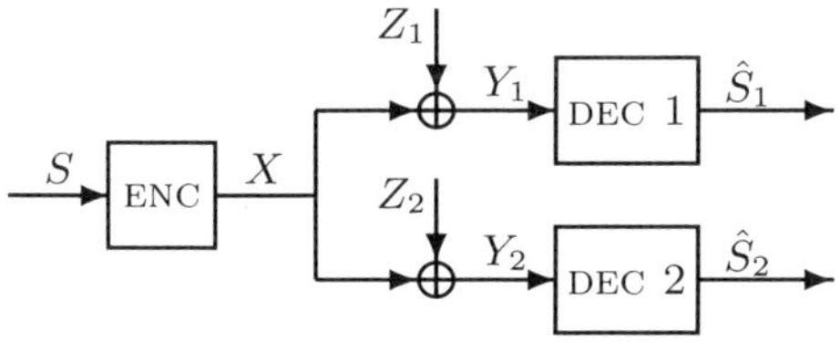

Fig. 8.7. Broadcasting a single source to two destinations.

For a separate source and channel code design, this means that we have to use a capacity-achieving channel code of rate C_2 to send to user 2. Can we "hide" (e.g., superimpose) another code for user 1? The answer is no: any superposition would be noise for user 2, and hence compromise D_2. For the example at hand, it is clear that decoder 1 can decode anything decoder 2 can, thus the smallest D_1 that can be achieved with a separate source and channel code design is $D_1 = D_2 = D(C_2)$.

However, it is quickly verified for this example that simply sending S without further coding achieves the distortions $D_1^* = D(C_1)$ and $D_2^* = D(C_2)$.

This has inspired several interesting constructions, see e.g. [26].

At a deeper level, this example concisely illustrates that information cannot be thought of as a *discrete* quantity in this context; it must be considered as analog.

8.4 Approximate Separation Theorems

In the previous section, we considered cases where no source/channel separation theorem applies. Specifically, as shown in Propositions 1 and 2, there can be an exponential difference (in the number of nodes) between the best separation-based code and the best code overall. However, for certain classes of source-channel network problems, one can give *approximate* separation theorems, meaning that in a certain limiting sense, a source-channel separation theorem applies, and a separation-based code performs close to the best code overall. Hence, in those cases, bits regain some of their universal significance.

8.4.1 The Large-Network Limit

Consider the model illustrated in Figure 8.8, and let us assume for simplicity that the all sources $S_1, S_2, \ldots, S_L$ are memoryless and Gaussian, with covariance matrix Σ_S, and all observation and communication noises are independent Gaussians of variance σ_W^2 and σ_Z^2, respectively. A is a real- or complex-valued matrix of dimensions $M \times L$, and B is a matrix of dimensions $N \times M$.

As before, we allow each encoder in Figure 8.8 to accumulate an arbitrary number of source observations before jointly encoding them into a suitable channel input sequence, but we assume that the number of channel uses equals the number of source observations. Moreover, whatever code is used, it must satisfy the power constraint

$$\sum_{m=1}^{M} P_m \leq P_{tot}, \tag{8.31}$$

where P_m is the average power consumed by encoder m.

Our goal is to understand the behavior of the resulting reconstruction quality in terms of the mean-squared error for the underlying sources $S_1, S_2, \ldots, S_L$, i.e.,

$$D = \frac{1}{L} \sum_{\ell=1}^{L} D_\ell, \tag{8.32}$$

where D_ℓ is the average mean-squared error in the reconstruction of S_ℓ.

We first consider the case where the number of underlying sources is of the same order of magnitude as the number of observations, and where these underlying observations are sufficiently independent. This could be thought of as an "expanding" sensor network: as more sensors are added, more ground needs to be covered. More precisely, we can get the following theorem [15]:

Proposition 3 ("expanding sensor network"). *For the general linear Gaussian sensor network, consider the case where $L = M = N$, that $A = B = I_M$, the M-dimensional identity matrix, and that the smallest eigenvalue of $\Sigma_S^{(M)}$ is strictly larger than zero for all M. Then, for* any *coding scheme, no matter how complex, the distortion D in Equation (8.32) can be at best a constant, independent of M. A separation-based code design can also attain constant distortion $D_{separation}$, independent of M.*

It is, of course, a matter of taste whether this should be considered an approximate separation theorem. All it says is that a separate source-channel code attains a distortion that does not blow up with M, but goes to a constant, and that even for the best possible scheme, the distortion also goes to a constant. However, the two constants may be arbitrarily different.

For the proof of the first fact, namely that no coding scheme, no matter how complex, can attain a distortion that vanishes as M increases, it suffices to note that the rate-distortion function for *jointly* encoding $S_1, \ldots, S_M$ is given by (see e.g. [5, Thm.10.3.3])

$$R_{\text{joint}} = \sum_{m=1}^{M} \frac{1}{2} \log \frac{M\lambda_m}{D} \geq \frac{M}{2} \log \frac{M\lambda_{min}}{D},$$

where λ_m are the eigenvalues of the covariance matrix of the observations, and λ_{min} is their smallest (by assumption strictly larger than zero). The maximum rate across the MIMO channel with inputs $(X_1, X_2, \ldots, X_M)$ and outputs $(Y_1, Y_2, \ldots, Y_M)$ scales no faster than $M \log_2(1 + MP_{tot}/\sigma_Z^2)$. Hence, we cannot ever hope to incur a distortion smaller than

$$D \geq \frac{M\lambda_{min}}{1 + MP_{tot}/\sigma_Z^2},$$

which tends to a constant as M increases. To see that a separation-based strategy can also attain a constant distortion, it suffices to note that in this simple example, there are really M separate, parallel channels, hence each of the observation sequences $U_1, U_2, \ldots, U_M$ can be decoded at a constant distortion, which implies the result.

As a second scenario derived from Figure 8.8, we now suppose that L is a small number, and that only M increases. We further suppose that there is no observation noise, i.e., $W_1 = W_2 = \ldots = W_M = 0$. Then, we obtain the following statement [15]:

Proposition 4 ("dense sensor network, noiseless observations"). *For the general linear Gaussian sensor network, consider the case where $L = N = 1$ and M increases. Let $A = B^T = \mathbf{1}$, where $\mathbf{1}$ is the all-ones vector, and let $\sigma_W^2 = 0$, i.e., there is no observation noise. Then, the distortion for* any *scheme, no matter how complex, is lower bounded by*

$$D \geq \frac{\sigma_S^2}{1 + MP_{tot}} \tag{8.33}$$

and a separation-based code design can attain

$$D_{separation} = \frac{\sigma_S^2}{1 + P_{tot}}. \tag{8.34}$$

This proposition can be understood as an "approximate" separation theorem when the total power P_{tot} increases at least linearly in M : In that case, both the separation-based code and the optimal code attain a distortion that convergence to zero as M increases, and the speed of this convergence is similar, at least in the sense that there is no exponential difference, by contrast to the case of noisy observations, discussed above in Propositions 1 and 2.

For the very particular case addressed in this proposition, it is also known that the lower bound of Equation (8.33) is attainable, simply by letting all the encoders use exactly the same code. Extensions of this simplistic lower bound beyond the specific case addressed in the proposition, however, are not known to be achievable. In fact, tighter bounds could be obtained using the lower bounding technique discussed in [9]. A further discussion of these two propositions can be found in [15].

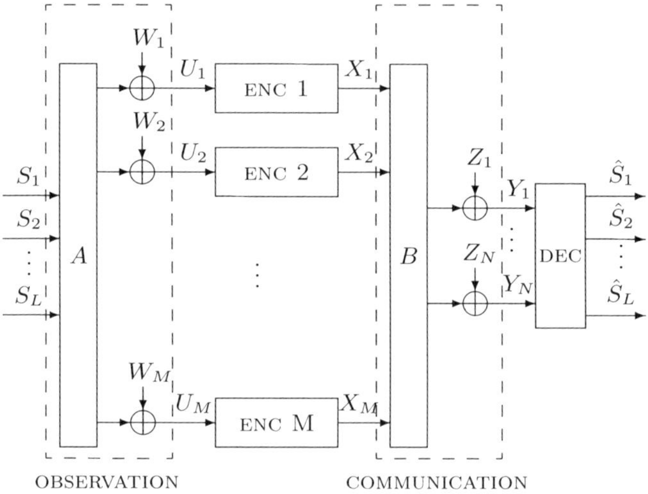

Fig. 8.8. The general linear Gaussian sensor network.

8.4.2 The High-SNR Limit

A second example of an approximate separation theorem can be developed for the standard additive white Gaussian noise M-user MAC as defined above in Section 8.2.5. For simplicity, let us consider a sum transmit power constraint as in Equation (8.31). Consider the problem of losslessly transmitting arbitrarily correlated sources across this channel, and suppose that the total rate required to encode the sources is R_{tot}. Then, a separation-based code works if

$$R_{tot} \leq \frac{1}{2}\log_2\left(1+\frac{P_{tot}}{\sigma_Z^2}\right), \tag{8.35}$$

which follows from the well-known capacity region of the Gaussian MAC (for independent sources), see e.g. [5, Sec.15.1.2]. Conversely, no code whatsoever works if

$$R_{tot} > \frac{1}{2}\log_2\left(1+\frac{MP_{tot}}{\sigma_Z^2}\right), \tag{8.36}$$

which follows by merging all M encoders into one "super-encoder." Now, suppose that M is a fixed number, and let the signal-to-noise ration (SNR) P/σ_Z^2 become large. Then,

$$\lim_{\frac{P}{\sigma_Z^2}\to\infty} \frac{\log_2\left(1+\frac{P_{tot}}{\sigma_Z^2}\right)}{\log_2\left(1+\frac{MP_{tot}}{\sigma_Z^2}\right)} = 1, \tag{8.37}$$

that is, in this relative sense, the two bounds become equivalent, which may again be interpreted as a type of an *approximate* separation theorem.

8.5 Elements of Partial Orderings

As explained in the introduction, the source-channel separation theorem has a double role; on the one hand, it provides architectural guidelines (at least so long as complexity and latency are not too constrained); on the other hand, it provides a universal (global) ordering of all (point-to-point) communication problems. To see this, suppose that a source is given, and one needs to choose between two channels. Clearly, the preferable channel is simply the one with higher capacity. Conversely, suppose that a channel is given, and one needs to decide which one of two candidate sources is less difficult to communicate across that channel. Clearly, the preferable source can be identified simply by comparing the respective rate-distortion functions of the two sources, evaluated at the capacity of the given channel. As discussed above (see e.g. Figure 8.5), there is no extension of these arguments to *general* networks. In the absence of such a global ordering, other techniques may provide at least partial orderings of communication problems. At one level, this can be interpreted to mean that the finer structure of the source and channel needs to be taken into account. Few such approaches have been taken in the literature, one notable exception being [24].

In line with the spirit of this chapter, in this section, we again argue in terms of a simple example.

8.5.1 Linear Gaussian Sensor Networks

Consider again the specific networked communication scenario illustrated in Figure 8.8. Suppose that the channel matrix B is fixed, and that we are given two different source matrices, A_1 and A_2 to choose from. This may model the sensor network scenario where by selecting the sensing modality or technology, the source observation process can be altered. A natural question is: Is the overall task, again captured by the trade-off between end-to-end distortion and sensor transmit power, easier if the source matrix is A_1 or if it is A_2? What we have shown in Section 8.3.3 is that there is *no* full answer available to date; we cannot universally tell which of the two matrices is preferable.

What *can* be given by exploiting mostly standard arguments is the following restricted partial ordering.

Proposition 5. *Let $L = N = 1$, and for simplicity, suppose that B is the all-ones vector. Then, if*

$$\|A_2\|^2 < \frac{\|A_1\|^2(1 - \frac{\sigma_Z^2}{\sigma_Z^2 + b(M)P_{tot}})}{1 - \frac{\sigma_Z^2}{\sigma_Z^2 + MP_{tot}} + \|A_1\|^2 \frac{\sigma_S^2 \sigma_Z^2}{\sigma_W^2} (\frac{1}{\sigma_Z^2 + b(M)P_{tot}} - \frac{1}{\sigma_Z^2 + MP_{tot}})}, \tag{8.38}$$

where

$$b(M) = \frac{(\sigma_W^2 + \sigma_S^2 \|A_1\|^2) \|A_1\|^2}{\sigma_S^2 \|A_1\|_4^4 + \sigma_W^2 \|A_1\|^2}, \tag{8.39}$$

then A_1 is strictly preferable over A_2.

As an explicit witness to the fact that this does not induce a universal ordering of all possible matrices A, one may note that the condition is not satisfied in the special case when $A_1 = A_2$.

8.6 Conclusions

In the absence of complexity and latency constraints, the potential offered by a noisy communication channel can be characterized by a single number, called its *capacity* and often measured in bits. This number is universal in the sense that via the so-called source/channel separation theorem, it applies to all (stationary, ergodic) communication problems, establishing that bits are a universal currency of information in point-to-point communication. No similarly universal notion is known for general communication networks: there is no general source/channel separation theorem, and bits are not a universal currency of information. In this chapter, we review some of the known partial results available to date. Lacking a more general theory, our exposition is mostly a collection of examples and illustrations, tentatively organized into four main sections. The first three cover, respectively, *exact* separation theorems (a collection of examples of networked communication problems for which bits are a universal currency of information), *counter-examples* where bits are far from being universal, and *approximate* separation theorems, where bits are somehow close to universal. The final section discusses some very tentative approaches of *partially* ordering networked communication problems according to their difficulty without using the notion of bits.

The underlying question that we do not attempt to answer in this chapter is which class typical *real-world* sensor networks fall into — are bits good enough? Eventually, only time can tell.

Acknowledgements

The author gratefully acknowledges stimulating discussions with K. Eswaran, B. Nazer, and A. D. Sarwate (University of California, Berkeley), as well as with P. L. Dragotti (Imperial College, London), G. Kramer (Bell Labs, Alcatel-Lucent), and M. Vetterli (EPFL/UC Berkeley). The comments of an anonymous reviewer were very helpful and are gratefully acknowledged. This work was supported in part by the National Science Foundation under awards CNS-0326503 and CCF-0347298 (CAREER).

References

[1] R. Ahlswede. Multi-way communication channels. In *Proc IEEE Int Symp Info Theory*, Tsahkadsor, Armenian S.S.R., September 1971.
[2] R. Ahlswede and T. S. Han. On source coding with side information via a multiple-access channel and related problems in multi-user information theory. *IEEE Transactions on Information Theory*, IT–29(3):396–412, May 1983.
[3] J. Barros and S. Servetto. Reachback capacity with non-interfering nodes. In *Proc IEEE Int Symp Info Theory*, Yokohama, Japan, July 2003.
[4] T. M. Cover, A. A. El Gamal, and M. Salehi. Multiple access channels with arbitrarily correlated sources. *IEEE Transactions on Information Theory*, 26(6):648–657, November 1980.
[5] T. M. Cover and J. A. Thomas. *Elements of Information Theory.* Wiley, New York, 2nd edition, 2006.
[6] I. Csiszár and J. Körner. *Information Theory: Coding Theory for Discrete Memoryless Systems.* Academic Press, New York, 1981.
[7] M. Gastpar. *To Code Or Not To Code.* PhD thesis, Ecole Polytechnique Fédérale (EPFL), Lausanne, Switzerland, 2002.
[8] M. Gastpar. On capacity under receive and spatial spectrum-sharing constraints. *IEEE Transactions on Information Theory*, 53(2):471–487, February 2007.
[9] M. Gastpar. Uncoded transmission is exactly optimal for a simple Gaussian sensor network. In *Proc. 2007 Information Theory and Applications Workshop*, San Diego, CA, February 2007.
[10] M. Gastpar, B. Rimoldi, and M. Vetterli. To code or not to code. In *Proc IEEE Int Symp Info Theory*, page 236, Sorrento, Italy, June 2000.
[11] M. Gastpar, B. Rimoldi, and M. Vetterli. To code, or not to code: Lossy source-channel communication revisited. *IEEE Transactions on Information Theory*, 49(5):1147–1158, May 2003.
[12] M. Gastpar and M. Vetterli. On the capacity of wireless networks: The relay case. In *Proc IEEE Infocom 2002*, volume 3, pages 1577 – 1586, New York, NY, June 2002.

[13] M. Gastpar and M. Vetterli. Source-channel communication in sensor networks. In Leonidas J. Guibas and Feng Zhao, editors, *2nd International Workshop on Information Processing in Sensor Networks (IPSN'03)*, pages 162–177. Lecture Notes in Computer Science, Springer, New York, NY, April 2003.
[14] M. Gastpar and M. Vetterli. Power, spatio-temporal bandwidth, and distortion in large sensor networks. *IEEE Journal on Selected Areas in Communications (Special Issue on Self-Organizing Distributive Collaborative Sensor Networks)*, 23(4):745–754, April 2005.
[15] M. Gastpar, M. Vetterli, and P. L. Dragotti. Sensing reality and communicating bits: A dangerous liaison. *IEEE Signal Processing Magazine*, 23(4):70–83, July 2006.
[16] J. Körner and K. Marton. How to encode the modulo-two sum of binary sources. *IEEE Transactions on Information Theory*, IT–25(2):219–221, March 1979.
[17] H. Liao. A coding theorem for multiple access communications. In *Proc IEEE Int Symp Info Theory*, Asilomar, CA, February 1972.
[18] J. L. Massey. Causality, feedback and directed information. In *Proc 1990 Int Symp Info Theory and Its Applications (ISITA'90)*, pages 303–305, Hawaii, U.S.A., November 1990.
[19] N. Merhav and S. Shamai. On joint source-channel coding for the Wyner-Ziv source and the Gel'fand-Pinsker channel. *IEEE Transactions on Information Theory*, IT–49(11):2844–2855, November 2003.
[20] B. Nazer and M. Gastpar. Reliable computation over multiple access channels. In *Proc 43rd Annual Allerton Conference on Communication, Control, and Computing*, Monticello, IL, September 2005.
[21] B. Nazer and M. Gastpar. Computing over multiple-access channels with connections to wireless network coding. In *Proceedings of the 2006 International Symposium on Information Theory (ISIT 2006)*, Seattle, WA, July 2006.
[22] Y. Oohama. The rate-distortion function for the quadratic Gaussian CEO problem. *IEEE Transactions on Information Theory*, IT–44(3):1057–1070, May 1998.
[23] L. H. Ozarow. The capacity of the white Gaussian multiple access channel with feedback. *IEEE Transactions on Information Theory*, IT–30(4):623–629, July 1984.
[24] S. S. Pradhan, S. Choi, and K. Ramchandran. An achievable rate region for multiple access channels with correlated messages. In *Proc IEEE Int Symp Info Theory*, Chicago, IL, June 27-July 2 2004.
[25] S. Ray, M. Effros, M. Médard, R. Koetter, and T. Ho abd D. Karger abd J. Abounadir. On separation, randomness and linearity for network codes over finite fields. March 2006. http://arxiv.org/abs/cs/0603022.
[26] Z. Reznic, R. Zamir, and M. Feder. Joint source-channel coding of a Gaussian mixture source over a Gaussian broadcast channel. *IEEE Transactions on Information Theory*, IT–48(3):776–781, March 2002.

[27] C. E. Shannon. A mathematical theory of communication. *Bell Sys. Tech. Journal*, 27:379–423, 623–656, 1948.
[28] D. Slepian and J. K. Wolf. Noiseless coding of correlated information sources. *IEEE Transactions on Information Theory*, IT–19:471–480, 1973.
[29] I. E. Telatar. Capacity of multi-antenna Gaussian channels. *Bell Labs Technical Memorandum*, June 1995. Also published in *European Transactions on Telecommunications*, 10(6):585-596, Nov.-Dec. 1999.

Part IV

Dynamics & Control over Communication Networks

9

Toward the Design of a Transport Layer for Networked Control Systems

C. L. Robinson[1] and P. R. Kumar[2]

[1] Department of Industrial and Enterprise Systems Engineering, and the Coordinated Science Lab, University of Illinois at Urbana-Champaign, clrobnsn@uiuc.edu

[2] Department of Electrical and Computer Engineering, and the Coordinated Science Lab, University of Illinois at Urbana-Champaign, prkumar@uiuc.edu

9.1 Introduction

In a 'classical' control system, such as the one shown in Figure 9.1, the system components (sensor, controller and actuator) are connected through dedicated and reliable communication links. As we move to deploy wireless networked control systems over a wireless communication medium, such as in Figure 9.2, we are faced with new problems. In particular, interference and channel fading result in unpredictable packet loss and delay. In data networks, where data integrity is paramount, the dropped packets may simply be retransmitted until they are successfully received. However, in control and sensor network systems, data has a *time value*, and transmitting outdated data has limited value. Instead of retransmitting the dropped data, it may be better to transmit any new data that is available.

A second characteristic that differentiates the embedded network control and sensor network environment from data networks is that the computational capacities of each node, and the bandwidth available between nodes, are typically small. Hence, from a state estimation and control system perspective,

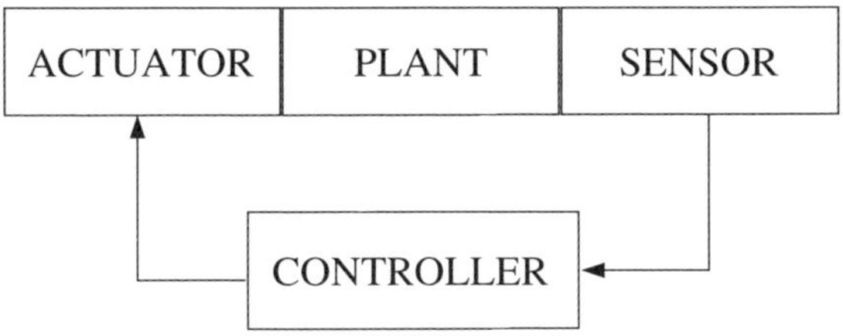

Fig. 9.1. A control system with the sensor, controller and actuator connected by dedicated reliable communication links.

it is useful to consider communication policies as well as system architectures that specifically take into account the time value associated with the data, as well as the communication and computation constraints.

In this chapter we will consider developing an appropriate *application level transport protocol* for optimal state estimation for a sensor network or networked control system, in which the communication channel is subject to random packet loss. We shall consider packet losses occurring between the sensor and controller in the control problem, and the sensor and the state estimator in the state estimation problem. Under this arrangement, we then pose the question:

Given that a measurement packet has been dropped and a new measurement is available, what information should be transmitted in the following packet?

Clearly, sending both old as well as new data would convey the maximal information and would yield the best possible state estimate. However, if multiple packets are dropped, *more* bandwidth is required to transmit the outdated data. At the level that we treat this problem, we ignore the difficulty involved in transmitting a real number and simply suppose that sending two real numbers is twice as difficult as sending just one. Thus, since older data is less valuable, sending both pieces of data represents an inefficient usage of the available bandwidth. Another option is to simply send the most recent observation, and ignore the information contained in the older dropped packets. This is the most common approach. However, there is a loss of information under this scheme. A third option, and the one we shall investigate in this chapter, is to form a *linear combination of the two measurements* and thus create a single measurement value which is then transmitted.

In Section 9.2 we discuss related literature and provide some background related to the issues at play in estimation and control over wireless channels. This section also places our results in context with regard to the many assumptions that can be made when dealing with such problems. We formally define the problem in Section 9.3 and then move on to present our main results in Section 9.4. To strengthen intuition we present several examples throughout the section. In Section 9.5 we extend the results to the case where the sensor sampling rate is several times faster than the allowable channel transmission rate. Section 9.6 presents some simulation results which illustrate the results

Fig. 9.2. Information sent from the sensor through the network to the estimation or control logic is subject to unpredictable loss.

and demonstrate the performance improvement achievable. A conclusion with directions for future work is given in Section 9.7. This work is a revision of earlier work presented in [37].

9.2 Background

9.2.1 Related Literature

Interest in networked control systems from both a research [2, 1, 22, 44, 12] and industry perspective [34, 24, 33] has been growing for several years. Just as there are many benefits for deploying such networked control systems, there are also several problems that are not addressed in classical control theory. These problems include primarily random packet delay and random packet loss. Other issues involve limited computational capacity and limited bandwidth. These problems become even more relevant as we move into *wireless* networked control systems and sensor networks.

In dealing with random packet loss, the stability of Kalman Filtering [40, 27], optimal LQG controllers [19, 16, 39] and some suboptimal controllers [4] have been examined. Performance issues and controller location is considered in [36]. The critical difficulty with the control problem revolves around the fact that packet or information loss may result in so called 'non-classical' information patterns, which have been shown to make the problem intractable [48, 32]. Generally, such information patterns arise whenever information upon which control actions are based is not consistent throughout the system. An example is when the controller is unaware that an actuation command was dropped enroute to the actuator. In the presence of such conditions the separation theorem breaks down.

In the context of networked control systems, avoiding this situation amounts to ensuring that the controller has knowledge of what control actions have been implemented by actuator. Such difficulties are not encountered in the estimation problem. Thus, packet delivery, or at least packet delivery status notification, between the *controller* and *actuator* is important. To realize such notification, a 'TCP' type information structure is assumed [19]. Under TCP, packet delivery notifications are enabled, whereas in UDP they are not.

Some transport layer protocols beyond TCP and UDP have also been developed which are of some use for networked control. TCP Friendly rate control (TFRC) [17] attempts to avoid abrupt send rate changes associated with TCP flows. The Datagram Congestion Control Protocol (DCCP) [21] foregoes reliable message delivery yet still achieves congestion control, and the Stream Control Transmission Protocol (SCTP) [41] treats transmissions as streams, as compared to individual packets.

There are also some protocols which have been designed specifically for control. The Controller Area Network (CAN) [9] protocol is an example is highly popular amongst automobile manufacturers. Under the CAN protocol, data from a component is assigned a unique identification code and transmitted on a simple two wire network. Transmission conflicts are resolved using message priorities and random backoffs. Thus, despite harsh operating environments, access to the communication medium can be strictly controlled and scheduled. As a result, the protocol provides reliable and timely message delivery. Other protocols developed for control systems include DeviceNet, BACnet, CEbus, LONWorks, ControlNet and Profibus [25, 38, 34].

The utility of these protocols, from a control and estimation perspective, is that they account for the time value of the data being transmitted, while addressing network considerations such as congestion control, collision avoidance and flow control. However, this importance is only addressed on a per-packet basis, i.e., ensuring the regular, timely or dependable delivery of an individual packet. Relationships between data transmitted at *different times* are not considered in the packet composition. The work we present here illustrates a way in which a transport layer protocol may be designed so as to *adapt packet contents* to meet both the control and estimation system requirements as well as deal with the network restrictions.

The other significant problem in networked control is random packet delay. Its effects (both network transmission and device computation) on performance have been investigated in [20, 25, 26, 10, 31, 3]. Stability issues are considered in [46, 29, 28]. Methodolgies for dealing with the delay include using estimators [8], jump linear systems [49], queues [11], scheduling transmissions [51, 18, 45, 50, 52], and others [44, 12].

Other work on networked control systems includes analyzing limited data rates [30, 14], using encoders on either side of the channel [42, 16] and developing appropriate software services and abstractions [6] and Middleware [5, 15, 7, 33, 43].

9.2.2 Background Material

Throughout this chapter we deal with the estimation of linear Gaussian systems. Consequently, we shall use the optimal Kalman Filter [35, 47] for prediction and state estimation. For ease of reference, we gather together the equations for a Kalman filter, for a system with state matrix A, control input matrix B, observation matrix C, state noise covariance Q_w and observation noise covariance R_v. The time update equations are:

$$\hat{x}_{k+1|k} = A\hat{x}_{k|k} + Bu_k, \tag{9.1}$$

$$P_{k+1|k} = AP_{k|k}A' + Q_w. \tag{9.2}$$

The measurement update equations are:

$$K_{k+1} = P_{k+1|k}C' \left(CP_{k+1|k}C' + R_v\right)^{-1}, \tag{9.3}$$

$$\hat{x}_{k+1|k+1} = \hat{x}_{k+1|k} + K_{k+1}\left(z_k - c\hat{x}_{k+1|k}\right), \tag{9.4}$$

$$P_{k+1|k+1} = \left(I - K_{k+1}C\right)P_{k+1|k}, \tag{9.5}$$

where the nomenclature $\hat{x}_{k+1|k}$ represents the state estimate at time $k+1$ given state observations up to time k. When combined, the Kalman filter measurement and prediction equations for the error covariance yield the discrete Ricatti equation:

$$P_{k+1} = AP_kA' - AP_kC'\left[CP_kC' + R_v\right]^{-1} CP_kA' + Q_w, \tag{9.6}$$

where by convention $P_{k+1} = P_{k+1|k}$. Throughout the remainder of the chapter we will deal with systems of dimension 2 and accordingly partition the covariance matrix as:

$$P_{k|k} = \begin{bmatrix} p_{k|k}^{1,1} & p_{k|k}^{1,2} \\ p_{k|k}^{1,2} & p_{k|k}^{2,2} \end{bmatrix}. \tag{9.7}$$

9.3 Problem Formulation

We consider linear Gaussian scalar systems described as follows:

$$\begin{aligned} x_{k+1} &= ax_k + bu_k + w_k, \\ y_k &= cx_k + v_k, \end{aligned} \tag{9.8}$$

where the system's state at time k is represented by x_k. A noisy observation at time k is represented by y_k and has associated Gaussian noise v_k with variance r_v. The variance of the Gaussian state noise w_k is given as q_w. Both w_k and v_k are independent. We have chosen scalar systems for simplicity and insight.

We have chosen to only consider packet loss between the sensor and controller or actuator to avoid intractability problems. As a consequence, the separation theorem is applicable, and the optimal control problem is separated into forming an optimal state estimate and an optimal certain equivalent controller. This can be explicitly illustrated by considering the standard quadratic cost criterion for the control problem:

$$J = E\sum_{k=0}^{N}\{x_k'Q_xx_k + u_k'R_uu_k\}, \tag{9.9}$$

where $Q_x \geq 0$ and $R_u > 0$. By resolving the state x_k into the state estimate $\hat{x}_k$, and state estimation error $\tilde{x}_k$, we can define $x_k = \hat{x}_k + \tilde{x}_k$. Conditioning the cost on the observations yields:

$$J = E\sum_{k=0}^{N-1}\left(\hat{x}_k'Q_x\hat{x}_k + u_k'R_uu_k\right) + \underbrace{E\sum_{k=0}^{N-1} Tr(Q_xP_k)}_{\text{Unaffected by controls}},$$

where P_k is the conditional covariance of the state estimation error. Hence, as a result of certainty equivalence, the control cost minimization problem reduces to that of minimizing the conditional state estimation error covariance, provided an optimal certain equivalent controller is used. Hence, for the remainder of this chapter we shall consider only the optimal *estimation* problem, and not per se the optimal *control* problem.

9.3.1 Formulating the Reaction to Dropped Packets

The system described in (9.8) represents only the discrete time system dynamics and does not explicitly account for the effect of dropped packets. We shall assume that that the communication link delivering observations to the estimator behaves as a Bernoulli erasure channel which randomly drops packets with probability λ. During periods of dropped packets, the state estimate will simply be a forward prediction of the most recent state estimate. This is the so called 'open loop' state prediction, and corresponds to the Kalman Filter time update equations (9.1) and (9.2). The state estimate produced by the Kalman filter has an associated state estimation error covariance which we are interested in minimizing so as to achieve lower cost. However, some complexity arises as a result of the random sequence of packet losses which result in a random state estimation covariance matrix. For our analysis we shall assume the following sequence of events:

1. A sensor observation is taken at time k and transmitted to the estimator through the network. The observation value is also stored at the sensor.
2. The packet may be dropped en-route to the estimator. The sensor is then made aware of the packet drop but cannot re-transmit anything before another observation is taken.
3. Another observation is taken at time $k+1$.
4. Given the two observations, the sensor must now construct a *single* measurement value to transmit to the estimator through the network. We will consider a strategy where this transmitted value is a linear combination of the old and new observations. This represents a *linear temporal coding* strategy.

For simplicity, we shall, for the meantime, restrict ourselves to the case where only a single packet is dropped. Later results extend this to the general case of multiple packet drops. With the above sequence of events in mind we define the following concatenated system:

$$\begin{aligned}
\begin{bmatrix} x_{k+1} \\ x_k \end{bmatrix} &= \begin{bmatrix} a & 0 \\ 1 & 0 \end{bmatrix} \begin{bmatrix} x_k \\ x_{k-1} \end{bmatrix} + \begin{bmatrix} 1 \\ 0 \end{bmatrix} u_k + \begin{bmatrix} w_k \\ 0 \end{bmatrix}, \\
y_k &= c x_k + v_k, \qquad (9.10) \\
z_k &= \begin{bmatrix} \alpha & \beta \end{bmatrix} \begin{bmatrix} y_k \\ y_{k-1} \end{bmatrix}, \\
&= \begin{bmatrix} \alpha & \beta \end{bmatrix} \begin{bmatrix} c x_k \\ c x_{k-1} \end{bmatrix} + \begin{bmatrix} \alpha & \beta \end{bmatrix} \begin{bmatrix} v_k \\ v_{k-1} \end{bmatrix}, \\
Q_w &= \begin{bmatrix} q_w & 0 \\ 0 & 0 \end{bmatrix}, \qquad (9.11) \\
R_v &= (\alpha^2 + \beta^2) r_v. \qquad (9.12)
\end{aligned}$$

Above we have explicitly separated the measurement taken at time k, denoted as y_k, from the information that is actually transmitted at time k, which is denoted as z_k. This allows us to compose the contents of the transmitted measurement by adjusting the relative weighting of the α and β terms in $C = (\alpha, \beta)$. Without loss of generality we will assume $c = 1$. In a similar vein, we note that it is only the *ratio* $\frac{\alpha}{\beta}$ that is of significance in constructing C, and not the individual values of α and β. The state noise covariance in this formulation is represented by Q_w, as shown in (9.11), and the observation noise is given in (9.12) as R_v. We have used the fact that $E[v_k v'_{k-1}] = 0$ and $E[v_k^2] = E[v_{k-1}^2]$, since they are i.i.d. with zero mean.

If there has *never* been a packet loss then there is no additional benefit obtained by retransmitting a previous observation. Hence the C matrix would be constructed with $\alpha = 1$ and $\beta = 0$, which corresponds to sending only the most recent observation. In general, any value of $\alpha \neq 0$ can be used so long as $\beta = 0$. We note that scaling (α, β) to $\gamma(\alpha, \beta)$ for $\gamma \neq 0$ yields an equally informative observation. Thus we only need to consider $(\alpha, \beta) = (0, 0)$ and all other (α, β) lying on the unit circle $\alpha^2 + \beta^2 = 1$. The former can be further ruled out since is corresponds to no information at all.

We now focus on deriving conditions for the optimal composition of C following the sequence of events described earlier. In Section 9.2 we have presented the Kalman filter equations and defined the state error covariance matrix in (9.7). Assume that $P_{k|k}$ represents the state estimation covariance at time k, immediately before the packet is dropped. We do not require any assumption regarding the arrival of an observation at or before that time, and simply state that $P_{k|k}$ represents the best state estimation covariance for time k. Thus, to move forward in time we perform a time update to create an open loop prediction of the state at time $k+1$ using (9.2) and (9.1). We focus here on the state covariance calculations (9.2), which yield:

$$\begin{aligned}
P_{k+1|k} &= A P_{k|k} A' + Q_w \\
&= \begin{bmatrix} a^2 p^{1,1}_{k|k} + q_w & a p^{1,1}_{k|k} \\ a p^{1,1}_{k|k} & p^{1,1}_{k|k} \end{bmatrix}. \qquad (9.13)
\end{aligned}$$

Note that these calculations are made *regardless* of the arrival of an observation. If the packet at time $k+1$ had been delivered, an 'Observation Update' calculation would have been done using (9.5). However, since it is dropped, another time update (9.2) is performed:

$$P_{k+2|k} = \begin{bmatrix} q_w + a^2(q_w + a^2 p^{1,1}_{k|k}) & a(q_w + a^2 p^{1,1}_{k|k}) \\ a(q_w + a^2 p^{1,1}_{k|k}) & q_w + a^2 p^{1,1}_{k|k} \end{bmatrix}. \tag{9.14}$$

In the sequence of events we are considering, the sensor has created a single combined measurement, via a linear temporal code, and transmitted it to the estimator. On arrival, a measurement update can be performed (9.5). Since the sensor has chosen a ratio of the 'new' and 'old' measurements, the update will be in terms of the α and β used to construct the C matrix. This will yield $P_{k+2|k+2}$. Recall the manner in which P can be split up as described in (9.7) and the description of the concatenated system in (9.10). The term in the upper left corner, $p^{1,1}_{k+2|k+2}$, represents the variance of the state estimate at the current time. The full expression for this term is:

$$p^{1,1}_{k+2|k+2} = q_w + a^2\left(a^2 p^{1,1}_k + q_w\right) - \frac{\left(q_w + \left(a^2 p^{1,1}_k + q_w\right)\left(a^2\alpha + a\beta\right)\right)\left(\left(a^2 p^{1,1}_k + q_w\right)\left(a\beta + a^2\alpha^2\right) + \alpha q_w\right)}{\left(a^2 p^{1,1}_k + q_w\right)\left(\alpha a + \beta\right)^2 + \alpha^2 q_w + r_v(\alpha^2 + \beta^2)}. \tag{9.15}$$

This term will incur a cost in the trace function in (9.9). We will investigate the optimal choice of α and β so as to minimize this term.

An Example

We first present an example to show that such a linear temporal coding can indeed provide a benefit, before proceeding further. Consider the system with $a = 2$, $r_v = 1$ and $q_w = 1$. Suppose that $p^{1,1}_{k|k} = 1$ at some time k. Then (9.15) yields:

$$p^{1,1}_{k+2|k+2} = \frac{21\alpha^2 + 26\beta^2}{22\alpha^2 + 20\alpha\beta + 6\beta^2}, \tag{9.16}$$

which can be minimized to a value of $\frac{13}{16}$ with $(\alpha, \beta) = \left(\frac{13}{5}, 1\right)$. In comparison, using only the most recent measurement corresponding to $(\alpha, \beta) = (1, 0)$ yields a covariance of $\frac{21}{22}$. A plot of the cost for various values of β is shown in Figure 9.3. This illustrates that by a linear combination of measurements one may indeed be able to reduce the cost in comparison to just sending a new measurement.

9.4 Analysis

The example in the previous section showed that there there can indeed be an optimal choice of α and β which yields a lower state estimation error than simply sending the most recent observation. However, before we proceed to compute the optimal ratio, we need to establish that improving the state estimate in the short term is optimal for the long term. In other words, we need to show that choosing an optimal α and β at time k is good for all future time, *regardless* of the policy implemented in the future. To do this we first restate a well known result [23] for convenience:

Lemma 1. *Monotonicity of Ricatti equation iterations. Let P_k and $\bar{P}_k$ represent the kth iteration of (9.6) corresponding to initial conditions P_0 and $\bar{P}_0$ respectively. Let P_∞ represent the steady state solution to (9.6). If $P_0 \geq \bar{P}_0 \geq P_\infty$ then $P_k \geq \bar{P}_k \geq P_\infty$ for all k.*

The Lemma below establishes that there is no trade-off between present and future estimates:

Lemma 2. *Minimizing $P^{1,1}_{k|k}$ is an optimal policy for all P_j for all $j > k$.*

Proof. Consider the structure of P after a Kalman time update in (9.2). The only term in $P_{k+1|k}$ from $P_{k|k}$ is $p^{1,1}_{k|k}$, as illustrated in (9.13). The consequence is that if the optimal (α^*, β^*) had been used to generate the minimal $p^{1,1\,*}_{k|k}$, then $P_{k+1|k}$ generated using $p^{1,1\,*}_{k|k}$ in (9.2) is also minimal.

Now consider a second system in which some other non-optimal choice of $(\tilde{\alpha}, \tilde{\beta})$ was used to compute some $\tilde{p}^{1,1}_{k|k}$, which is then used to compute $\tilde{P}_{k+1|k}$. Since $\tilde{p}^{1,1}_{k|k} > p^{1,1\,*}_{k|k}$, we must have $\tilde{P}_{k+1|k} > P^*_{k+1|k}$. Then, any subsequent choice of (α, β) by the second system can also be used by the first system, and hence by the dominance Lemma 1, $\tilde{P}_j > P^*_j$ for all $j > k$. □

This Lemma enables us to proceed directly and minimize the covariance of the state estimate obtained at each step from the Kalman Filter. We note however, that forming the optimal estimate for $p^{1,1}_{k+2|k+2}$ does *not* yield the optimal estimate for $p^{2,2}_{k+2|k+2}$. An example of this is given later in the chapter in Figure 9.4.

Our first result deals with the case when only a single packet may be dropped and there are no consecutive packet drops. Later we will deal with the more general case of multiple consecutive dropped packets. We will assume throughout that $q_w > 0$ and $r_v > 0$.

Theorem 1. *Following a single dropped observation in a system described by (9.8), with noise process described in (9.12), the optimal linear combination of the past (dropped) measurement with the latest (new) sampled measurement is:*

$$\frac{\alpha^*}{\beta^*} = a + \frac{q_w}{a r_v} + \frac{q_w}{a(a^2 p^{1,1}_k + q_w)}. \tag{9.17}$$

Proof. The burden of the proof is to find conditions under which $p^{1,1}_{k+2|k+2}$ is minimized by a choice of α and β. Note that $P^{1,1}_{k+2|k+2}$ is continuous everywhere except at $(\alpha, \beta) = (0, 0)$. We can exclude this point since it represents no information transfer. For all $(\alpha, \beta) \neq (0, 0)$ we also note that scaling to $\gamma(\alpha, \beta)$ where $\gamma \neq 0$ yields no change. We shall first find the stationary points and then check which of them represent a minimum. Necessary conditions for a stationary point $\left(\frac{dp^{1,1}_{k+2|k+2}}{d\alpha} = \frac{dp^{1,1}_{k+2|k+2}}{d\beta} = 0\right)$ are:

$$\frac{-1}{\beta}\frac{dp^{1,1}_{k+2|k+2}}{d\alpha} = \frac{1}{\alpha}\frac{dp^{1,1}_{k+2|k+2}}{d\beta} = -\frac{2\Theta\Gamma}{\Phi^2} = 0, \tag{9.18}$$

where

$$\begin{aligned}
\Theta &= \left(a^4\, p_k^{1,1} + q_w + a^2\, q_w\right)\, \alpha + a\, \left(a^2\, p_k^{1,1} + q_w\right)\beta, \\
\Gamma &= \left(a^2 p_k^{1,1} + q_w\right)\left(a r_v \alpha - a^2 r_v \beta - q_w \beta\right) - q_w r_v \beta, \\
\Phi &= \left(a^2 p_k^{1,1} + q_w\right)(a\alpha + \beta)^2 + (q_w + r_v)\,\alpha^2 + r_v \beta^2.
\end{aligned}$$

Hence, for a stationary point we require either the numerator terms Θ or Γ to be equal to zero, since $\Phi \neq 0$. So, $\Theta = 0$ or $\Gamma = 0$ is a necessary condition. Figure 9.3 illustrates the stationary points for a particular example.

To establish if the stationary points are minimizing, consider the positive definiteness of the Hessian $\mathcal{H}$ which is guaranteed when all the determinant of the principal minors are all positive. Taking the upper left term in $\mathcal{H}$, and substituting $\Gamma = 0$ yields the following condition for positive definiteness:

$$\frac{2\,\Psi\beta^4\left(\Delta_1 + \Delta_2 + \Delta_3 + \Delta_4\right)^2\left(a^4\, p_k^{1,1} + q_w + a^2\, q_w\right)^6}{a^2\left(a^2\, p_k^{1,1} + q_w\right)^2 r_v^2\beta^6\left(\Delta_1 + \Delta_2 + \Pi_1 + \Pi_2\right)^3} \geq 0, \tag{9.19}$$

where

$$\begin{aligned}
\Psi &= \left(a^4\, p_k^{1,1}\, r + (q_w + r_v)^2 + a^2\left(q_w\, r_v + p_k^{1,1}\,(q_w + r_v)\right)\right), \\
\Delta_1 &= a^8\, {p_k^{1,1}}^2\, r_v + {q_w}^2\,(q_w + r_v), \\
\Delta_2 &= a^6\, p_k^{1,1}\left(2\, q_w\, r_v + p_k^{1,1}\,(q_w + r_v)\right), \\
\Delta_3 &= a^4\, q_w\left(q_w\, r_v + 2\, p_k^{1,1}\,(q_w + 2\, r_v)\right), \\
\Delta_4 &= a^2\, {q_w}^2\left(p_k^{1,1} + q_w + 3\, r_v\right), \\
\Pi_1 &= a^4\, q_w\left(q_w\, r_v + 2\, p_k^{1,1}\,(q_w + 2\, r_v)\right), \\
\Pi_2 &= a^2\, {q_w}^2\left(p_k^{1,1} + q_w + 3\, r_v\right).
\end{aligned}$$

Closer examination of the terms in (9.19) shows that the expression is always positive definite and only in some special cases, such as $a = r_v = q_w = 0$, disallowed by our assumption, is the expression positive semi-definite. Hence, $\Gamma = 0$ is a minimizing stationary point. In a similar manner, it can be shown that the condition $\Theta = 0$ yields a negative definite expression, and is hence is a maximizing solution. Note further that the local minimum corresponding to $\Gamma = 0$ is actually a global minimum since attention can be restricted to the circle $\alpha^2 + \beta^2 = 1$. Hence, we conclude that $\Gamma = 0$ is necessary and sufficient condition for the global minimizer. □

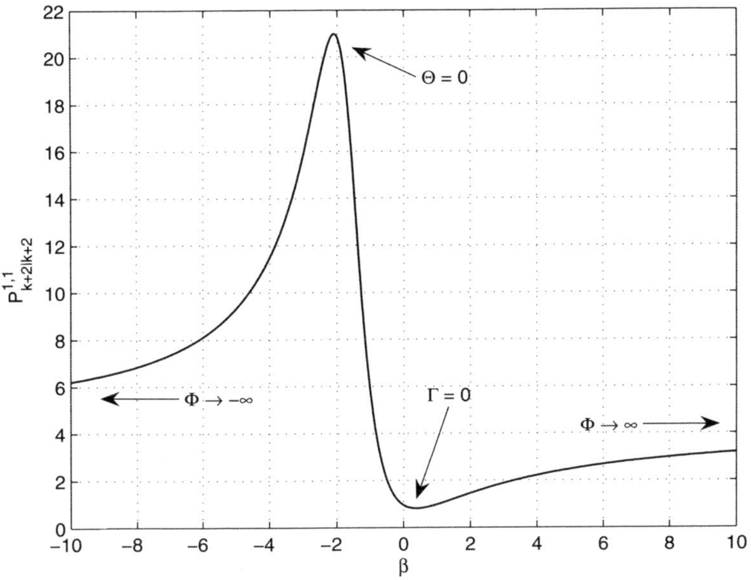

Fig. 9.3. The magnitude of the error covariance $p^{1,1}_{k+2|k+2}$ as a function of β with $\alpha = 1$, $a = 2$, $p^{1,1}_{k|k} = 1$, $r_v = 1$ and $q_w = 1$. Note that $\lim_{\Phi \to \pm\infty} P^{1,1}_{k+2|k+2} = q_w + a^2 \left(a^2 p^{1,1}_k + q_w\right)$, as one expects.

The previous theorem dealt with the case where only a single packet is dropped. The following Theorem addresses the more general case of *multiple* successive packet drops.

Theorem 2. *Following D successive dropped observations in a system described by (9.8), with noise process described in (9.12), the optimal linear combination of the most recently dropped measurement with the latest sampled measurement is given by:*

$$\frac{\alpha^*}{\beta^*} = a + \frac{q_w}{a r_v} + \frac{q_w}{a(a^{2D} p^{1,1}_k + \sum_{i=1}^{D} a^{2(i-1)} q_w)}. \tag{9.20}$$

Proof. This follows directly from the previous proof once it is noted that $a^{2D}p_k^{1,1} + \sum_{i=1}^{D} a^{2(i-1)}q_w$ is the state estimation error covariance projected forward by D drops, and that it simply replaces the expression for a single drop state error covariance in (9.13).

This result establishes that regardless of the number of drops there is an optimal combination of the two most recent measurements that can be sent in order to minimize the estimation error covariance. We have not dealt with the topic of creating a single measurement to be transmitted from *all* of the dropped packets. This is more complex to analyze, as well as to implement, and may not be desirable. (See also the comparison with the lower bound in Section 9.6).

9.4.1 Interpretation of Results

The results that we have presented can be examined for a number of special cases and interpreted in a variety of ways to provide some insight. In the simplest case, consider a kinematic system with no process noise. There is only noise on the observations. In this case the state will remain constant and the optimal policy is simply to average over the two observations, which will create a single observation with a lower variance. To see this with our results, set $q_w = 0$ and $a = 1$ in (9.17) and note that $\alpha^* = \beta^*$, indicating equal weighting in the combined measurement. As we extend this example and add system dynamics and model noise we still observe this coupling and reduction in variance, although the past measurements have less significance. Considering Theorem 1 we observe:

r_v Taking the limit as $r_v \to 0$ represents receiving noiseless, perfect observations. In the limit, the term $\frac{\alpha}{\beta} \to \infty$. This ratio indicates we should place no importance on the dropped observation, and the transmitted measurement should comprise only the most recent observation. This is intuitive since the observation is perfect and hence there is no need to include past information. On the other hand, for systems that have r_v very large (i.e., high observation noise), the ratio is in fact lower bounded by $\frac{\alpha}{\beta} \geq a + \frac{q_w}{a(a^2 p_k^{1,1} + q_w)}$. So, regardless of how 'noisy' an observation is, it still has value.

a As the system dynamics disappear and the system state is continually forced to zero ($a \to 0$), any present control action or system noise completely determines the future state of the system, irrespective of the current or past states. Thus, there is no correlation between the states at observation instants, and thus there is no advantage in communicating any information from a past observation. This is represented by $\frac{\alpha}{\beta} \to \infty$ as $a \to 0$.

q_w The case where $q_w \to 0$ represents noiseless state equation updates. It is intriguing that in this case $\frac{\alpha}{\beta} \to a$, for which we still have no intuitive explanation. The case where $q_w \to \infty$ represents a very poor system model, and as one expects, more emphasis should be placed on the more recent observation, i.e., $\frac{\alpha}{\beta} \to \infty$.

$p_k^{1,1}$ The fact that the ratio depends on the state estimation error covariance before the packet drop is the first point of interest. The second is that the ratio is bounded by below by $a + \frac{q_w}{ar_v}$ for large values of $p_k^{1,1}$, and from above by $a + \frac{q_w}{ar_v} + \frac{1}{a}$ for small values of $p_k^{1,1}$. Interestingly, the tightness of these bounds depends on the system dynamics, a.

9.5 Oversampling

In contrast to the scenarios presented earlier where packets are randomly dropped, in this section we consider the case where a sensor can take measurements *more rapidly* than the communication network can transmit them. We shall start by considering the case where the state is sampled at twice the transmission frequency and analyze what the optimal combination of the two measurements should be to minimize the estimation error. We shall, in the end, arrive at the same result as in the previous section. However, here we use a slightly different approach, exploiting the fact that packets are not dropped under this oversampling formulation. Using the model in (9.8) and system formulation in (9.10), we consider the state equation for the *next* time step:

$$\begin{aligned}
\begin{bmatrix} x_{k+2} \\ x_{k+1} \end{bmatrix} &= \begin{bmatrix} a & 0 \\ 1 & 0 \end{bmatrix} \begin{bmatrix} x_{k+1} \\ x_k \end{bmatrix} + u_{k+1} + \begin{bmatrix} w_{k+1} \\ 0 \end{bmatrix} \\
&= \begin{bmatrix} a^2 & 0 \\ a & 0 \end{bmatrix} \begin{bmatrix} x_k \\ x_{k-1} \end{bmatrix} + \begin{bmatrix} a & 1 \\ 1 & 0 \end{bmatrix} \begin{bmatrix} u_k \\ u_{k+1} \end{bmatrix} \\
&\quad + \begin{bmatrix} a & 1 \\ 1 & 0 \end{bmatrix} \begin{bmatrix} w_k \\ w_{k+1} \end{bmatrix}, \\
z_k &= \alpha y_k + \beta y_{k-1} = \begin{bmatrix} \alpha & \beta \end{bmatrix} \begin{bmatrix} y_k \\ y_{k-1} \end{bmatrix} \\
&= \begin{bmatrix} \alpha & \beta \end{bmatrix} \begin{bmatrix} x_k \\ x_{k-1} \end{bmatrix} + \begin{bmatrix} \alpha & \beta \end{bmatrix} \begin{bmatrix} v_k \\ v_{k-1} \end{bmatrix}.
\end{aligned}$$

This represents the evolution of the system between successive transmitted observations. Since measurements are delivered regularly, we can restrict attention to even k by setting $k = 2n$, and defining the following new terms:

$$
\begin{aligned}
X_n &:= \begin{bmatrix} x_{2n} \\ x_{2n-1} \end{bmatrix}, \\
\bar{W}_n &:= \begin{bmatrix} w_{2n} \\ w_{2n+1} \end{bmatrix} \rightsquigarrow N\left(\begin{bmatrix} 0 \\ 0 \end{bmatrix}, \begin{bmatrix} q_w & 0 \\ 0 & q_w \end{bmatrix}\right), \\
\bar{V}_n &:= \begin{bmatrix} v_{2n} \\ v_{2n-1} \end{bmatrix} \rightsquigarrow N\left(\begin{bmatrix} 0 \\ 0 \end{bmatrix}, \begin{bmatrix} r_v & 0 \\ 0 & r_v \end{bmatrix}\right), \\
Z_n &:= z_{2n}.
\end{aligned}
$$

This can be used to specify the system:

$$
\begin{aligned}
X_{n+1} &= \begin{bmatrix} a^2 & 0 \\ a & 0 \end{bmatrix} X_n + \begin{bmatrix} a & 1 \\ 1 & 0 \end{bmatrix} \bar{W}_n, \\
Q_w &= cov\left(\begin{bmatrix} a & 1 \\ 1 & 0 \end{bmatrix} \bar{W}_n\right) = \begin{bmatrix} a^2+1 & a \\ a & 1 \end{bmatrix} q_w, \\
R_v &= cov\left(\begin{bmatrix} \alpha & \beta \end{bmatrix} v_n\right) = \left(\alpha^2 + \beta^2\right) r_v, \\
Z_n &= \begin{bmatrix} \alpha & \beta \end{bmatrix} X_n + \begin{bmatrix} \alpha & \beta \end{bmatrix} \bar{V}_n.
\end{aligned}
$$

This representation is in a more familiar linear system form, for which we can immediately investigate the Kalman filter and Ricatti equations. A first consideration is stability of the estimate which is guaranteed by existence and uniqueness of the solution to the difference Riccati equation. This requires stabilizability of the pair (A, Q_w) and detectability of the pair (A, C), where $C = [\alpha, \beta]$, which is true for all (a, q_w) stabilizable, and (a, α) or (a, β) detectable. Thus, there is a unique positive definite solution to the Riccati equation.

Since packets are not dropped, a time and observation update using the Kalman equations is equivalent to a single update of the discrete Ricatti Equation (9.6). Note that if $P_{n+1|n} \to P$ as $n \to \infty$, then $P_{n|n} \to \tilde{P}$ where $\tilde{P}$ is related to $P_{n+1|n}$ through (9.5). Computing $p^{1,1}_{n+1|n+1}$ and taking the derivative with respect to β and α yields:

$$
\frac{-1}{\beta}\frac{dp^{1,1}_{n+1|n+1}}{d\alpha} = \frac{1}{\alpha}\frac{dp^{1,1}_{n+1|n+1}}{d\beta} = -\frac{2\Theta\Gamma}{\Phi^2}.
$$

We have seen this expression before in the proof of Theorem 1, and thus, following the same logic as in the proof we find that $\Gamma = 0$ is the minimizer yielding the following condition for the minimizer:

$$
\frac{\alpha^*}{\beta^*} = a + \frac{q_w}{a r_v} + \frac{q_w}{a(a^2 p^{1,1}_n + q_w)}. \tag{9.21}
$$

This is exactly the expression obtained in Theorem 1.

9.5.1 Intermediate State Estimation

This section examines the optimal ratio for estimating the state at an *intermediate time*, when an observation was taken but not transmitted, $p^{1,1}_{k+1|k+2}$

(or equivalently $p^{2,2}_{k+2|k+2}$), is of importance. As would be expected, the optimal (α, β) ratio differs from the one for estimating the state at the most recent time. Computing once again the first order necessary conditions for a minimum of $p^{1,1}_{k+1|k+2}$ yields:

$$\frac{\alpha}{\beta} = \frac{2(a^2 p_k^{1,1} + q_w)^2 \overbrace{\alpha(a\alpha + \beta)}^{\Gamma_1} \overbrace{(-(q_w + r_v)\alpha + ar_v\beta)}^{\Gamma_2}}{\Phi^2}.$$

Hence, potential minima are obtained at Γ_1 or Γ_2 equal zero. Intuition suggests that any minima must depend on the observation noise, and since Γ_1 is independent of r_v it can be ignored. The intuition is confirmed by computing the second derivative of the above function and observing that $\Gamma_1 = 0$ yields a maximizing solution. Hence, the condition $\Gamma_2 = 0$ yields the ratio for optimal estimation of the state at the 'missed' sample instant

$$\frac{\alpha}{\beta} = \frac{ar_v}{q_w + r_v}. \tag{9.22}$$

This ratio represents the optimal choice for the 'smoothing' problem, where future observations are used to generate a better state estimate. The difference between the minima associated with previous estimation problem, and this smoothing problem are shown in Figure 9.4. The analysis in this section has focused on the case where sampling frequency was double that of the transmission frequency. In a similar manner to the extension to multiple packet drop in earlier sections, we now extend Theorem 2 to the general case of multiple sample instants occurring between transmission times.

Corollary 1. *When sensor observations are transmitted once every D samples, the optimal linear combination of current and previous measurements is given by*

$$\frac{\alpha}{\beta} = a + \frac{q_w}{ar_v} + \frac{q_w}{a(a^{2D} p_k^{1,1} + \sum_{i=1}^{D} a^{2(i-1)} q_w)}.$$

Proof. Follows directly from Theorem 2.

9.6 Simulations

We demonstrate our results, with a simulation study of the example cited throughout the chapter, i.e., $a = 2$, $r_v = 1$, $q_w = 1$ and $\alpha = 1$. We use average estimation error covariance per step as a performance metric, i.e., $\frac{1}{N} \sum_{k=0}^{N} p_k^{1,1}$. One million steps were used. The results are shown in Figure 9.5 where λ represent the packet drop probability. We assume that packet drops are i.i.d. Bernoulli random variables. In Figure 9.5 we have used a baseline for

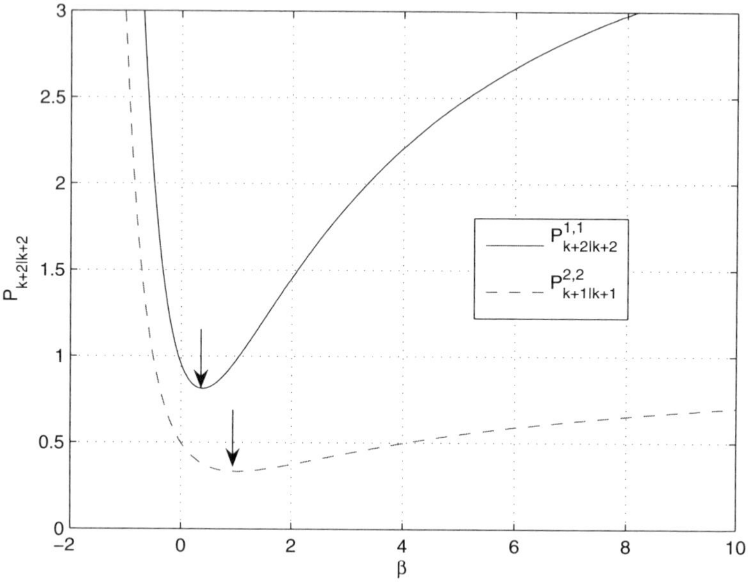

Fig. 9.4. Figure showing the minimum points of the error covariance $p^{1,1}_{k+2|k+2}$ and $p^{2,2}_{k+2|k+2}$ as a function of β, with $\alpha = 1$, $a = 2$, $p^{1,1}_{k|k} = 1$ and $q_w = 1$. The arrows indicate the minimum point. As expected, more weighting is given to the 'missed' observation (larger β) at time $k + 1$ when the state covariance at that time is to be minimized.

comparison from [36], represented with a value of 0 cost for all drop probabilities. The lower bound derived in [36] represents the absolute best performance that is achievable under *any* information sharing structure or drop sequence. Consequently, the costs illustrated in Figure 9.5 represent a deviation from achievable optimality. The linear temporal coding scheme clearly performs very well.

9.7 Conclusions

In this chapter we have examined the problem of retransmissions in a networked control or sensor network environment where sensor observations are occasionally lost enroute to the controller or state estimator. In particular, we address what single value should be sent following a packet drop in order to best estimate the state. We show that sending the most recent observation is not optimal, and derive necessary and sufficient conditions for the existence of a combination of past and present measurements that minimizes the state error covariance. We extend the results in several directions including

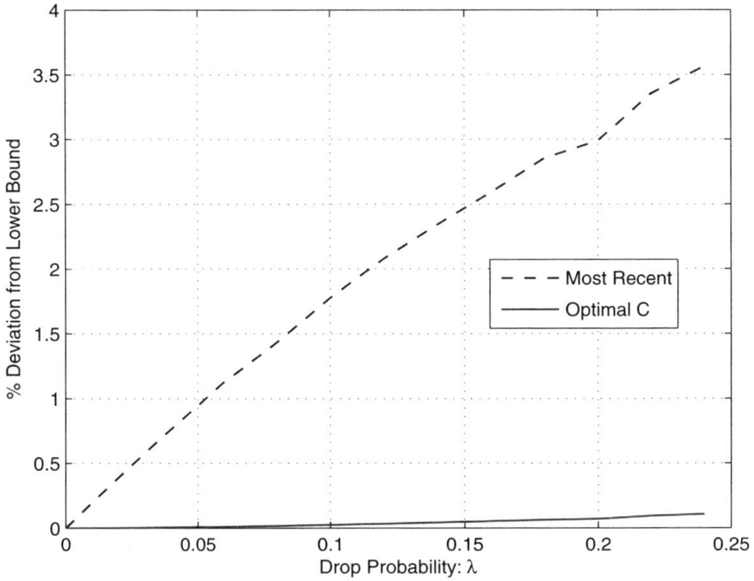

Fig. 9.5. Simulation results showing the percentage deviation in cost from the absolute lower bound achievable as described in [36]. λ is the packet loss probability.

multiple packet drops, rapid sampling and slow transmission, and estimate smoothing. Simulation results are presented which illustrate the performance improvement.

These results highlight the utility of *linear temporal coding*, which suggests several potential directions of further work. First, modifiying the contents of packets before transmission, represents a viable strategy for congestion control compared to traditional TCP type methods. This might form the basis of a networked control specific transmission protocol that modifies packet contents based on previous message disposition. The significance of the approach is that it recognizes the time value of the transmitted data, as compared to TCP which values only data integrity.

Second, the results enable better performance to be achieved in environments with limited bandwidth. This could even be achieved after system deployment without requiring the network be modified. For example, transmitting sensors capable of faster sampling can still use their scheduled transmission time to transmitt a combined sensor observation. Further, if bandwidth is a limiting constraint, the sensor sampling rate could be selected so as to achieve the performance required.

More generally, these results also show that, for network coding, realized in this context as linear temporal codes, is indeed valuable for networked control systems. However, in contrast to "in-network information processing" in sensor networks [13], this would be done at the transport layer, rather than the network layer.

References

[1] *Special edition on Networks and Control*, volume 21. IEEE Control Systems Magazine, Feb 2001.

[2] *Special Issue: Technology of Networked Control Systems.* Proceedings of the IEEE, Jan 2007.

[3] M. Adès, P. E. Caines, and R. P. Malhamé. Stochastic optimal control under poisson-distributed observations. *IEEE Transactions on Automatic Control*, 45(1), January 2000.

[4] B. Azimi-Sadjadi. Stability of networked control systems in the presence of packet losses. In *Proc. of the IEEE Conf. on Decision and Control.*, 2003.

[5] G. Baliga. *A Middleware Framework for Networked Control Systems.* PhD thesis, University of Illinois at Urbana-Champaign, 2005.

[6] G. Baliga, S. Graham, and P. R. Kumar. Middleware and abstractions in the convergence of control with communication and computation. In *Proc. of the IEEE Conf. on Decision and Control.*, pages 4245–4250, 2005.

[7] G. Baliga, S. Graham, Lui Sha, and P. R. Kumar. Etherware: Domainware for wireless control networks. In *Proceedings of The 7th IEEE International Symposium on Object-oriented Real-time Distributed Computing*, May 2004.

[8] O. Beldiman, G. C. Walsh, and L. Bushnell. Predictors for networked control systems. In *Proc. of the American Controls Conf.*, 2000.

[9] BOSCH. *CAN Specification*, 1991. http://www.semiconductors.bosch.de/pdf/can2spec.pdf.

[10] A. Cervin, D. Hendriksson, B. Lincoln, J. Eker, and K. Arzen. How does control timing affect performance? *IEEE Control Systems Magazine*, pages 16–30, June 2003.

[11] H. Chan and U. Ozguner. Closed-loop control of systems over a communication network with updates. In *Proc. of the American Controls Conf.*, 1994.

[12] M. Chow and Y. Tipsuwan. Network-based control systems: A tutorial. In *The 27th Annual Conference of the IEEE Industrial Electronics Society.*, pages 1593–1602, 2001.

[13] A. Giridhar and P. R. Kumar. Towards a theory of in-network computation in wireless sensor networks. *IEEE Communications Magazine*, 44(4):98–107, April 2006.

[14] G. C. Goodwin, H. Haimovich, D. E. Quevedo, and J. S. Welsh. A moving horizon appproach to networked control system design. *IEEE Transactions on Automatic Control*, 2004.
[15] S. Graham, G. Baliga, and P. R. Kumar. Issues in the convergence of control with communication and computing: proliferation, architecture, design, services, and middleware. In *Proc. of the IEEE Conf. on Decision and Control.*, pages 1466–1471, 2004.
[16] V. Gupta, D. Spanos, B. Hassibi, and R. R. Murray. On lqg control across a stochastic packet-dropping link. In *Proc. of the American Controls Conf.*, June 2005.
[17] M. Handley, S. Floyd, J. Pahdye, and J. Widmer. TCP friendly rate control (TFRC): Protocol specification. Technical report, ftp://ftp.isi.edu/in-notes/rfc3448.txt, 2003.
[18] D. Hristu-Versakelis. Feeback control systems as users of a shared network: Communication sequences that guarantee stability. In *Proc. of the IEEE Conf. on Decision and Control.*, 2001.
[19] O. C. Imer, S. Yuksel, and T. Basar. Optimal control of dynamical systems over unreliable communication links. In *NOLCOS*, 2004.
[20] A. Khan, N. Agarwal, D. Tilbury, and J. Moyne. The impact of random device processing delays on networked control system performance. In 42nd *Annual Allerton Conference on Communication, Control, and Computing.*, 2004.
[21] E. Kohler, M. Handley, and S. Floyd. Designing DCCP: Congestion control without reliability. In *Proceedings of ACM SIGCOMM*, 2006.
[22] P. R. Kumar. New technological vistas for systems and control: the example of wireless networks. *IEEE Control Systems Magazine*, 21(1), Feb 2001.
[23] P. R. Kumar and P. Varaiya. *Stochastic Systems: Estimation, Identification and Adaptive Control*, volume 1. Englewood Cliffs, 1986.
[24] G. Leen, D. Heffernan, and A. Dunne. Digital networks in the automotive vehicle. *Computing & Control Engineering Journal*, 10(6), December 1999.
[25] F Lian, J Moyne, and D Tilbury. Performance evaluation of control networks: Ethernet, ControlNet, and DeviceNet. *IEEE Control Systems Magazine*, 21(1):66–83, 2001.
[26] F Lian, J Moyne, and D Tilbury. Optimal controller design and evaluation for a class of networked control systems with distributed delays. In *Proc. of the American Controls Conf.*, 2002.
[27] X. Liu and A. Goldsmith. Kalman filtering with partial observation losses. In *Proc. of the IEEE Conf. on Decision and Control.*, pages 4180–4186, Atlantis, Bahamas, December 2004.
[28] L. A. Montestruque and P. J. Antsaklis. Model-based networked control systems-necessary and sufficient conditions for stability. In *Proceedings of the* 10th *Mediteranean Conference on Control and Automation*, Lisbon, Portugal, July 2002.

[29] G. N. Nair and R. J. Evans. Stabilization with data-rate-limited feedback: tightest attainable bounds. *Systems and Control Letters*, 41, 2000.
[30] G. N. Nair and R. J. Evans. Exponential stabilisability of finite dimensional linear systems with limited data rates. *Automatica*, 39(4), April 2003.
[31] J. Nilsson, B. Bernhardsson, and B. Wittenmark. Stochastic analysis and control of real-time systems with random time delays. *Automatica*, 34(1):57–64, 1998.
[32] C. H. Papadimitriou and J. Tsitsiklis. Intractable problems in control theory. *SIAM Journal on Control and Optimization*, 24(4):639–654, July 1986.
[33] J. Quillinan. The connected building [internet connected building control networks]. *IEE Review*, 51(4):44–47, April 2005.
[34] R. S. Raji. Smart networks for control. *IEEE Spectrum*, 31(6):49–55, 1994.
[35] I. B. Rhodes. A tutorial introduction to estimation and filtering. *IEEE Trans. on Automatic Control*, AC-16(6), December 1971.
[36] C. L. Robinson and P. R. Kumar. Control over networks of unreliable links: Location of controllers, control laws and bounds on performance. In *Proceedings of Control over Communication Channels (ConCom) - To Appear*, Cyprus, April 2007.
[37] C. L. Robinson and P. R. Kumar. Sending the most recent observation is not optimal in networked control: Linear temporal coding and towards the design of a control specific transport layer. In *Proc. of the IEEE Conf. on Decision and Control.*, page Submitted, 2007.
[38] M. Santori. A tale of three buses: Devicenet, profibus-dp, foundation fieldbus. *EDN*, 42(22), 1997.
[39] P. Seiler and R. Sengupta. Analysis of communication losses in vehicle control problems. In *Proc. of the American Controls Conf.*, June 2001.
[40] B. Sinopoli, L. Schenato, M. Franceschetti, K. Poolla, M. I. Jordan, and S. S. Sastry. Kalman filtering with intermittent observations. In *IEEE Transactions on Automatic Control*, volume 49, 2004.
[41] R. Stewart and Q. Xie. *Stream Control Transmission Protocol (SCTP): A Reference.* Addison-Wesley, 1 edition, 2001.
[42] A. Tatikonda, S. Sahai and S. Mitter. Control of lqg systems under communication constraints. In *Proc. of the IEEE Conf. on Decision and Control.*, pages 1165–1170, Tampa, Fl., December 1998.
[43] Y. Tipsuwan and M. Chow. Gain scheduler middleware: A methodology to enable existing controllers for networked control and teleoperation part i: Networked control. *IEEE Transactions on Industrial Electronics*, 51(6), 2004.
[44] Y. Tipsuwan and M Chow. Control methodologies in networked control systems. *Control Engineering Practice*, 11:1099–1111, 2003.
[45] G. C. Walsh and H. Ye. Scheduling of networked control systems. *IEEE Control Systems Magazine*, 2001.

[46] G. C. Walsh, H. Ye, and L. G. Bushnell. Stability analysis of networked control systems. *IEEE Transactions on control systems technology*, 10(3):438–446, May 2002.
[47] G. Welch and G. Bishop. An introduction to the kalman filter. http://www.cs.unc.edu/~welch/media/pdf/kalman_intro.pdf, April 2004.
[48] H. S. Witsenhausen. A counter example in stochastic optimal control. *Siam J. Control*, 6:131–147, 1968.
[49] L. Xiao, A. Hassibi, and J. P. How. Control with random communication delays via a discrete-time jump system approach. In *Proc. of the American Controls Conf.*, 2000.
[50] Y. Xu and J. P. Hespanha. Optimal communication logics in networked control systems. In *Proc. of the IEEE Conf. on Decision and Control.*, December 2004.
[51] Y. Xu and J. P. Hespanha. Estimation under uncontrolled and controlled communications in networked control systems. In *Proc. of the IEEE Conf. on Decision and Control.*, pages 842–847, 2005.
[52] J. K. Yook, D. M. Tilbury, and N. R. Soparkar. Trading computation for bandwidth: Reducing communication in distributed control systems using state estimators. *IEEE Transactions on Control Systems Technology*, 10(4):503–518, July 2002.

10

Reliable Distributed Estimation with Intermittent Communications *

Venkatesh Saligrama and David A. Castañón[1]

Department of Electrical and Computer Engineering, Boston University, Boston, MA 02215 {srv,dac}@bu.edu

10.1 Introduction

Consider the problem of tracking multiple objects using passive sensors mounted on the sensor heads of mobile robots as illustrated in Figure 10.1. We are currently investigating multi-target tracking techniques on such platforms, specifically, the so called REDOWL platform testbed (Robot Enhanced Detection Outpost with Lasers [15]).

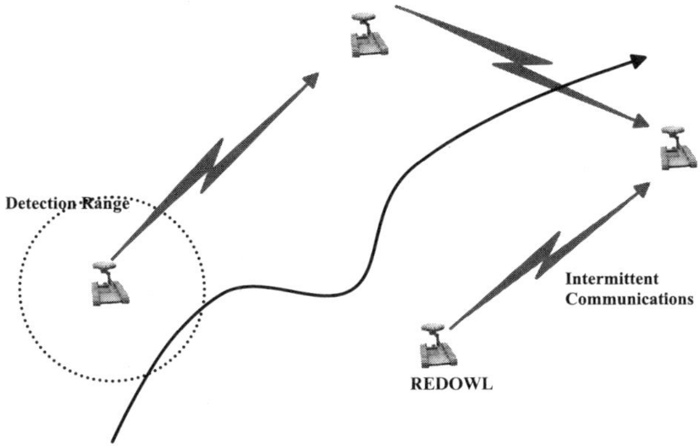

Fig. 10.1. Target Tracking with REDOWLs

Each REDOWL platform has acoustic arrays that can detect and analyze individual acoustic emissions to identify the directions of sounds of interest over time. By combining information from multiple platforms, the objective

* This work was supported by PECASE grant no. N00014-02-100362, NSF CAREER award ECS 0449194, Army grant W911NF-06-2-0040 and NSF Grants CCF 0430983, CNS 0435353, DMI 0330171 and AFOSR FA9550-04-1-0133

is to rapidly localize and track objects. The messages transmitted would be broadcast using an 802.11b protocol to other REDOWLs in the vicinity. The 802.11b protocol limits the potential range of these transmissions, and introduces an element of unreliability of communications and delay that needs to be addressed. Intermittent communications may also be employed by design. Indeed, since communication-energy/bit far outstrips computational as well as sensing energy expenditures, we can expect that in such networked systems successful communications will occur less frequently than measurements.

In this context a mathematical question that arises is that of distributed tracking with a collection of such sensing platforms connected to a fusion center by means of communication links that are subject to packet losses. Our problem is to design efficient encoding and fusion rules to optimally track moving objects under the constraints imposed by intermittent communications. Our emphasis in this chapter is on designing lossless protocols, i.e., protocols which recover centralized performance.

The general problem has received significant attention (see [2, 3, 5, 7, 1, 9, 10, 8, 6, 11] and references therein). Of particular relevance to our scenario is the recent work on Kalman Filtering with intermittent observations [9]. In their setup the sensors directly send their local observations to the fusion center but the arrival of these messages are subject to random packet losses. The fusion center fuses these intermittent observations through Kalman Filtering techniques. They study the statistical convergence properties of the estimation error covariance, showing the existence of a critical value for the arrival rate of the observations, beyond which a transition to an unbounded error occurs. Moreover, beyond this critical regime the likelihood that the error remains bounded goes to zero as well.

This strategy in the context of our problem would imply that each REDOWL would intermittently broadcast raw measurements, as and when the communication link can be established. Nevertheless, in most sensor network applications a second option is usually available. A REDOWL could encode its past observations and transmit the encoded observation. The dramatic improvements in computational speed makes complex encoding schemes routine and can be practically implemented on a REDOWL system. Encoding of measurements holds several potential advantages. It can help mitigate the effect of packet losses. For instance, one possible encoding of measurements is the local state estimate. In a sporadic communication environment a raw angle measurement does not provide sufficient information to narrow down the track (for example at least two measurements are required to narrow down position and velocity) while a state estimate provides accurate track state. A subtle point here is that encoding observations has the same effect as having access to all the raw measurements (not just the transmitted measurement).

Motivated by these advantages we will investigate efficient encoding and decoding rules in this chapter. Suboptimal schemes have been proposed in the target tracking literature for fusion of tracks from multiple sensors; Chong *et al* [11] provide a thorough review of these approaches. We show through examples in Section 10.3.2 that these schemes can lead to poor performance. We then develop approaches that achieve optimal centralized performance and quantify communication cost for different network topologies in Section 10.3.2. Our problem setup elaborated in Section 10.2 is schematically illustrated in Figure 10.2. The main difficulty arises from optimally fusing the intermittent local statistics received at the fusion center. These difficulties arise due to the lack of conditional independence and packet losses. In

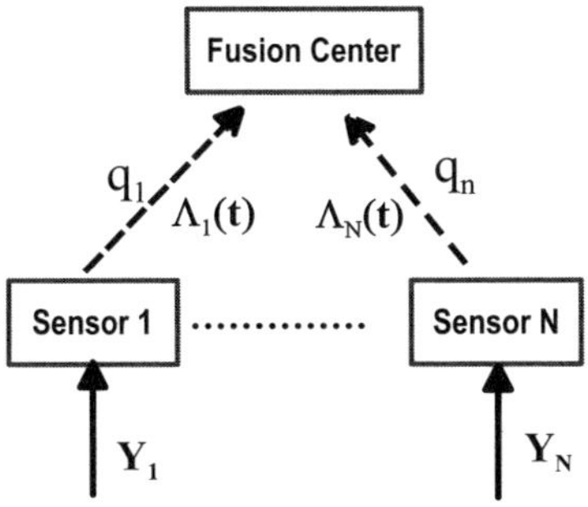

Fig. 10.2. Schematic illustration of the setup a sensor network. Any sensor or all sensors can serve as fusion centers. Sensors are connected through lossy communications links.

Gaussian estimation problems, if the local estimates are conditionally independent when conditioned on the underlying state, the optimal fusion rule is a linearly weighted average of the optimal local estimates. However, when the underlying state (or parameter) has random evolution the local estimates are no longer conditionally independent. In the synchronous setting, when all the sensor transmissions arrive simultaneously at each time, this difficulty can be overcome ([2, 3], others) by either fixing a decoder (fusion) structure and optimizing for the encoders (i.e., what each sensor must transmit) or vice versa. However, for asynchronous settings, where the transmissions are intermittent and do not arrive simultaneously these techniques do not apply. In such scenarios, the fusion center does not have local estimates from each sensor corresponding to a common time, so the local encoding and fusion rules have to be modified. To address this issue we propose joint optimization of both the encoder and decoder. We discuss different network architectures and information structures for which centralized performance can be recovered. We show in Section 10.4 that if packet arrival times are globally revealed to each sensor then optimal centralized performance can be achieved but communication message complexity is not scalable. On the other hand if optimal estimates are desired only at appropriately chosen time instants, we show that communication message complexity is scalable. We then discuss asymptotic stability of the track estimates in Section 10.5 and provide necessary and sufficient conditions on communication link availability to guarantee asymptotic stability.

10.2 Problem Statement

Consider the discrete-time system

$$X_{t+1} = AX_t + W_t, \quad X_0 \sim N(0, \Sigma_0), \tag{10.1}$$

where A is an $n \times n$ matrix, assumed invertible to simplify the exposition, and $W = (W_t : t = 0, 1, 2, \cdots)$ is an IID sequence such that $W_t \sim N(0, \Sigma_W)$, independently of X_0.

We wish to estimate the sequence $(X_t : t = 0, 1, 2, \cdots)$ causally based on measurements taken by a collection V of sensors. The measurement of sensor $v \in V$ taken at time t is denoted by $Y_t(v) \in \mathbf{R}^m$ and it satisfies

$$Y_t(v) = C_t(v)X_t + U_t(v), \quad v \in V,$$

where $C_t(v)$ is an $m \times n$ matrix, and $(U_t(v) : t = 0, 1, 2, \cdots)$ is an IID sequence such that $U_t(v) \sim N(0, \Sigma_U)$. We assume that the sequences $U(v)$ are mutually independent for different sensors $v \in V$. When all measurements are immediately available to a central processor then the MMSE estimator is a Kalman filter. Specifically, the MMSE estimate $X_{t|t} = E[X_t | Y_\tau(v) : v \in V, \tau \leq t]$ of X_t is

$$X_{t|t} = X_{t|t-1} + P_{t|t} \sum_{v \in V} C_t^T(v) \Sigma_U^{-1} (Y_t(v) - C_t(v) X_{t|t-1}) \tag{10.2}$$

where $P_{t|t} = E(X_t - X_{t|t})(X_t - X_{t|t})^T$ is the conditional error covariance matrix at time t and is given by the recursion:

$$P_{t|t}^{-1} = P_{t|t-1}^{-1} + \sum_{v \in V} C_t^T(v) \Sigma_U^{-1} C_t(v)$$

The Kalman filter prediction is:

$$X_{t+1|t} = AX_{t|t}, \;\; P_{t+1|t} = AP_{t|t}A^T + \Sigma_W \tag{10.3}$$

with the initial conditions $P_{0|-1} = \Sigma_0, X_{0|-1} = 0$.

Communication Model

To complete the problem setup we need to describe the communication connectivity model. We explore two models where the connections at each instant of time form an IID process. In the Poisson clock model each node has an independent Poisson clock with rate $\gamma(v)$. At each arrival of its local clock, node i broadcasts its messages to its neighbors. Two simultaneous transmissions lead to collisions; such packet losses can be incorporated in this model by modifying the clock rate. The other model is the Bernoulli model; connections between each sensor, v, and the fusion center are independent Bernoulli processes each with parameter p.

For convenience we define the indicator function $I_v(t)$ to denote whether or not the sensor, v, communicates at time t and $N_v(t) = \max\{k \leq t \mid I_v(k) = 1\}$ to denote the last communication time of sensor v up to time t. Let $\mathcal{Y}_v^k$ denote sensor v's measurement up to time k and $\mathbf{N}(t) = (N_1(t), N_2(t), ...N_{|V|}(t))$ the vector of arrival times. We deal with two versions of optimality.

Definition 1. *A decentralized fusion strategy (i.e. encoding/decoding strategy) is said to be* **anytime optimal** *if the fusion center at any time, t, can realize the following estimate:*

$$X_{t|\mathbf{N}(t)} = E(X(t) \mid \mathcal{Y}_v^{N_v(t)}, \forall v \in V)$$

By convention if no arrival occurs before t the conditioning event $\mathcal{Y}_v^{N_v(t)}$ is omitted.

Definition 2. *Let T be a stopping time with respect to the random process, $\mathbf{N}(t)$. A decentralized fusion strategy is* **weakly optimal** *with respect to T if the fusion center can realize the following estimate at times $t > T$:*

$$X_{t|\mathbf{N}(T)} = E(X(t) \mid \mathcal{Y}_v^{N_v(T)}, \forall v \in V)$$

The above objectives formalize the notion of optimal performance in that it characterizes recovery of centralized performance through decentralized processing.

10.3 Encoding and Decoding Algorithms

In this section we outline two basic fusion algorithms for distributed estimation that achieve anytime optimality. The first algorithm is designed for dealing with the no process noise case (conditionally deterministic systems), which is often encountered in parameter estimation problems. Unfortunately, if process noise is not negligible these algorithms can lead to poor results. We develop joint encoding and decoding rules to deal with situations when process noise is not insignificant. Process noise arises naturally as a consequence in modeling the uncertain dynamics of target tracks.

10.3.1 Conditionally Deterministic Systems

The simplest case to consider is where the process noise $W(t) = 0$ in Equation 10.1, so the resulting problem becomes a conditionally deterministic system. Parameter estimation problems fall into the class of systems with A

matrix equal to identity. In conditionally deterministic systems, the time histories of past observations, i.e., $\mathcal{Y}^{N_v(t)}$, $v \in V$ are conditionally independent given the initial state X_0.

In this case, it is well-known that there are protocols that communicates the local estimates $X^v_{t|t}$ (as in [2, 3, 8]) to the fusion center. Note that each of these estimates has in common the prior information $X_0 \sim N(0, \Sigma_0)$. The local estimates are computed as

$$\begin{aligned}(P^v_{t|t})^{-1}X^v(t|t) &= (P^v_{t|t-1})^{-1}X^v(t|t-1) + C_t^T(v)\Sigma_U^{-1}Y_t(v)\\ (P^v_{t|t})^{-1} &= (P^v_{t|t-1})^{-1} + C_t^T(v)\Sigma_U^{-1}C_t(v)\end{aligned} \tag{10.4}$$

and the local predict rule as

$$X^v(t+1|t) = AX^v(t|t);\ P^v_{t+1|t} = AP^v_{t|t}A^T + \Sigma_W \tag{10.5}$$

In addition, define the open-loop predicted covariance (corresponding to sensor $v = 0$) as

$$P^0_{t|t} = AP^0_{t-1|t-1}A^T;\quad P^0_{0|0} = \Sigma_0$$

Assume now that the fusion center has received the most recent local estimates from the different sensors at the times in $\mathbf{N}(t)$. Let $X_{t|\mathbf{N}(t)}$ and $P_{t|\mathbf{N}(t)}$ denote the centralized estimate and error covariance of $X(t)$ given the information $\{Y_s(v), s = 0, \ldots, N_v(t), v \in V\}$. Define $X^v(t|N_v(t)), P^v_{t|N_v(t)}$ as the local predicted estimate and error covariance of $X(t)$ given local observations up to time $N_v(t)$. Note that these local estimates can be computed at the fusion center from the communications $X^v(N_v(t)|N_v(t))$ and knowledge of the local covariance $P^v_{N_v(t)|N_v(t)}$ by applying eq. 10.5 repeatedly. Our first result specifies the optimal algorithm at the fusion center:

Theorem 1. *Assume that the fusion center has received communications $X^v(N_v(t)|N_v(t))$ from sensors $v \in V$ at times $N_v(t) \leq t$. Then the following decentralized fusion rule achieves anytime optimality, i.e.,*

$$X(t|\mathbf{N}(t)) = P_{t|\mathbf{N}(t)} \sum_{v \in V} (P^v_{t|N_v(t)})^{-1} X^v(t|N_v(t)) \tag{10.6}$$

$$P^{-1}_{t|\mathbf{N}(t)} = \sum_{v \in V} (P^v_{t|N_v(t)})^{-1} - (|V| - 1)(P^0_{t|t})^{-1} \tag{10.7}$$

The proof follows from the conditional independence of $Y_s(v), s \leq t$ given $X(t)$ and the prior mean of X_0 is 0, and is a straightforward extension of the results in [2, 3, 8].

To highlight the advantages of the above protocol, consider two scenarios: (a) a single sensor with guaranteed connection to a fusion center, (b) $|V|$ sensors with unreliable links with Bernoulli parameter, $p = 1/|V|$. For Case (b) the average number of samples in time T is $p|V|T = T$. For the latter scenario consider two possibilities: transmitting raw observations vs. transmitting local MMSE estimates.

Theorem 2. *Denote the average MMSE errors (where the average is taken w.r.t. number of samples) in these three cases are denoted as* $MMSE_{single}$, $MMSE_{raw}$, $MMSE_{local}$. *Then:*

$$E(P_{t|\mathbf{N}(t)}) = MMSE_{local} = \frac{1}{|V|} MMSE_{raw}$$
$$\xrightarrow{t\to\infty} \frac{1}{|V|} MMSE_{single}$$

This establishes that the fusion protocols are efficient, in that they recover the performance of having all the information centrally located.

10.3.2 Encoding & Decoding Rules with Process Noise

We first present an example to illustrate that, when there is process noise in $X(t)$, transmitting only the local estimates from each sensor, as in the case of conditionally deterministic systems can lead to significant performance loss compared to when measurements are centrally available. Several suboptimal methods discussed in [10, 11], are based on some form of local state estimate transmission. The example will justify our emphasis on developing lossless encoding and decoding fusion rules.

Example 1. Consider a set of N sensors with identical measurement model and observing a random walk:

$$\begin{aligned} X(t+1) &= X(t) + W(t) \\ Y_v(2t) &= X(2t) + V_v(2t) \\ Y_v(2t+1) &= X(2t+1), \ t = 1, 2, \ldots \end{aligned}$$

where $W(t)$ and $V_v(t)$ are i.i.d. Gaussian noise sequences with mean zero and variance σ^2 and $x(0)$ is Gaussian with mean zero and variance σ^2. At even time instants, the local KF estimate at sensor v is:

$$X^l(2t|2t) = \frac{1}{2}(Y_v(2t) + Y_v(2t-1))$$

and the centralized KF estimate is:

$$X(2t|2t) = \frac{X(2t-1)}{N+1} + \frac{1}{N+1}\sum_v Y_v(2t)$$

Since local KF estimates are Gaussian, the optimal fused estimate is a linear superposition of the local estimates. By symmetry the fusion weights are identical. Therefore,

$$X_{fuse}(2t|2t) = \frac{1}{N}\sum_v X^l(2t|2t) = \frac{X(2t-1)}{2} + \frac{1}{2N}\sum_v Y_v(2t)$$

The error in the fused estimate is

$$X(2t) - X_{fuse}(2t|2t) = \frac{1}{2}W(2t-1) - \frac{1}{2N}\sum_v V_v(2t)$$

whereas the error in the centralized estimate is

$$X(2t) - X(2t|2t) = \frac{1}{N+1}W(2t-1) - \frac{1}{N+1}\sum_v V_v(2t)$$

which shows that

$$MMSE_{centralized} \leq \frac{2}{N+1}MMSE_{local}$$

We consider the following simple architecture. There is a single remote sensor, l, transmitting messages to a fusion center, which has side information from a sensor, f, located in situ as shown in Figure 10.3.

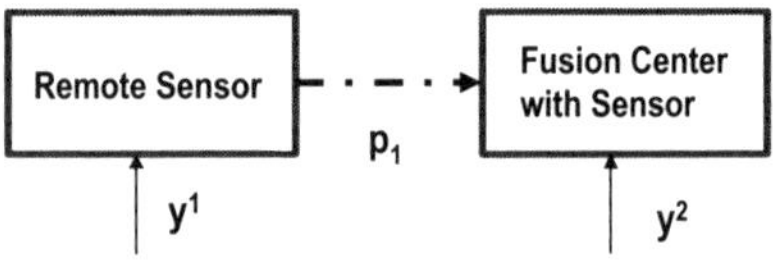

Fig. 10.3. One way transmission from remote sensor. y^1, y^2 are sensor measurements respectively and p_1 is communication-link rate

One can extend results from [2, 3] to construct a compensated estimate at the remote sensor to achieve anytime optimality as we do in [13]. However, a better approach is to use ideas from [4, 8] and we follow our approach in [14] here. The global KF estimate can be written in the information form:

$$\begin{aligned} P_{t|t}^{-1}X_{t|t} &= P_{t|t-1}^{-1}X_{t|t-1} + \sum_{v\in\{l,f\}} C_t^{\top}(v)(R^v)^{-1}Y_t^v \\ &= P_{t|t-1}^{-1}AP_{t-1|t-1}P_{t-1|t-1}^{-1}X_{t-1|t-1} + \sum_{v\in\{l,f\}} C_t^{\top}(v)(R^v)^{-1}Y_t^v \end{aligned}$$

Let $S(t) = P_{t|t}^{-1}X_{t|t}$ and $\tilde{A}(t) = P_{t|t-1}^{-1}AP_{t-1|t-1}$ to get:

$$S(t) = \tilde{A}(t)S_{t-1} + \sum_{v\in\{l,f\}} C_t^{\top}(v)(R^v)^{-1}Y_t^v \tag{10.8}$$

The main observation is that the state evolution of the information, $S(t)$, is linear. Therefore, the output can be realized through superposition of each local processed sensor output, i.e., $S^v(t) = \tilde{A}(t)S^v(t-1) + C_t^\top(v)(R^v)^{-1}Y_t^v$ for sensors $v \in \{l, f\}$, where the matrix $\tilde{A}(t)$ uses the global statistics in the problem.

When packet losses occur, the processing is as follows: When no messages are received from the local sensor, the fusion center can use its measurements to update the state estimate $X(t|\mathbf{N}(t))$, and also computes the fusion statistic S_t^f. Whenever the local sensor succeeds, it transmits S_t^l to the fusion center. The fusion center then simply computes, $P_{t|t}(S_t^f + S_t^l)$ to obtain its current state estimate. Observe that computing S_t^l, S_t^f incorporates global covariance estimates in $\tilde{A}(t)$. This protocol exploits the fact that the local sensor is aware of the fusion center having uninterrupted observations.

The packet loss scenario highlights that encoding at the local sensor has to be combined with a dynamic fusion rule in contrast to the no packet loss case. Specifically, for time instants with $N_l(t) < t$ the fusion center maintains one set of recursive measurement update and predictor equations and another set of time instants $N_l(t) = t$.

Due to linearity, the local sensor can reset its state $S^l(t)$ to zero after each successful transmission. The fusion center in this case will set the initial condition of its compensator to the the global state. Consequently, each transmission only encodes measurements between two successful transmissions. This type of encoding is related to the idea of tracklets of [12, 11]. Let $\sigma_1, \ldots, \sigma_n$ be a sequence of time instants of successful transmissions. Then, for $t \in (\sigma_j, \sigma_{j+1})$:

$$
\begin{aligned}
S^l(t) &= \tilde{A}(t)S^l(t-1) + C_t^\top(l)(R^l)^{-1}Y_t^l, \; S^l(\sigma_j) = 0 \\
S^f(t) &= \tilde{A}(t)S^f(t-1) + C_t^\top(f)(R^f)^{-1}Y_t^f \\
S^f(\sigma_j) &= P_{\sigma_j|\sigma_j}^{-1} X_{\sigma_j|\sigma_j}
\end{aligned}
\tag{10.9}
$$

The above discussion is summarized below:

Theorem 3. *The decentralized fusion strategy outlined in Equation 10.9 for a single remote sensor achieves anytime optimality.*

A Note on Channel Noise: We have primarily focused our attention on intermittent communication links. However, the Shannon capacity of such a channel in the absence of noise is infinite. Therefore, any meaningful technique with wireless communication channels must account for an additive noise distortion at the fusion center. In other words, the channel operates as follows: whenever communication is successful a noisy encoded signal is received at the fusion center. Note that in the absence of noise one could presumably transmit all the raw data instantaneously at the time the communication link is active. Noise precludes this possibility and to be realistic one must look for techniques that gracefully degrade in performance with increasing channel noise. The encoding and decoding rules presented above have this characteristic. Specifically, if a noisy version of the compensated estimate $S^l(t)$ in Equation 10.9 is made available at the fusion center, the error covariance degrades gracefully as a function of channel noise variance.

10.4 Protocols for Networks

In this section we present protocols for optimally tracking moving objects in a networked setting with intermittent communications. We consider two types of networks: completely connected and star networks. Our encoding and decoding rules are based on the algorithm developed in Section 10.3.2. Our emphasis is on designing lossless protocols, i.e., protocols which recover centralized performance either in the sense of anytime optimality or stopping time optimality.

10.4.1 Anytime Optimality

Two Sensor Two Way Network

This network consists of two sensors with two way communication. In this case the decentralized algorithm of Section 10.3.2 applies directly. Concretely, each sensor serves as fusion center and updates according to the fusion rule outlined for the single sensor case exactly as in Equation 10.9. Both sensors achieve anytime optimality; unfortunately, this property is hard to extend to completely connected multi sensor networks. The following analysis for Star Networks will clarify the underlying issues in achieving anytime optimality.

Star Networks

Assume that the central node of the star network serves as the fusion center and that successful communication times instants for every sensor are globally revealed by the fusion center.

We illustrate the protocol and the accompanying difficulties here by means of a timing diagram for two sensors as in Figure 10.4. Assume that the system starts at time zero. Let σ_i, τ_i with $\sigma_i < \tau_i$ be time instants of successful transmission from Sensor 1 and Sensor 2 respectively and s, t be two time instants as shown in Figure 10.4. For achieving anytime optimality we need

Fig. 10.4. Two sensor timing diagram

to ensure that optimal "centralized" estimates can be formed at any time by the fusion center. At time s Sensor 1 is the only sensor that has successfully communicated. Therefore, it must transmit the statistic that allows the fusion center to predict the track based only on data observed at Sensor 1. However, since there is a possibility that Sensor 2 transmission will arrive next, at time τ_1, at the fusion center. Consequently for time instant t, Sensor 1 must also transmit a compensated statistic that allows the fusion center to optimally combine data from Sensor 1 up to time σ_1 and up to time τ_1 from Sensor 2. Generally, these two statistics require different encodings of the observation. This can be observed from Equation 10.9 where one notes that Sensor 1 must process its information using two different $\tilde{A}(t)$ matrices. For track estimation at time s the $\tilde{A}(t)$ matrix is based only on the local error covariance (because this is the global covariance matrix as well) matrix. For time t, however, the $\tilde{A}(t)$ matrix should be based on information received from both sensors.

The above discussion establishes the existence of an anytime optimal protocol, but with some limitations. Specifically, for an N sensor star network, the number of statistics computed and communicated by each sensor scales as 2^{N-1}, the number of possible combinations of successful transmission from other sensors.

Completely Connected Network:

Here we consider a completely connected network in a federated architecture with each sensor serving as a fusion center. We assume a broadcast protocol, i.e., if any sensor communicates it communicates the same information to all its neighbors. The main advantage of a completely connected network is that: (a) Estimates from each sensor are globally known; (b) the event times of successful transmissions are globally known. We can utilize the second aspect in designing a scalable protocol but with increased delay. In particular we consider a round robin protocol where the order in which sensors attempt to communicate is fixed. After a successful transmission, each sensor node waits until all sensors have transmitted in the correct order before transmitting again. For this protocol the number of statistics that each sensor must compute scales as $\mathcal{O}(N)$ to achieve anytime optimality. This is because the order of successful communications is fixed. Let $t_1 < t_2 < t_3 \ldots < t_N$ correspond to message arrivals from the N different sensors. At time t_1, sensor 1 transmits N different estimates for fusion: One for times $t \in (t_1, t_2)$, one for $t \in [t_2, t_3)$, $\ldots$, one for $t \in [t_{N-1}, t_N)$ and one for $t \geq t_N$. The fusion center propagates these estimates for fusion at the appropriate times assuming no further measurements from Sensor 1, achieving anytime optimality.

10.4.2 Weak Optimality at Stopping Times

We consider weaker versions of optimality for star and completely connected networks as a means to obtain scalable protocols. We are specifically interested in optimality at stopping times. Recall from Section 2 that a protocol is weakly optimal for a stopping time T if the centralized performance is achievable at T.

Star Networks

As before assume that the successful communication instants are revealed to sensor nodes by the fusion center. Let $L(t),\ t > 0$ be a renewal process with renewal periods corresponding to an interval where at least one message has been received from each sensor and the last transmit time corresponds to the only message for some sensor in that interval.

Denote the renewal times as $0 < T_1 < T_2 < \ldots$. These renewal instants are stopping times where we will achieve weak optimality. First, we determine the mean renewal time. This problem is the so called *coupon-collector's problem*. The problem here is that there are N coupons which arrive i.i.d. and the problem is to estimate the minimum time when all coupons have arrived. It turns out [16] that

$$E(T_{j+1} - T_j) = \frac{N \log N}{N\lambda}$$

where λ is the mean arrival rate for each coupon. We have the following result for message scaling per sensor:

Theorem 4. *There is a protocol with a maximum of $N-1$ estimates per sensor message such that centralized performance can be achieved at each renewal time T_j. Furthermore, the expected number of messages transmitted by each sensor in a renewal period is N.*

The proof is as follows: Suppose Sensor 1 is the first to arrive in the interval (T_j, T_{j+1}), it encodes a single estimate by accounting for all future $N-1$ arrivals from all other sensors. The second sensor if distinct from the first transmits two estimates: one is for fusion of $N-2$ future arrivals and the second estimate accounts for the possibility that the first sensor arrives again before the next renewal time. If the first sensor arrives again after the second but before the third sensor, it becomes the second sensor and again transmits two estimates. The protocol is extended so that the $N-1$ sensor in the sequence transmits $N-1$ estimates. To find the expected number of estimates transmitted per sensor we proceed as follows: Divide the renewal period in epochs, where epoch i begins with i-th (distinct) sensor arrival and ends with $(i+1)$th distinct sensor arrival. Suppose R_i denotes number of sensor arrivals in the ith epoch. Observe that each sensor arriving in the ith epoch must transmit i estimates. This implies that the expected number of estimates, M, per sensor transmission in a renewal interval is:

$$E(M) = \frac{1}{N} E \sum_{i=1}^{N-1} i R_i \approx \Theta(N)$$

Thus weak optimality provides exponential reduction in number of estimates transmitted per message over anytime optimality.

Completely Connected Network

As before we consider the case where sensors control transmission of messages, i.e., the messages are transmitted in order in a round robin protocol. In this case one obtains $\mathcal{O}(1)$ scaling to achieve weak optimality.

Theorem 5. *For the round robin protocol weak optimality adapted to the time instant of arrival of last sensor can be achieved with a single estimate in each sensor transmission.*

10.5 Asymptotic Results

In the previous sections we focused on encoding and decoding rules along with various protocols for achieving anytime and weak optimality. In this section we present asymptotic results for error covariances.

First, we consider weak optimality for Bernoulli channels. The result is similar in spirit to the asymptotic results established in [9] for time-invariant, extended to the multi-sensor case:

Theorem 6. *Consider an ad-hoc protocol with the Bernoulli channel model. Suppose the stopping time are time instants when all sensors transmit synchronously. Then there is a weakly optimal algorithm satisfying:*

$$Prob\{\liminf P_{t|\mathbf{N}(t)} = P_\infty\} = 1$$

where, P_∞ is the asymptotic covariance when all the sensor measurements are centrally available at all instants of time.

The proof of this result follows from several properties: For Bernoulli communication channels, there is positive probability that all sensors communicate at the same time. At those times, the error covariance is equivalent to the centralized error covariance, as this corresponds to the synchronous communications protocol. These time instants can be thought of as renewal periods, and the probability that an infinite number of such time instants exist is 1, which establishes the $\liminf$ property claimed above.

The above result implies that irrespective of the message loss probability there exists time instants when the error covariance of the decentralized scheme is arbitrarily close to that of a fully centralized filter. In contrast if local raw measurements are transmitted and are subject to random losses, the error covariance is always arbitrarily large and may not approach the centralized covariance. However, this does not guarantee that the expected value of the fusion center error covariance is bounded, since the expected length of the renewal intervals can be large. Consider first the case of a single remote sensor communicating its statistics: the following result characterizes the behavior of the expected error covariance

Theorem 7. *Assume that the $[A, C]$ is detectable and $[A, Q^{1/2}]$ is stabilizable. Then, for the ad-hoc protocol with the Bernoulli channel model and a single remote sensor,*

$$\lim_{t \to \infty} E[\|P_{t_{\mathbf{N}}(t)}\|] < \infty \tag{10.10}$$

if and only if $(1-p)^{1/2}|\lambda_{max}(A)| < 1$.

To show this, assume t is large, so that the local estimate generated by a sensor has steady state error P_∞. If t was a successful communication time,

$$P_{t|\mathbf{N}(t)} = P_\infty \tag{10.11}$$

If t was not a successful communication time, let $\tau(t)$ denote latest successful communication time prior to time t. Then,

$$P_{t|\tau(t)} = A^{t-\tau(t)} P_\infty (A^T)^{t-\tau(t)} + \sum_{k=0}^{t-\tau(t)-1} A^k Q (A^T)^k$$

Note that $Prob(t - \tau(t) = n) = p(1-p)^n$ because of the Bernoulli model. Then,

$$\begin{aligned} E[P_{t|\sigma(t)}] &= pP_\infty \\ &+ p \sum_{j=1}^{\infty} (1-p)^j [A^j P_\infty (A^T)^j + \sum_{k=0}^{j-1} A^k Q (A^T)^k] \end{aligned}$$

Thus,

$$\begin{aligned} \| E[P_{t|\mathbf{N}(t)}] \| &\leq p \| P_\infty \| \\ &+ p \| \sum_{j=1}^{\infty} [((1-p)^{1/2} A)^j P_\infty ((1-p)^{1/2} A^T)^j \\ &+ \sum_{k=0}^{j-1} ((1-p)^{1/2} A)^k Q ((1-p)^{1/2} A^T)^k \| \end{aligned}$$

Note that detectability and stabilizability guarantee that $0 < \| \Sigma_\infty \| < \infty$. The second term is the limiting solution of the linear equation

$$P(t+1) = ((1-p)^{1/2} A) P(t) ((1-p)^{1/2} A^T) + Q, P(0) = P_\infty$$

which has a bounded solution whenever $|\lambda_{max}((1-p)^{1/2} A)| < 1$ and has an unbounded solution whenever $|\lambda_{max}((1-p)^{1/2} A)| \geq 1$ because of the detectability assumption on $[A, Q^{1/2}]$, establishing the theorem.

The above results can be generalized to the N sensor case as follows.

Theorem 8. *Assume that the* $[A, C(v)]$ *is detectable for each sensor* v *and* $[A, Q^{1/2}]$ *is stabilizable. Then, for the ad-hoc protocol with* N *sensors,and the Bernoulli channel model with identical success probabilities* p

$$\lim_{t\to\infty} E[\| P_{t_{\mathbf{N}(t)}} \|] < \infty \tag{10.12}$$

if and only if $(1-p)^{N/2}|\lambda_{max}(A)| < 1$.

The proof follows the above discussion. Assuming t is large, the errors associated with the local estimates at each sensor v approach steady state error covariance $P^v_\infty > 0$. Note that, for any successful communication time t where sensor v communicates,

$$P_{t|\mathbf{N}(t)} \leq P^v_\infty$$

because the fusion center will fuse the estimate from sensor v optimally with additional information from other sensors.

If t is not a successful communication time for any sensor, let $\tau(t)$ denote latest successful communication time prior to time t, and let $j(\tau(t))$ denote the index of a sensor with successful communication at time $\tau(t)$. Then,

$$P_{t|\mathbf{N}(t)} = A^{t-\tau(t)} P_{\tau(t)|\mathbf{N}(t)} (A^T)^{t-\tau(t)} + \sum_{k=0}^{t-\tau(t)-1} A^k Q (A^T)^k$$

Note that both $\tau(t)$ and $P_{\tau(t)|\mathbf{N}(t)}$ are event-dependent. Note that $Prob(t - \tau(t) = n) = (1-(1-p)^N)(1-p)^{Nn}$ because of the Bernoulli model and the independent access across all sensors. Note also that

$$P_{t|\mathbf{N}(t)} \leq A^{t-\tau(t)} P^v_\infty (A^T)^{t-\tau(t)} + \sum_{k=0}^{t-\tau(t)-1} A^k Q (A^T)^k$$

for any v that communicated successfully at $\tau(t)$. Then, conditioned on knowing v,

$$\begin{aligned} E[P_{t|\mathbf{N}(t)}|v] &\leq (1-(1-p)^N) P^v_\infty(t) \\ &\quad + (1-(1-p)^N) \sum_{j=1}^{\infty} (1-p)^{Nj} [A^j \Sigma_{opt}(t-j)(A^T)^j + \sum_{k=0}^{j-1} A^k Q (A^T)^k] \end{aligned}$$

$$\begin{aligned} \| E[P_{t|\mathbf{N}(t)}|v] \| &\leq (1-(1-p)^N) \| P^v_\infty \| \\ &\quad + (1-(1-p)^N) \| \sum_{j=1}^{\infty} [((1-p)^{N/2} A)^j \Sigma^i_\infty ((1-p)^{N/2} A^T)^j \\ &\quad + \sum_{k=0}^{j-1} ((1-p)^{N/2} A)^k Q ((1-p)^{N/2} A^T)^k \| \end{aligned}$$

Detectability and stabilizability guarantee that $0 <\| P_\infty^v \|< \infty$. The second term is the limiting solution of the linear equation

$$P(t+1) = ((1-p)^{N/2}A)P(t)((1-p)^{N/2}A^T) + Q, P(0) = P_\infty^v$$

which has a bounded solution whenever $|\lambda_{max}((1-p)^{N/2}A)| < 1$. Since the existence of a bounded solution does not depend on the choice of v, this establishes the sufficiency of the condition.

To show necessity, let P_∞ denote the steady-state error of a centralized estimator. For t not a successful communication time, we have

$$P_{t|\mathbf{N}(t)} \geq A^{t-\tau(t)}P_\infty(A^T)^{t-\tau(t)} + \sum_{k=0}^{t-\tau(t)-1} A^kQ(A^T)^k$$

and for a successful communication time,

$$P_{t|\mathbf{N}(t)} \geq P_\infty$$

Hence,

$$\begin{aligned} \| E[P_{t|\mathbf{N}(t)}] \| &\geq (1-(1-p)^N) \| P_\infty \\ &+ (1-(1-p)^N) \sum_{j=1}^{\infty} [((1-p)^{N/2}A)^j \Sigma_\infty ((1-p)^{N/2}A^T)^j \\ &+ \sum_{k=0}^{j-1} ((1-p)^{N/2}A)^k Q((1-p)^{N/2}A^T)^k \| \end{aligned}$$

The second term sum is the limiting solution of

$$P(t+1) = ((1-p)^{N/2}A)P(t)((1-p)^{N/2}A^T) + Q, P(0) = P_\infty$$

which has an unbounded solution whenever $|\lambda_{max}(\rho^{N/2}A)| \geq 1$ because of the detectability assumption on $[A, Q^{1/2}]$. This establishes the necessity of the condition.

The above result characterizes the effect of having additional sensors in the distributed system. The mean value of the error covariance can be kept bounded provided that the probability that at least one sensor gets through is sufficiently high relative to the maximum unstable eigenvalue of the system dynamics.

10.6 Simulation Results

To illustrate the behavior of our asynchronous distributed estimation algorithms, we conducted experiments using a target tracking model with 2-dimensional motion. The target dynamics were defined as

$$\begin{pmatrix} x \\ vx \\ y \\ vy \end{pmatrix}(t+1) = \begin{pmatrix} 1 & \Delta & 0 & 0 \\ 0 & 1 & 0 & 0 \\ 0 & 0 & 1 & \Delta \\ 0 & 0 & 0 & 1 \end{pmatrix} \begin{pmatrix} x \\ vx \\ y \\ vy \end{pmatrix} + w(t) \tag{10.13}$$

where Δ was set to 1 as the sampling time, and $w(t)$ was an iid zero-mean Gaussian process with covariance

$$Q = q * \begin{pmatrix} \Delta^3/3 & \Delta^2/2 & 0 & 0 \\ \Delta^2/2 & \Delta & 0 & 0 \\ 0 & 0 & \Delta^3/3 & \Delta^2/2 \\ 0 & 0 & \Delta^2/2 & \Delta \end{pmatrix}$$

The first set of experiments included two sensors, each of which observed orthogonal subspaces. The measurement model was

$$\begin{pmatrix} y_1 \\ y_2 \end{pmatrix}(t) = \begin{pmatrix} x(t) \\ y(t) \end{pmatrix} + \begin{pmatrix} v_1(t) \\ v_2(t) \end{pmatrix}$$

where $v_1(t), v_2(t)$ are iid, zero-mean Gaussian process with variance Σ. In our experiments, we set the process noise variance at $q = 0.01$, and the measurement noise variance at $\Sigma = 900$, so that accurate estimation requires integration of measurements at multiple different times. This is typical of systems where local sensors are observing phenomena at faster rates but with less accuracy.

Figure 10.5 shows the results of an experiments with two sensors communicating to a fusion center. Sensor 1 observes only $y_1(t)$ and sensor 2 observes only $y_2(t)$. The orthogonal observability subspaces make this case simple. In this case, each sensor has a probability p of communicating successfully independently, which is varied over a range of values from 0.05 to 1. Figure 10.5 shows the trace of the expected steady state error covariance, averaged over 50 Monte Carlo simulations, for the case where sensors communicate their most recent measurements and the case where they communicate their local estimates. Note the reduction in error as the probability of successful communication decays. Since the state dynamics are not exponentially unstable, the filters reach an average steady state error for low p.

The second set of experiments used the same two sensor case, but using the Poisson communication model where one and only one sensor communicates at a time, and we use the anytime optimality protocol. In this case, Figure 10.6 shows the trace of the fusion center error covariance as a function of time. for both measurement communications and estimate communications, averaged over 30 Monte Carlo simulations. The results in Fig. 10.6 illustrate the advantages of communicating estimates versus observations: a reduction in 33% on the sum of the eigenvalues of the average error covariance. The last set of experiments consists of a set of two identical sensors, both of which measure both x and y with independent additive measurement noises. Again, we use the

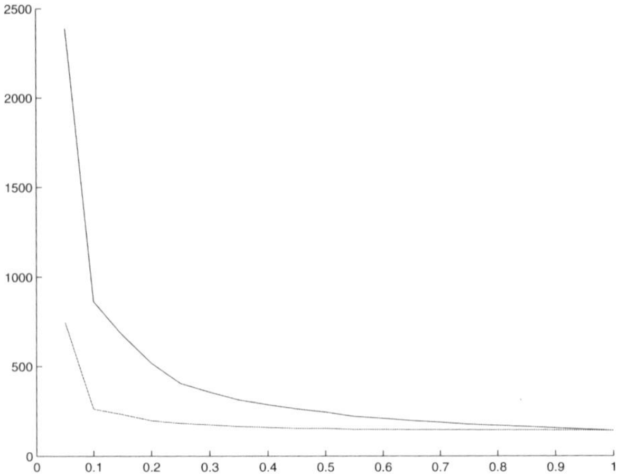

Fig. 10.5. Trace of steady state error covariance versus communication success probability. Upper curve is measurement transmission, lower curve is estimate transmission.

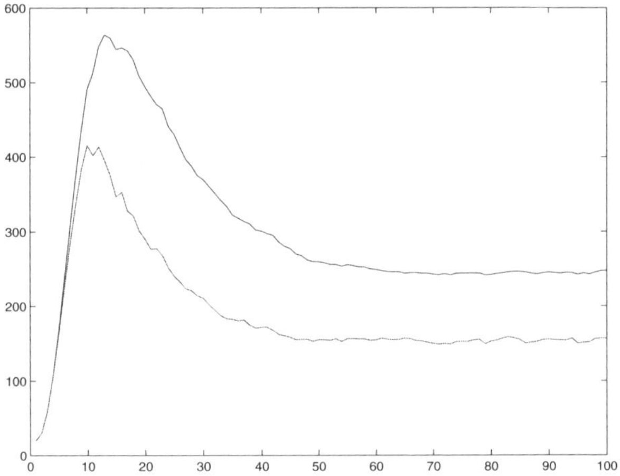

Fig. 10.6. Trace of average error covariance versus time. Upper curve is measurement transmission, lower curve is estimate transmission.

anytime optimal protocols that allow the fusion center to reconstruct optimal estimates from asynchronous communications . Figure 10.7 shows the trace of the average steady state covariance as a function of the Bernoulli probability of communication p for each sensor, for the case where individual measurements are communicated and the case where our distributed estimation protocols are used. The results illustrate the advantage of our distributed protocols in situations with unreliable communications. As noted before, the expected steady state covariance remains bounded, as the maximum eigenvalue of A is 1.

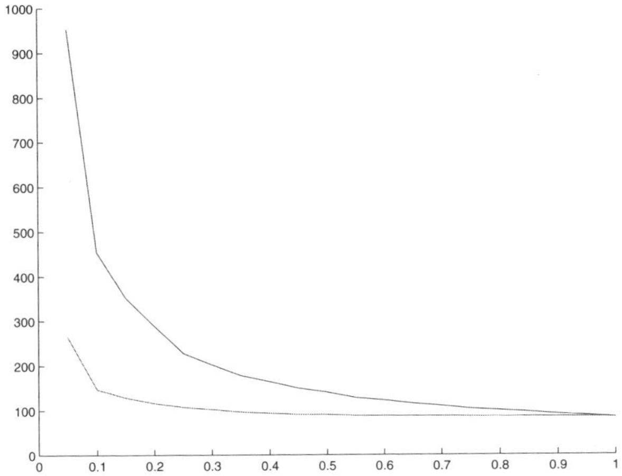

Fig. 10.7. Trace of steady state error covariance versus communication success probability for two identical sensors. Upper curve is measurement transmission, lower curve is estimate and compensator transmission.

10.7 Conclusions

In this chapter, we have developed algorithms for optimal distributed estimation for linear Gaussian models in networks of sensors subject to intermittent, random communications. We developed local processing strategies, communication protocols and fusion algorithms that guarantee that the fusion center is able to reconstruct the optimal estimate of the state, given all of the available measurements at each sensor up to their last communication times. However, unlike the case where all sensors communicate synchronously, the presence of asynchronous communication requires increased local processing and communications as the number of sensors increases. Our theoretical results and experiments show the advantages of including local estimates instead of raw measurements when unreliable communication is present.

The focus of this chapter was on the development of optimal distributed estimation algorithms in asynchronous communication environments. It is possible to generate alternative fusion algorithms that trade off computation and communication complexity versus estimation performance. Furthermore, for nonlinear estimation models, the asynchronous algorithms discussed in this chapter will require approximations, as the local compensation processes will no longer have access to the required statistics. Such problems remain as challenges for future investigations.

References

[1] M. Alanyali and V. Saligrama, *Distributed Tracking over Multi-hop Networks*, IEEE Workshop on Statistical Signal Processing, Bordeaux, France 2005.
[2] J. Speyer, "Computation and transmission requirements for a decentralized linear-quadratic-Gaussian control problem," *IEEE Transactions on Automatic Control*, vol. 24, no. 2, pp. 266–269, 1979.
[3] A. Willsky, M. G. Bello, D. A. Castañón, B. C. Levy, G. C. Verghese "Combining and Updating of Local Estimates and Regional Maps Along Sets of One Dimensional Tracks," *IEEE Transactions on Automatic Control*, vol. 24, no. 4, pp. 799–813, 1982.
[4] B. C. Levy, D. A. Castañón, G. Verghese and A. S. Willsky, "A Scattering Framework for Decentralized Estimation Problems," *Automatica,* V 19, July 1983, pp. 373-384.
[5] T. H. Chung, V. Gupta, J. W. Burdick, R. M. Murray, "On a Decentralized Active Sensing Strategy using Mobile Sensor Platforms in a Network," *Proc. Conf. Decision and Control,* Bahamas, Dec 2004.
[6] A. Makarenko and H. F. Durrant-Whyte, "Decentralized Data Fusion and Control in Active Sensor Networks," in *INt. Conf. on Info. Fusion,* Stockholm, Sweden, 2004.
[7] D. Spanos, R. Olfati-Saber, R. Murray, "Distributed Kalman Filtering in Sensor Networks with Quantifiable Performance," In *Proc. of Information Processing in Sensor Networks*, April 2005
[8] V. Gupta, T. Chung, B. Hassibi and R. M. Murray, "On a Stochastic Algorithm for Sensor Scheduling," *Proc. 16th IFAC World Congress,* Prague, July 2005.
[9] B. Sinopoli, L. Schenato, M. Franceschetti, K. Poolla, M. Jordan and S. Sastry "Kalman Filtering with Intermittent Observations," *IEEE Transactions on Automatic Control*, September 2004.
[10] Y. Bar-Shalom, "On the Track-to-Track Correlation Problem," *IEEE Transactions on Automatic Control*, AC-25, pp. 802-807, Aug., 1980.
[11] C. Y. Chong, F. Zhao, S. Mori, and S. Kumar, "Distributed tracking in wireless ad hoc sensor networks," in *Proc. 6th Int. Conf. Information Fusion,* 2003, pp. 431-438.

[12] O. E. Drummond, "A Hybrid Sensor Fusion Algorithm Architecture and Tracklets," in *Proc. SPIE, Signal and Data Processing of Small Targets,* vol. 3163, 1997.
[13] V. Saligrama and D. Castañón, "Reliable Tracking with Unreliable Communication Links" ICASSP, 2006
[14] V. Saligrama and D. Castañón, "Reliable Distributed Estimation with Intermittent Communications" Proc. of IEEE Conference on Decision and Control, Dec. 2006
[15] "Unmanned Systems Magazine," *Association for Unmanned Vehicles Systems International*, vol. 23, no. 6, Nov. 2005
[16] S. Ross, "Stochastic Processes," *John Wiley & Sons*, 2nd Edition, 1996.

11

Smart Sleeping Policies for Energy-Efficient Tracking in Sensor Networks

Jason A. Fuemmeler and Venugopal V. Veeravalli

Department of Electrical and Computer Engineering and Coordinated Science Laboratory
University of Illinois at Urbana-Champaign
{fuemmele,vvv}@uiuc.edu

11.1 Introduction

Advances in technology are enabling the deployment of vast sensor networks through the mass production of cheap wireless sensor units with small batteries. Such sensor networks can be used in a variety of application areas – indeed any application where there are signals to be detected. Our focus in this chapter is on applications of sensor networks that involve *tracking*, e.g., surveillance, wildlife studies, environmental control, and health care. More specifically, due to the use of battery power in these networks we concern ourselves with energy efficiency in tracking applications.

Many researchers have examined energy efficiency in the context of tracking using sensor networks. In this research, the problem is typically to decide which sensors should participate in the tracking procedure at each time step. The rest of the sensors in the network are then allowed to conserve energy by entering a sleep mode. Some previous research has focused on the design of protocols for sensor networks (e.g., [1, 3, 2, 5, 4]). Other research has approached the problem more analytically (e.g., [7, 8, 6, 9, 11, 10, 13, 12]), however the analysis often takes a myopic view where it is assumed that the actions at one time step do not impact future time steps.

We wish to formulate the problem of energy-efficient tracking in sensor networks in a systematic framework that takes into account the impact of current actions on future time steps. There has been some recent research that has posed the problem in this manner (e.g., [14, 16, 17, 18, 15, 20, 19]). However, whereas previous research has assumed it is possible to wake up sleeping sensors externally, our formulation does not make this assumption. The result is a more realistic, albeit more complicated, framework for energy efficient tracking in sensor networks.

A straightforward sleeping strategy when external wake-ups are not allowed is to have each sensor enter and exit the sleep mode using a fixed or a random duty cycle. A more intelligent, albeit more complicated, approach is to use information about the object trajectory that is available to the sensor from the network to determine the sleeping strategy. In particular, it is easy to see that the location of the object (if known) at the time when the sensor is put to sleep would be useful in determining the sleep duration of the sensor; the closer the object, the shorter the sleep duration should be. We take this latter approach in designing sleeping policies for the sensors.

11.1.1 Notation

In this chapter, we will use the following notational conventions:

- Scalars are written in lower case (e.g. c).
- Matrices are written in upper case (e.g. P).
- All vectors are row vectors and are written in bold face (e.g. $\boldsymbol{p}$).
- The vector $\boldsymbol{e}_i$ is a vector with a one in the ith position and zeros elsewhere.
- The vector $\boldsymbol{e}$ is a vector with a one in every position.
- The indicator function of the set $\mathcal{A}$ is written as $\mathbb{1}_{\mathcal{A}}$.
- Let $\boldsymbol{p}$ be a probability vector length n, let $\mathcal{S} \subseteq \{1 \ldots n\}$ be a *set* of integers, and suppose $p_i > 0$ for some $i \in S$. Then define $[\boldsymbol{p}]_{\mathcal{S}}$ to be a probability vector formed by setting all components p_i such that $i \notin S$ to zero and then normalizing the vector so that the sum of the components is 1.
- We will also have occasion to reference $[\cdot]_i$ where i is an *integer* and the expression in brackets evaluates to a vector. In this case, we are simply referring to the ith component of the vector in brackets.

11.2 A General Problem Formulation

The system of interest is a sensor network designed to track one or more objects in discrete time. Such a system is depicted in Figure 11.1. We consider a sensor network with n sensors. These sensors are equipped with hardware that allow them to perform two primary functions. First, each sensor is able to make observations about objects according to a sensing model. Second, each sensor is able to communicate with other sensors (or perhaps with another controller node, defined in subsequent sections) in order to combine observations and exchange control information.

The sensor nodes typically need to operate on limited energy budgets. In order to conserve energy, the sensors may be put into a sleep mode. A sensor that is sleeping cannot make observations about objects or communicate. The inability to communicate implies that a sleeping sensor cannot be awoken by external means. Note that this is in contrast to some previous research that has assumed the presence of a low-power wakeup radio used for waking the sensor (see, e.g., [7, 9, 8, 1, 12]). Our reason for not using the wakeup radio approach is that we understand it is not feasible to design a receiver that requires negligible power for operation. Our method for waking sensors

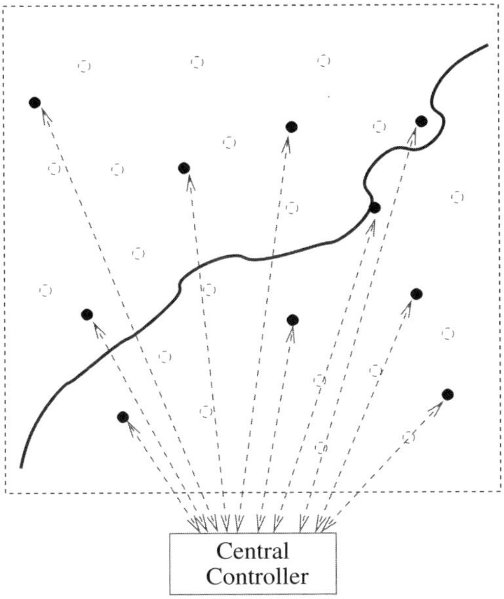

Fig. 11.1. Depiction of system operation. Sensors in active mode are opaque while are sensors in sleep mode are not opaque. The curve indicates a possible object trajectory through the network.

involves the use of a sleep timer at each sensor. A sensor entering sleep mode sets its sleep timer to an initial value (a sleep time). The sleep timer counts down as the sensor sleeps and the sensor comes awake only when this timer reaches zero.

It is clear that having sleeping sensors in the network that cannot be woken up could result in reduced tracking performance. Thus there is a tradeoff between energy savings and tracking performance that results from the sleeping at the sensors. The selections of the sleep times (i.e., the sleeping policies) at the sensors should be designed to optimize this tradeoff.

Although we have defined our design problem quite generally, we can state that this problem fits quite naturally into the framework of a partially observable Markov decision process (POMDP). Such problems are typically easy to formulate mathematically but difficult to solve either analytically or numerically. The ease of solution depends on the assumptions we make about the particulars of our system. Examples of these particulars include the following:

- Whether we have distributed or centralized control.
- The number of objects being tracked.
- The sensing model.
- The model for object movement.
- The tracking cost structure.
- The energy cost structure.

In what follows, we first examine a formulation where these particulars are chosen to simplify the analysis. As we shall see, difficulties are still encountered. We then go on to discuss how with less restrictive formulations we can continue to design effective policies.

11.3 A Particular Problem Formulation

We now turn our attention to the system depicted in Figure 11.2. In this

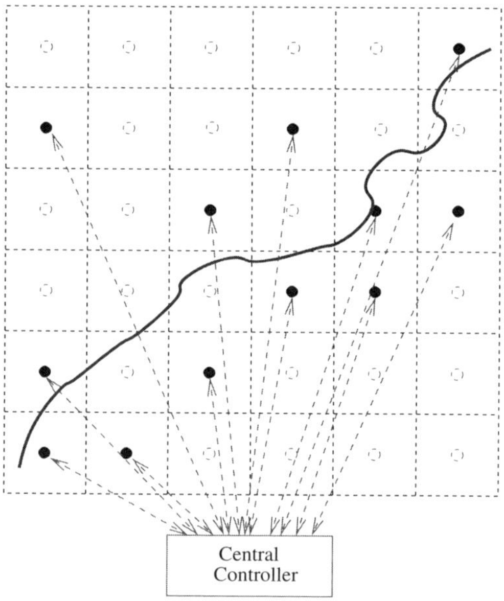

Fig. 11.2. Depiction of system operation. Sensors in active mode are opaque while are sensors in sleep mode are not opaque. The curve indicates a possible object trajectory through the network.

setup, we have for simplicity assumed that the sensing ranges of the sensors completely cover the region of interest with no overlap. In other words, the region can be divided into n cells with each cell corresponding to the sensing range of a particular sensor. Moreover, we also constrain the movement of the object so that the object can take on exactly one possible location per cell. Note that the sensors and cells need not follow a regular topology as in Figure 11.2; this depiction is only for convenience.

While we discuss the main results of our research on this problem in this section, we omit some details for the sake of brevity. For a more complete treatment, the reader is invited to examine [21].

11.3.1 Mathematical Description

The state of our system consists of two parts. The first part of the state is the current location of the object at time k, denoted b_k. The object location can take on one of n possible locations corresponding to the number of cells. However, we also append a special absorbing terminal state, denoted as $\mathcal{T}$, that occurs when the object leaves the network. Thus, there are $n+1$ possible states for the object and we will refer to the terminal state as both $\mathcal{T}$ and $n+1$. The statistics for the object movement are described by a $(n+1)\times(n+1)$ probability transition matrix P such that P_{ij} is the probability of the object being in state j at the next time step given that it is currently in state i. Since the problem remains in the terminal state once the object leaves the network, we can write P as

$$P = \left[\begin{array}{c|c} Q & 1 - Q\boldsymbol{e}^T \\ \hline 0\,0\,\ldots\,0 & 1 \end{array}\right] \tag{11.1}$$

where Q is a $n \times n$ matrix. We assume there is a path from every state to the terminal state, which is equivalent to having $\lim_{k\to\infty} Q^k = 0$. Let b_k denote the location of the object at time k. Having described the evolution of the object location in words, we note that we can write the evolution more compactly as

$$b_{k+1} \sim \boldsymbol{e}_{b_k} P \tag{11.2}$$

The second part of the state of our system consists of the residual sleep times. Let $r_{k,\ell}$ denote the value of the sleep timer of sensor ℓ at time k. Then the n-vector $\boldsymbol{r}_k$ is the residual sleep times of the sensors at time k. Also, let $u_{k,\ell}$ denote the sleep time input supplied to sensor ℓ at time k. We can describe the evolution of the residual sleep times as

$$r_{k+1,\ell} = (r_{k,\ell} - 1)\mathbf{1}r_{k,\ell} > 0 + u_{k,\ell}\mathbf{1}r_{k,\ell} = 0 \tag{11.3}$$

We see that we have a discrete-time dynamical model that describes our system, with state $x_k = (b_k, \boldsymbol{r}_k)$ and control input $\boldsymbol{u}_k$. The state evolution equations are given in (11.2) and (11.3). Unfortunately, not all of x_k is known to the central unit at time k since b_k is known only if the object is currently being tracked perfectly. Thus we have a dynamical system with incomplete (or partially observed) state information.

At each time k, an n-vector $\boldsymbol{y}_k$ of observations is generated by the sensors. The sensing model for these observations is described by the rule

$$y_{k,\ell} = \begin{cases} \ell & \text{if } b_k = \ell \text{ and } r_{k,\ell} = 0 \\ \mathcal{E} & \text{else} \end{cases} \tag{11.4}$$

where $\mathcal{E}$ denotes an erasure. However, we also assume that if the object leaves the network that this becomes known to the central controller immediately. We therefore specify that there should be a unique value for $\boldsymbol{y}_k$ that occurs if the object has left the network.

As stated earlier, our state x_k is only partially observable. We therefore need to know what information at each time step is a sufficient statistic for optimal decision making. It can be shown that a sufficient statistic for such decision making can be written as $v_k = (\boldsymbol{p}_k, \boldsymbol{r}_k)$, where

$$p_{k,\ell} = \mathsf{P}(b_k = \ell | \boldsymbol{y}_0, \ldots, \boldsymbol{y}_k) \tag{11.5}$$

The task of recursively computing $\boldsymbol{p}_k$ for each k is a problem in nonlinear filtering. Although in many contexts nonlinear filtering is a difficult problem, in the present context the solution is quite straightforward. One reason for this is our simple model for the movement of the object which allows us to update a discrete-space probability vector rather than a continuous-space probability distribution. A second reason is our simple, deterministic sensing model. We can write the evolution of $\boldsymbol{p}_k$ as

$$\begin{aligned} \boldsymbol{p}_{k+1} = \boldsymbol{e}_{\mathcal{T}} \mathbf{1} b_{k+1} = \mathcal{T} + \boldsymbol{e}_{b_{k+1}} \mathbf{1} r_{k+1, b_{k+1}} &= 0 \\ + [\boldsymbol{p}_k P]_{\{j : r_{k+1,j} > 0\}} \mathbf{1} r_{k+1, b_{k+1}} &> 0 \end{aligned} \tag{11.6}$$

where $\boldsymbol{r}_{k+1}$ is defined through (11.3) and b_{k+1} (conditioned on $\boldsymbol{p}_k$) is distributed as

$$b_{k+1} \sim \boldsymbol{p}_k P \tag{11.7}$$

To understand (11.6), note that if the object is observed at time $k+1$, $\boldsymbol{p}_{k+1}$ becomes a point-mass distribution with all the probability mass concentrated at b_{k+1}. If the object is not observed, we eliminate all probability mass at sensors that are awake (since the object is known to not be at these locations) and renormalize. Thus, all information from observations is incorporated.

We now identify the two costs present in our tracking problem. The first is an energy cost of $c \in (0, 1]$ for each sensor that is awake. The second is a tracking cost of 1 for each time unit that the object is not observed by the sensor at the current object location. Note that this means that we may incur a tracking cost even if the object location is known due to all other possible object locations being eliminated by other sensors. Despite this drawback, we use our definition of tracking cost to simplify the formulation, with the understanding that it is a good approximation to the actual tracking cost. If the object leaves the network (i.e., b_k enters the terminal state), we assume the problem terminates and no further cost is incurred. In reality, there will be n additional wakeups that occur once the object leaves the network but since this energy cost is fixed, we do not include it in the formulation. Thus, the total cost at time k is given by

$$g(x_k) = \mathbf{1} b_k \neq \mathcal{T} \left(\mathbf{1} r_{k,b_k} > 0 + \sum_{\ell=1}^{n} c \mathbf{1} r_{k,\ell} = 0 \right) \tag{11.8}$$

Note that c is the parameter used to tradeoff energy consumption and tracking errors.

The sleep time inputs k are allowed to be a function of v_k, i.e.,

$$\boldsymbol{u}_k = \mu_k(v_k) \tag{11.9}$$

The vector-valued function μ_k is the sleeping policy at time k. The total cost (over a possibly infinite horizon trajectory) for the system is given by

$$J(v_0, \mu_0, \mu_1, \ldots) = \mathsf{E}\left[\sum_{k=1}^{\infty} g(x_k) \middle| v_0\right] \tag{11.10}$$

The goal is to compute the solution to

$$J^*(v_0) = \min_{\mu_0, \mu_1, \ldots} J(v_0, \mu_0, \mu_1, \ldots) \tag{11.11}$$

The solution to this optimization problem for each value of c yields an optimal sleeping policy.

11.3.2 Performance Gains of Our Approach

We comment now on what can be gained through our approach. For the purposes of comparison, we consider a sleeping policy that does not use information about the location of the object, which we call the *duty cycle* scheme. In this sleeping policy, each sensor is awake for a fixed fraction π of the time slots. Whether the time slots where a particular sensor is awake are chosen deterministically or randomly is immaterial since the resultant performance is the same. For a duty cycle scheme, we see that in order to ensure zero tracking errors (an "Always Track" sleeping policy) we must use a value of $\pi = 1$, which means that every sensor is awake in every time slot. In contrast, our scheme uses information about the location of the object. Thus, for a sparse P matrix, our Always Track policy will allow many sensors in the network to remain asleep since it will be known that the object could not be at those locations. Note that P will in fact be sparse if the range of the object movement over a single time step is limited. Thus we surmise that using location information in this fashion will result in a tradeoff curve between energy and tracking costs that is significantly better than those for a duty cycle scheme. These suppositions will be confirmed in Section 11.3.5.

We also note that our approach can result in better asymptotic behavior as the size of the network becomes large. Note that as n becomes large, the number of sensors awake per unit time for an Always Track duty cycle scheme grows as $O(n)$. In contrast, it can be shown that when the movement of the object per time step is bounded, the number of sensors awake per unit time for our Always Track policy grows at most as $O(\log(n))$ for one-dimensional networks and at most as $O(\sqrt{n})$ for two-dimensional networks [21].

11.3.3 Suboptimal Solutions to the Problem

Having formulated our optimization problem, we seek a solution using the tools of dynamic programming. It can be shown via standard arguments (see, e.g. [22]) that there exists a stationary optimal policy for our problem (i.e. $\mu_0 = \mu_1 = \cdots = \mu^*$). Such a policy and the optimal cost J^* can be found by solving the Bellman equation given as

$$J(v) = \min_{\mu} \mathsf{E}\left[g(x_1) + J(v_1) | v_0 = v, \boldsymbol{u}_0 = \mu(v)\right] \tag{11.12}$$

with μ^* being the minimizing value of μ in this equation. Note that there are multiple functions J^* that satisfy this equation since adding a constant to a particular J^* yields another solution. We are interested in J^* such that $J^*(v) = 0$ when $v = (\boldsymbol{e}_{\mathcal{T}}, \boldsymbol{r})$.

There are two main approaches to solving (11.12). The first approach is to solve (11.12) analytically. However, we are unable to do this due to the complexity of the expressions involved. The second approach is to solve the equation using an iterative technique such as value iteration or policy iteration. However, even if we truncate the state space to a finite set, the number of states still grows exponentially with the number of sensors. The result is that such numerical methods are intractable except for the most trivial cases. We conclude that we cannot find an optimal solution to our problem and that we need other approaches to finding near-optimal solutions.

Much of the complexity of our problem stems from the complicated evolution of $\boldsymbol{p}_k$ given in (11.6). In deriving suboptimal solutions to our problem, we will make assumptions about the observations that will be available in the future. These assumptions will allow us to simplify the evolution in (11.6) considerably. In fact, the evolution of $\boldsymbol{p}_k$ will no longer be affected by the sleeping actions of the sensors. Furthermore, each sensor will only be able to affect the energy and tracking costs that occur at its location. The result is that the optimization problem easily separates into n simpler problems, one for each sensor. In each of these simpler problems, we will be able to eliminate the residual sleep times $\boldsymbol{r}_k$ from the state since the only times of interest will be those when the sensor comes awake. It will then be possible to solve each of the n simpler problems to find a cost function and policy.

The assumptions we make will be inaccurate. However, the usefulness of our assumptions must be measured in terms of how well the resultant solutions approximate optimal performance. Of course, we have no idea as yet what optimal performance may be. Fortunately, in the course of our derivations we will obtain a lower bound on optimal performance that will be useful in later performance analysis.

Our first suboptimal solution is the first cost reduction (FCR) solution. Here we assume that we will have no future observations. In other words, we are replacing (11.6) with

$$\boldsymbol{p}_{k+1} = \boldsymbol{p}_k P \tag{11.13}$$

Note that this does not mean that it will be impossible to track the object; we are simply making an assumption about the future state evolution in order to generate a sleeping policy. The Bellman equation to generate a sleeping policy for sensor ℓ is

$$J^{(\ell)}(\boldsymbol{p}) = \min_u \left(\sum_{j=1}^{u} \left[\boldsymbol{p}P^j\right]_\ell + \sum_{i=1}^{n} c \left[\boldsymbol{p}P^{u+1}\right]_i + J^{(\ell)}(\boldsymbol{p}P^{u+1}) \right) \tag{11.14}$$

To understand this equation note that within the main parentheses we have three terms. The first term represents the expected tracking cost for times 1 to u, the second term represents the expected energy cost at time $u+1$, and the third term represents the expected future cost that will be incurred when the sensor comes awake at time $u+1$.

It is easy to verify that

$$J^{*(\ell)}(\boldsymbol{p}) = \sum_{j=1}^{\infty} \min \left\{ \left[\boldsymbol{p}P^j\right]_\ell, \sum_{i=1}^{n} c \left[\boldsymbol{p}P^j\right]_i \right\} \tag{11.15}$$

is indeed a solution to (11.14). In other words, at each time step we incur a cost that is the minimum of the expected tracking cost at sensor ℓ and the expected energy cost at sensor ℓ. The policy is to come awake at the first time such that the expected tracking cost exceeds the expected energy cost. This is why this solution is called the first cost reduction solution.

Our second suboptimal solution is the Q_{MDP} solution. In the POMDP literature (see, e.g., [23]), a Q_{MDP} solution is one in which it is assumed that the partially observed state becomes fully known after a control input has been chosen. In our problem, this means assuming that we will have perfect future observations, i.e., the location of the object will be known in the future. In other words, we are replacing (11.6) with

$$\boldsymbol{p}_{k+1} = \boldsymbol{e}_{b_{k+1}} \tag{11.16}$$

Note that this does not mean that it will be impossible to incur tracking errors; we are simply making an assumption about the future state evolution in order to generate a sleeping policy. The Bellman equation to generate a sleeping policy for sensor ℓ is

$$J^{(\ell)}(\boldsymbol{p}) = \min_u \left(\sum_{j=1}^{u} \left[\boldsymbol{p}P^j\right]_\ell + \sum_{i=1}^{n} c \left[\boldsymbol{p}P^{u+1}\right]_i + \sum_{i=1}^{n} \left[\boldsymbol{p}P^{u+1}\right]_i J^{(\ell)}(\boldsymbol{e}_i) \right) \tag{11.17}$$

This equation is identical to that in (11.14) except the term representing the future cost has changed due to the different state evolution.

Unfortunately, we are unable to find an analytical solution to (11.17). However, note that if we can solve (11.17) for $\boldsymbol{p} = \boldsymbol{e}_b$ for all $b \in \{1, \ldots, n\}$, then it is straightforward to find the solution for all other values of $\boldsymbol{p}$. We therefore concern ourselves with finding values of $J^{*(\ell)}(\boldsymbol{e}_b)$ and $\mu^{*(\ell)}(\boldsymbol{e}_b)$ that satisfy (11.17) for all $b \in \{1, \ldots, n\}$. This can be achieved through policy iteration.

Policy iteration proceeds as follows:

1. Set $\mu^{(\ell)}(\boldsymbol{e}_b) = \infty$ and $J^{(\ell)}(\boldsymbol{e}_b) = \sum_{j=1}^{\infty} \left[\boldsymbol{p}P^j\right]_\ell$ for all $b \in \{1, \ldots, n\}$.
2. Compute a new value for $\mu^{(\ell)}(\boldsymbol{e}_b)$ as

$$\mu^{(\ell)}(\boldsymbol{e}_b) = \arg\min_u \left(\sum_{j=1}^{u} \left[\boldsymbol{e}_b P^j\right]_\ell + \sum_{i=1}^{n} c \left[\boldsymbol{e}_b P^{u+1}\right]_i + \sum_{i=1}^{n} \left[\boldsymbol{e}_b P^{u+1}\right]_i J^{(\ell)}(\boldsymbol{e}_i) \right) \tag{11.18}$$

for all $b \in \{1, \ldots, n\}$, with the additional caveat that if there are multiple minimizing values of u, the smallest should be chosen.
3. Solve a set of linear equations to find new values for $J^{(\ell)}(\boldsymbol{e}_b)$ for all $b \in \{1, \ldots, n\}$. Using the shorthand $u_b = \mu^{(\ell)}(\boldsymbol{e}_b)$, the linear equations are given as

$$J^{(\ell)}(\boldsymbol{e}_b) = \sum_{j=1}^{u_b} \left[\boldsymbol{e}_b P^j\right]_\ell + \sum_{i=1}^{n} c \left[\boldsymbol{e}_b P^{u_b}\right]_i + \sum_{i=1}^{n} \left[\boldsymbol{e}_b P^{u_b}\right]_i J^{(\ell)}(\boldsymbol{e}_i) \tag{11.19}$$

for all $b \in \{1, \ldots, n\}$.
4. If $\mu^{(\ell)}(\boldsymbol{e}_b)$ is different from the previous value for $\mu^{(\ell)}(\boldsymbol{e}_b)$ for at least one value of b, return to step 1. Otherwise, terminate and set $J^{*(\ell)} = J^{(\ell)}$ (of course, we will then have that $\mu^{*(\ell)} = \mu^{(\ell)}$).

There are portions of this algorithm that warrant further discussion. Note that the minimization in step 2, although well defined, is nontrivial since we are minimizing a non-convex function over a countably infinite set. Although we could restrict the set of sleep times to a finite set, this could lead to loss in optimality. A better approach is to start with an initial guess of $u = \infty$ for the minimizing u and a minimum value equal to the limit of the function to be minimized as $u \to \infty$. We then start at $u = 0$ and search for a minimum by repeatedly increasing u by 1. At each step, we can compute a lower bound on the function to be minimized over all values of u' such that $u' \geq u$. If the minimum found so far is less than or equal to this lower bound, then a global minimum has been found and the search terminates. This procedure will work as long as the lower bound we compute becomes appropriately tight as $u \to \infty$. It is frequently possible to find such lower bounds, so this is an attractive approach. Turning our attention to step 3 of the policy iteration algorithm, it is easy to establish that the set of linear equations described does have a unique and nonnegative solution. It is also clear that if the algorithm terminates, a solution to the Bellman equation has been found. Although we can apply policy iteration to any particular tracking problem and hope for termination, we would like to know if there are any conditions under which termination is assured. It can be shown that one such condition is for Q (the previously defined submatrix of P in (11.1)) to be primitive, i.e., a square matrix with nonnegative elements that has a unique maximal eigenvalue (see [24]). Note that constructing the lower bounds discussed for step 2 of the policy iteration algorithm is made relatively simple if Q is primitive because the cost function to be minimized becomes an exponential function asymptotically.

Note that for the Q_{MDP} solution, we are assuming more information than is actually available. Thus, the cost function obtained under the Q_{MDP} is a lower bound on optimal performance. We will use this lower bound when we present our numerical results.

11.3.4 Point Mass Approximations

The suboptimal policies derived in the preceding sections are considerably easier to compute than the optimal policy and can be computed on-line after some initial off-line computation has been completed. However, such on-line computation requires sufficient processing power and could introduce delays. It would be convenient if the suboptimal μ^* could be pre-computed and stored either at the central controller or distributed across the sensors themselves. The latter option is particularly attractive since it could potentially allow for a distributed implementation of the sleeping strategy without the need for a central controller. But the set of possible distributions $\boldsymbol{p}$ is potentially quite large — even if quantization is performed — and could make the storage requirements prohibitive.

To make the storage requirements feasible, we consider approximating $\boldsymbol{p}$ with a point mass distribution. The number of sleep times to be stored is then only n per sensor. We consider two options for the placement of the unit point mass when computing the sleep time for sensor ℓ: (i) the centroid of $\boldsymbol{p}$, and (ii) the nearest point to sensor ℓ on the support of $\boldsymbol{p}$. Note that the latter option allows for the implementation of policies without detailed information about the statistics of the random walk – only the support of the random increment at each time step is required.

11.3.5 Numerical Results

In this section, we give some sample simulation results that illustrate the performance of the policies we derived in previous sections. We will present simulation results for a one-dimensional sensor network with 41 sensors where the object moves with equal probability either one to the left or one to the right in each time step. In each simulation run, the object was initially placed at the center of the network and the location of the object was made known to each sensor. A simulation run concluded when the object left the network. The results of many simulation runs were then averaged to compute an average tracking cost and an average energy cost. To allow for easier interpretation of our results, we then normalized our costs by dividing by the *expected* time the

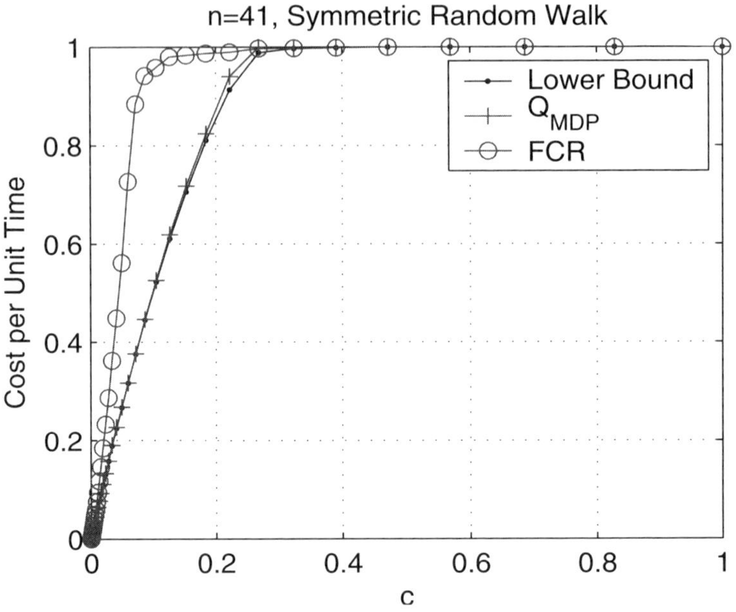

Fig. 11.3. Cost per unit time comparisons as a function c.

object spends in the network. We refer to these normalized costs as costs per unit time, even though the true costs per unit time would use the *actual* times the object spent in the network (the difference between the two was found to be small).

In Figure 11.3 we plot cost curves as a function of c for our network. In this figure three curves are shown. The first curve is the lower bound on optimal performance discussed in Section 11.3.3. The second and third curves are the costs for the Q_{MDP} and FCR policies respectively. From these data we can see that the Q_{MDP} policy consistently outperforms the FCR policy. Moreover, the cost for the Q_{MDP} policy is extremely close to the lower bound on optimal performance except at a few data points. We therefore suspect that the Q_{MDP} policy is a near-optimal policy.

In Figure 11.4, we now examine the tradeoff curves between energy cost and tracking cost for our network. From these data, we again see that the Q_{MDP} policy outperforms the FCR policy, although the difference does not appear as large. This does not contradict our previous results since it is possible that if one policy achieves certain values of energy and tracking costs at a particular value of c, it is possible for another policy to achieve these same energy and tracking costs at a somewhat different value of c. Note that the difference between the performance of the Q_{MDP} policy and the lower bound on optimal performance becomes small as the number of tracking errors becomes small. This makes sense since when there are few tracking errors, the Q_{MDP}

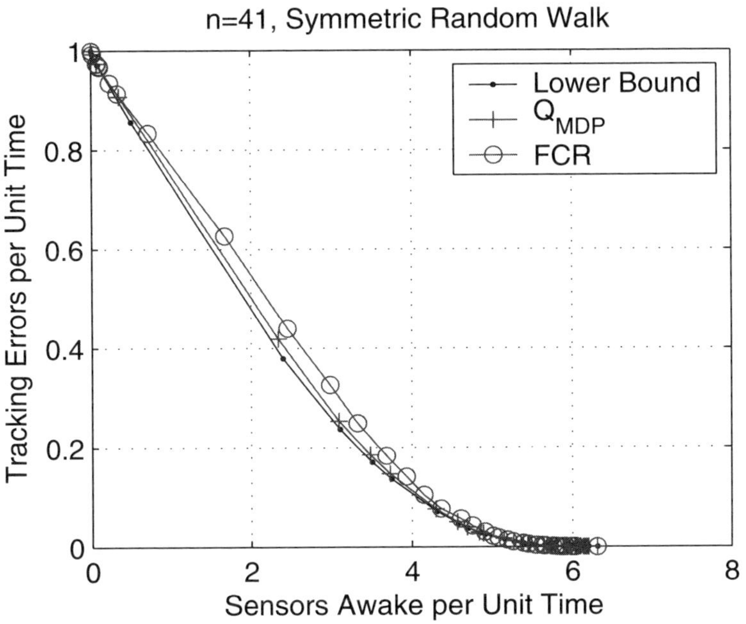

Fig. 11.4. Tradeoff curves.

assumption (that we will know the object location in the future) becomes realistic.

For the moment, consider the duty cycle scheme discussed earlier where each sensor is awake in a fraction π of the time slots. As π is varied, we achieve a tradeoff curve that is a straight line between the points $(0, 1)$ and $(n, 0)$ (where n is the appropriate number of sensors) in the coordinate systems used in Figures 11.4. When compared with this policy, the schemes we have proposed result in significant improvement.

In Figure 11.5 we explore the impact of using the point mass approximations discussed in Section 11.3.4 on the performance of the Q_{MDP} policy for our network. Four curves are shown in the figure. The first two are the lower bound and Q_{MDP} tradeoff curves already seen. The third and fourth curves are the tradeoff curves for the Q_{MDP} policy using the centroid and nearest point mass approximations respectively. It can be seen that there is indeed some loss in performance when using point mass approximations, but this loss becomes small as the number of tracking errors becomes small. This makes sense since when tracking errors are infrequent, the object location is usually known exactly and so the distribution is usually already a point mass distribution.

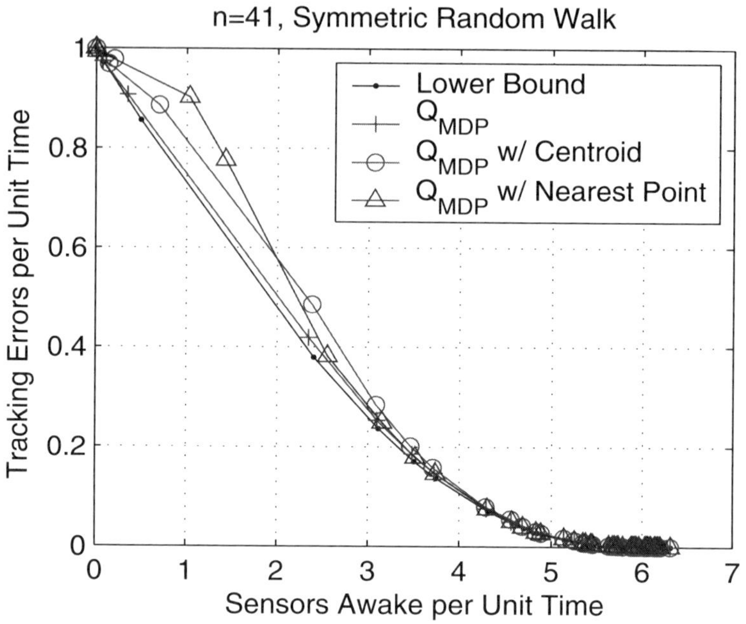

Fig. 11.5. Tradeoff curves using Q_{MDP} and point mass approximations.

11.4 Removing the Restrictions

Having met with some success in our treatment of the problem of Section 11.3, we wish to examine problems with less restrictive assumptions. In particular, we wish to make more general assumptions about the particulars mentioned at the end of Section 11.2.

11.4.1 A New Formulation

In [25], we consider a modified version of the problem of Section 11.3. In this formulation, we dispense with the notion of cells around each sensor and thus return to the setup in Figure 11.1. Although there are still n senors, we now allow for m possible object locations with m possibly not equal to n. The object still moves according to a probability transition matrix at each time step, although the matrix is now of size $(m+1) \times (m+1)$. However, a major change to the formulation is that the sensing model for the observations is generalized to be described by a conditional distribution as

$$\boldsymbol{y}_k \sim p_{\boldsymbol{Y}|X}(\boldsymbol{y}|x_k) \tag{11.20}$$

with the addition of two restrictions. First, we dictate that if $b_k = \mathcal{T}$ then the observation vector takes on a unique value that signals that the object has left the network. Second, we dictate that if the object has not left the network and a sensor is not awake at time k, its observation is an erasure. Mathematically, we say that $b_k \neq \mathcal{T}$ and $r_{k,\ell} > 0$ imply $y_{k,\ell} = \mathcal{E}$.

In addition to changing the sensing model, we also change the tracking cost structure. At each time step k, we make an estimate of the object location. Denoting this estimate as $\hat{b}_k$, we write the tracking cost at time k as $d(b_k, \hat{b}_k)$. If the object leaves the network (b_k enters the terminal state), we assume the problem terminates and no further cost is incurred. Thus, the total cost at time k is given by

$$g(x_k) = \mathbf{1} b_k \neq \mathcal{T} \left(d(b_k, \hat{b}_k) + \sum_{\ell=1}^{n} c \mathbf{1} r_{k,\ell} = 0 \right) \tag{11.21}$$

Although we can think of $\hat{b}_k$ as a new control input that only affects the cost, the selection of the optimal choice of $\hat{b}_k$ is known *a priori*. It can be shown that the optimal choice is

$$\hat{b}_k = \arg\min_{b} \mathsf{E}_{\boldsymbol{p}_k} \left[d(b_k, b) \right] \tag{11.22}$$

where $\mathsf{E}_{\boldsymbol{p}_k}$ denotes an expectation with respect to the distribution $\boldsymbol{p}_k$. Thus we do not need to add a new control input to our design problem.

In generating suboptimal solutions to this new problem formulation, we cannot directly apply the previously used methods. The main reason for this is the generalization of the sensing model. Since it is now possible for multiple sensors to simultaneously make nonerasure observations about the object, it is necessary to combine these observations in a meaningful way to help pinpoint the object. This process is referred to as cooperative localization. Note that the nonlinear filtering framework still provides the optimal method for performing this localization. The reason for the additional complexity is that since there is now cooperation among the sensors, the problem does not nicely decouple into a simple problem for each sensor.

We choose to generate suboptimal policies by performing two key steps. The first step is to assume that after the sleep times have been selected, the object location becomes known. Furthermore, we assume that at each subsequent time step the object location becomes known after the observations have been made and the costs have been incurred. Note that this is similar to the assumption made in the Q_{MDP} solution discussed in Section 11.3.3. The second key step taken to simplify the problem is to *artificially* separate the problem into a set of n subproblems, one for each sensor. In each of these subproblems, the task is to select the sleep time for that sensor as a function of $\boldsymbol{p}_k$ alone; in other words, $\boldsymbol{r}_k$ is ignored. We can think of this artificial separation of the problem as restricting the class of policies considered. While there will likely be loss in optimal performance due to this restriction, we can still seek an optimal within this class.

Because our tracking costs do not in general separate due to cooperation among the sensors, we define two matrices that will act as tracking costs in our suboptimal solutions. The performance of our suboptimal solutions will depend on the selection of these matrices. We define T^S to be an $m \times n$ matrix such that $T^S_{b,\ell}$ is the expected tracking cost given that sensor ℓ is forced to be asleep and the object was known to have been in location b in the previous time step. Similarly, we define T^W to be an $m \times n$ matrix such that $T^W_{b,\ell}$ is the expected tracking cost given that sensor ℓ is forced to be awake and the object was known to have been in location b in the previous time step. We use T^S_ℓ and T^W_ℓ to denote the ℓth columns of T^S and T^W respectively. We can write the Bellman equation for the subproblem for sensor ℓ as

$$J^{(\ell)}(\boldsymbol{p}) = \min_u \left(\sum_{j=1}^{u} \boldsymbol{p}P^{j-1}T^S_\ell + \boldsymbol{p}P^u T^W_\ell + \right. \\ \left. \sum_{i=1}^{m} c\left[\boldsymbol{p}P^{u+1}\right]_i + \sum_{i=1}^{m} \left[\boldsymbol{p}P^{u+1}\right]_i J^{(\ell)}(\boldsymbol{e}_i) \right) \tag{11.23}$$

Note that if we can solve this equation for $\boldsymbol{p} = \boldsymbol{e}_b$ for all $b \in \{1, \ldots, m\}$, then we can solve it for arbitrary $\boldsymbol{p}$. These solutions can be found via policy iteration as in Section 11.3.3 .

Note that in computing $T^S_{b,\ell}$ or $T^W_{b,\ell}$, the action of sensor ℓ is fixed. The values of these elements therefore depends on the *assumed* actions of the other sensors in the network. We identify three possible assumptions for these actions in [25] and the result is three candidate policies. In Policy 1, when we compute a particular element of T^S or T^W, we assume that all other sensors are asleep. Thus, this policy assumes no cooperation among the sensors in reducing the tracking cost. In Policy 2, when we compute a particular element of T^S or T^W, we assume that other sensors will be awake if they belong to a particular set. This set is found by searching over all 2^n possible sets of sensors that could be awake and finding the set that minimizes the expected cost. Note that finding this optimal set could be difficult. In our simulation results, however, we will consider a case where this set is easy to find. This policy does model cooperation among the sensors but the cooperation it assumes may not be representative of the behavior of the other sensors. In fact, this policy in effect assumes that other sensors can be woken up externally, which is of course not the case. Finally, in Policy 3 we iteratively solve for the cooperation among the sensors. We start with an initial estimate of T^S and T^W and compute an initial policy. For an initial distribution for the object location of $\boldsymbol{p}_0$, it is possible to use this policy to compute the probability that a given sensor will be awake given that the object was at a location b at the previous time step. New values for the T^S and T^W matrices can then be computed by conducting Monte Carlo simulations in which the behavior of the other sensors is governed by these probabilities of being awake. We then continue to generate new policies and values of T^S and T^W until convergence is achieved in the cost functions defined in (11.23).

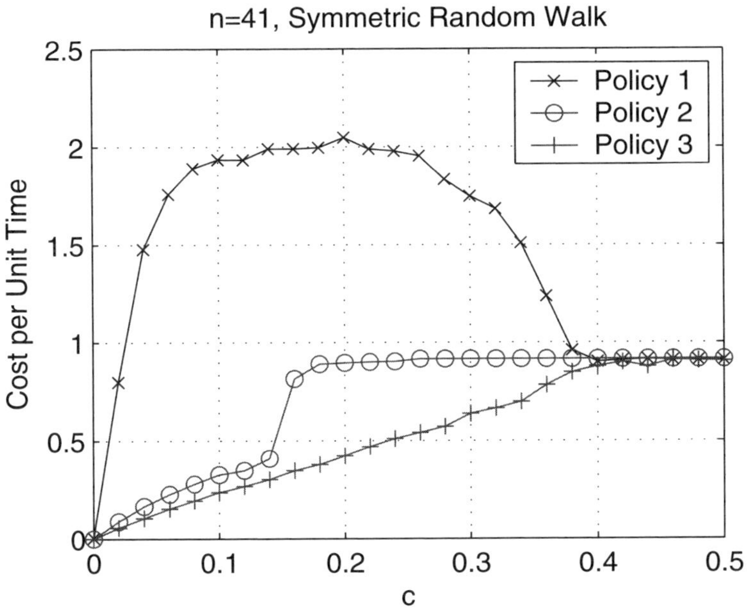

Fig. 11.6. Cost per unit time as a function of c for each policy.

We now give some numerical results. We again consider a 41-sensor line network with an object following a symmetric random walk (in this case $m = n$). The sensing model used was a lognormal shadowing model. The tracking cost function was a Hamming cost with a cost of 0 being incurred if the object location was tracked correctly and 1 otherwise. We plot cost curves and tradeoff curves for our three policies in Figures 11.6 and 11.7, respectively.

We see that there is a strict ordering in the performance of the three policies. We therefore conclude that modeling the cooperation among the sensors is crucial to policy design. However, further study is needed to determine if further gains in performance can be achieved or if any bound on performance can be proved.

11.4.2 Further Extensions

Further modifications could be made to our problem formulation to make it even more general. There are three major areas for exploration.

The first area is dealing with distributed control. In other words, we would like to eliminate the need for a central controller, and have the sensors that are awake communicate directly with each other. An example of previous research that has examined distributed control can be found in [17], where it is assumed that sensor can be woken up externally. In this work, one of the

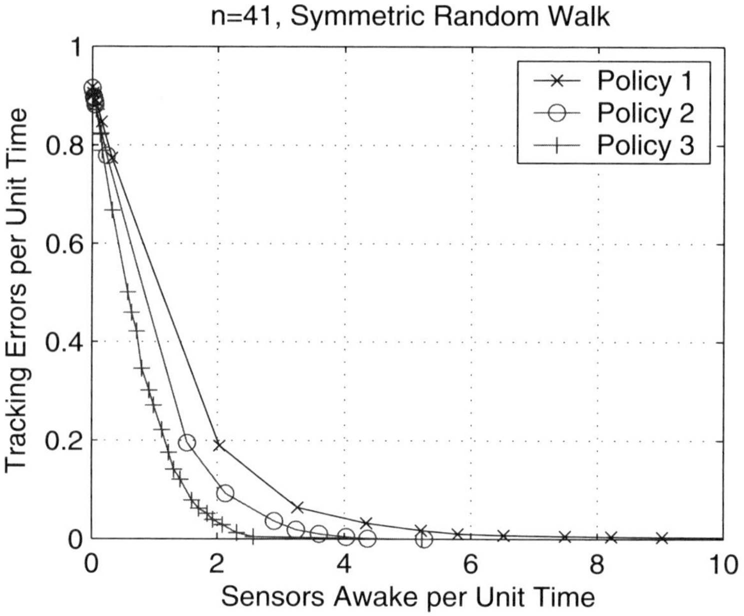

Fig. 11.7. Tradeoff curves for each policy.

sensors that is awake during each time step is designated the leader node and other sensors forward their observations to this leader node. The leader node is then responsible for determining which sensors should come awake at the next time step, which sensor should be the leader node, and also for forwarding the state of the system to the next leader node. A similar framework could be adopted in our formulation, and in particular, techniques such as the point-mass approximation described in Section 11.3.4 would facilitate the exchange of state information between the sensors. However, our assumption that sensors cannot be woken up externally adds a complication. In particular, it may be possible that there are no sensors in common among the ones that awake during one time step and the ones that are awake during the next. Whether the additional constraints needed to retain this system state can arise naturally from the problem formulation or whether more ad hoc constraints are needed remains to be seen.

A second area for further research is the setting where the statistics for the object movement are unknown or partially known. In this case, learning approaches such as one described in [26] for the multi-armed bandit problem with unknown statistics may be useful in generating useful solutions. We also noted in Section 11.3.4, for our simple sensing and object movement model, that the nearest-point point mass approximation to the system state required only information about the support of the random increment at each time step. It may be possible to extend such schemes to more general tracking problems.

The third major area for exploration is the tracking of multiple objects. Multi-target tracking (MTT) is a nontrivial extension of single target tracking since a sensor making an observation may not necessarily know which object(s) contributed to that observation. In practice, this additional complexity becomes most problematic when objects cross paths. A recent survey of MTT with resource constraints is provided in [27]. It is of interest to see how some of the MTT techniques described in [27] can be extended to the case where the sensors are allowed to sleep and cannot be woken up externally.

11.5 Conclusions

In this chapter we have examined the problem of energy-efficient tracking in sensor networks. A key feature of our work that distinguishes it from other research on the problem is the realistic assumption that it is not possible wake up sleeping sensors externally. We have formulated a basic version of the problem as a POMDP and found that while finding optimal solutions is intractable, good suboptimal policies can be designed. We have also discussed extensions of our formulation to accommodate more general assumptions about the underlying sensor network and the objects we wish to track. Further research on this problem should result in policies that will allow real-world sensor networks to track objects in an energy efficient manner.

References

[1] H. Yang and B. Sikdar, "A protocol for tracking mobile targets using sensor networks," *Proc. IEEE Int. Workshop on Sensor Network Protocols and Applications*, 2003, pp. 71–81.
[2] C. Gui and P. Mohapatra, "Virtual patrol: a new power conservation design for surveillance using sensor networks," in *IEEE/ACM Information Processing in Sensor Networks (IPSN) 2005*, 2005.
[3] X. Du and F. Lin, "Efficient energy management protocol for target tracking sensor networks," *International Symposium on Integrated Network Management*, May 2005, pp. 45–58.
[4] N. Vasanthi and S. Annadurai, "Energy saving schedule for target tracking sensor networks to maximize the network lifetime," in *First International Conference on Communication System Software and Middleware*, Jan. 2006, pp. 1–8.
[5] T. He et al., "Achieving real-time target tracking using wireless sensor networks," *Real-Time and Embedded Technology and Applications Symposium*, Apr. 2006, pp. 37–48.

[6] J. Liu, J. Reich, and F. Zhao, "Collaborative in-network processing for target tracking," *EURASIP Journal on Applied Signal Processing*, Mar. 2003, pp. 378–391.

[7] R. R. Brooks, P. Ramanathan, and A. M. Sayeed, "Distributed target classification and tracking in sensor networks," *Proc. of the IEEE*, vol. 91, no. 8, Aug. 2003, pp. 1163–1171.

[8] R. Gupta and S. R. Das, "Tracking moving targets in a smart sensor network," *Proceedings of the 58th IEEE Vehicular Technology Conference*, vol. 5, Oct. 2003, pp. 3035–3039.

[9] S. Balasubramanian et al., "Distributed and collaborative tracking for energy-constrained ad-hoc wireless sensor networks," in *IEEE WCNC*, vol. 3, Mar. 2004, pp. 1732–1737.

[10] W. Zhang and G. Cao, "DCTC: Dynamic convoy tree-based collaboration for target tracking in sensor networks," *IEEE Trans. on Wireless Communications*, vol. 3, no. 5, Sep. 2004, pp. 1689–1701.

[11] B. S. Malhotra and A. A. Aravind, "Energy efficient on-site tracking of mobile target in wireless sensor network," *Intelligent Sensors, Sensor Networks and Information Processing Conference*, Dec. 2004, pp. 43–48.

[12] L. Yang et al., "A multi-modality framework for energy efficient tracking in large scale wireless sensor networks," *IEEE Conf. on Networking, Sensing and Control*, Apr. 2006, pp. 916–921.

[13] Y. Xu and W.-C. Lee, "DTTC: Delay-tolerant trajectory compression for object tracking sensor networks," *IEEE Conf. on Sensor Networks, Ubiquitous, and Trustworthy Computing*, vol. 1, Jun. 2006, pp. 436–445.

[14] Y. He and E. K. P. Chong, "Sensor scheduling for target tracking in sensor networks," *IEEE Conf. on Decision and Control (CDC)*, vol. 1, Dec. 2004, pp. 743–748.

[15] W.-L. Yeow, C.-K. Tham, W.-C. Wong, "A novel target movement model and energy efficient target tracking in sensor networks," *IEEE Vehicular Technology Conf.*, vol. 5, May 2005, pp. 2825–2829.

[16] A. S. Chhetri, D. Morrell, and A. Papandreou-Suppappola, "Energy efficient target tracking in a sensor network using non-myopic sensor scheduling," *International Conf. on Information Fusion*, vol. 1, Jul. 2005, pp. 558–565.

[17] J. L. Williams, J. W. Fisher III, and A. S. Willsky, "An approximate dynamic programming approach to a communication constrained sensor management problem," *IEEE Conf. on Information Fusion*, vol. 1, Jul. 2005, pp. 582–589.

[18] W.-L. Yeow, C.-K. Tham, W.-C. Wong, "Energy efficient multiple target tracking in sensor networks," *IEEE Global Telecommunications Conf.*, vol. 1, Nov. 2005, pp. 168–172.

[19] W. Xiao et al., "Multi-step adaptive sensor scheduling for target tracking in wireless sensor networks," *IEEE ICASSP*, vol. 4, May 2006, pp. 705–708.

[20] S. Aeron, V. Saligrama, and D. A. Castañón, "Energy efficient policies for distributed target tracking in multihop sensor networks," in *IEEE Conf. on Decision and Control (CDC)*, Dec. 2006.
[21] J. Fuemmeler and V.V. Veeravalli, "Smart Sleeping Policies for Energy Efficient Tracking in Sensor Networks," to appear in *IEEE Transactions on Signal Processing*, 2007.
[22] D. Bertsekas, *Dynamic Programming and Optimal Control*, Athena Scientific, Belmont, MA, 2007.
[23] D. Aberdeen, "A (revised) survey of approximate methods for solving POMDP's," *Technical Report*, Dec. 2003, `http://users.rsise.anu.edu.au/~daa/papers.html`.
[24] R. Horn and C. Johnson, *Matrix Analysis*, Cambridge University Press, New York, 1985.
[25] J. Fuemmeler and V.V. Veeravalli, "Smart Sleeping Strategies for Localization and Tracking in Sensor Networks," *Asilomar Conference on Signals, Systems, and Computers*, Oct. 2006.
[26] P. Auer, N. Cesa-Bianchi, Y. Freund, and R.E. Schapire, "Gambling in a rigged casino: The adversarial multi-armed bandit problem," *Proceeding of 36th Annual Symposium on Foundations of Computer Science (FOCS'95)* p. 322, 1995.
[27] J. Liu, M. Chu, and J.E. Reich, "Resource-Aware Multi-Target Tracking in Distributed Sensor Networks," *IEEE Signal Processing Magazine Special Issue on Resource-Constrained Signal Processing, Communications, and Networking*, May 2007.

12

Distributed coverage of nonconvex environments

Anurag Ganguli[1], Jorge Cortés[2], and Francesco Bullo[3]

[1] Coordinated Science Laboratory, University of Illinois at Urbana-Champaign, IL 61801, USA, and the Department of Mechanical Engineering, University of California, Santa Barbara, CA 93106, USA, aganguli@uiuc.edu
[2] Department of Applied Mathematics and Statistics, University of California, Santa Cruz, CA 95064, USA, jcortes@ucsc.edu
[3] Department of Mechanical Engineering, University of California, Santa Barbara, CA 93106, USA, bullo@engineering.ucsb.edu

12.1 Introduction

Sensor networks and multi-agent robotic systems have been receiving increasing attention in recent times. This is due in no small part to the remarkable advances made in recent years in the development of small, agile, relatively inexpensive sensor nodes with mobile and networking capabilities. These sensor nodes are envisioned to be the basic components of complex networks intended to perform a wide variety of tasks. These include search and rescue, exploration, environmental monitoring, location-aware computing, and the maintaining of structures. The potential advantages of employing arrays of robotic sensors are numerous. For instance, certain tasks are difficult, if not impossible, when performed by a single agent. Further, a group of agents inherently provides robustness to failures of single agents or communication links.

The existence of such motion-enabled sensing devices and the anticipated development of still more advanced versions raise compelling questions. A particularly important issue is whether large numbers of such small autonomous devices will be successfully deployed as a search team to cooperatively carry out a prescribed task reliably, robustly and adaptively, without a centralized controller and with limited communications among its members.

Motivated by these future scenarios, this chapter focuses on algorithms for **visually-guided agents,** i.e., mobile robotic agents with line-of-sight sensing and communication capabilities, to solve a distributed version of the **Art Gallery Problem**. In the remainder of the introduction, we describe the problem in its original context, broadly highlight the characteristics of visually-guided agents and reformulate the original problem with respect to visually-guided agents.

Art Gallery and Illumination Problems

The classic Art Gallery Problem, was introduced by Klee and first analyzed by Chvátal, see [1, 2]. This combinatorial and geometric problem is stated as follows:

> Imagine placing guards inside an art gallery in the shape of a nonconvex polygon with n vertices: how many guards are required and where should they be placed in order for each point in the gallery to be visible by at least one guard?

The Art Gallery Theorem [1] states that $\lfloor n/3 \rfloor$ guards are sufficient and sometimes necessary to guard any polygon with n vertices. An elegant "triangulation + coloring" proof was proposed by Fisk [3]. The proof is constructive, i.e., it includes an efficient placement algorithm; an illustration is provided in Figure 12.1.

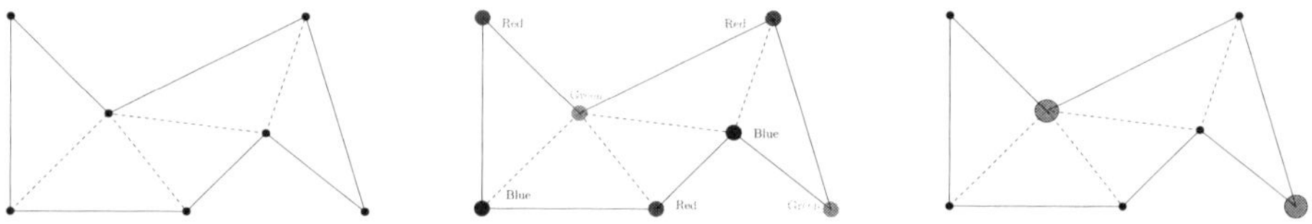

Fig. 12.1. Fisk's Algorithm: 1: triangulate the polygon (see dashed lines). 2: three-color the vertices so that each triangle has all three colors (possible because the "dual graph" is a tree). 3: select the color with smallest cardinality and place guards at the corresponding vertices (see the two guards in right picture).

Fisk's solution is, however, **centralized**, that is, it assumes that a central processor has global knowledge of the environment and that guards can be placed in desired locations without accounting for sensor-based and/or communication-based deployment.

Networks of visually-guided agents

Taking the Art Gallery Problem as a starting point, we consider a novel scenario where the guards are robotic agents in a simple nonconvex environment and are equipped with "line-of-sight" sensing and communication capabilities. In other words, our version of the Art Gallery Problem is different from its classic counterpart by the use of distributed feedback and communication protocols, rather than open-loop centralized computation.

We consider agents moving in a nonconvex planar or spatial environment, and make the following assumptions: (A1) Each agent is equipped with an "omnidirectional sensor." By this we mean a device or combination of devices (omnidirectional cameras, range and proximity sensors)

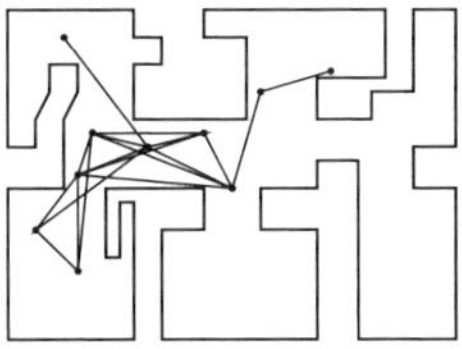

Inter-agent communication graph

that sense distance to the environment boundaries and to other agents within unobstructed lines of sight; (A2) The agents do not know the entire environment and their positions in it; (A3) Depending on the problem at hand, the guards are also allowed to exchange information with agents within line-of-sight through an asynchronous communication channel with delays and packet losses. This communication graph is depicted on the side; (A4) The agents are assumed to evolve asynchronously, i.e., a different sensing/communication/control schedule is allowed for each agent; (A5) For simplicity's sake, we model these agents as point masses with first-order dynamics. Assumptions (A1) through (A5) characterize what we refer to as **visually-guided agents**.

Illuminating art galleries via incremental partition and deployment

Combining the discussion in the earlier subsections, we obtain the following version of the Art Gallery Problem: starting from arbitrary positions, how should the agents move (and what should they communicate) in order to reach final positions such that each point of the environment is visible to at least one agent. This is what we refer to as the **distributed art-gallery deployment problem.** Remarkably, the difficulty of this problem is inherently due to the communication and sensing constraints: the agents are not given a map of the environment and no central entity controls them.

The proposed algorithms allow for sensor-based, distributed, asynchronous execution and guaranteed visibility is achieved when the number of agents is at least $\lfloor n/2 \rfloor$. The algorithm is organized in three steps:

[Geometric Structure]: first, we show that any simple nonconvex polygon can be partitioned into star-shaped polygons in an incremental distributed way. This induces a graph, the *vertex-induced tree*, as follows: every star-shaped polygon in the partition is a node and edges between nodes exist only when the corresponding polygons are contiguous;

[Distributed Information Processing]: second, we design appropriate distributed algorithms to manage the geographic information obtained by the network of agents. This entails deciding what information needs to be stored by what agent and how it needs to be transmitted and updated;

[Local Navigation and Global Exploration]: third and final, we devise navigation algorithms for two purposes: (i) to traverse edges of the vertex-induced tree, i.e., to move individual agents between contiguous polygons, and (ii) to explore and deploy a group of agents over the nodes of the vertex-induced tree.

This combination of "geometric structure + information management + navigation algorithms" is the key idea that allows individual agents to explore and traverse the nonconvex polygon only based on local sensing and communication. We refer to solutions of this form as **incremental partition and deployment algorithms**.

The rest of the chapter is organized as follows. In Section 12.2, we present the literature related to our current work. Section 12.3 contains preliminaries and notation. In Section 12.4, we present the algorithm details. Finally, we present our conclusions in Section 12.5.

12.2 Related Work

The content of this chapter is related to the works on map building and exploration, deployment of robotic networks, illumination and geometric optimization problems, and distributed algorithms. In the following, we cite the works that are relevant, by subject or by the tools therein, to either the problem or the approach in this chapter or both.

Map building and exploration

The robotics literature is abound in works on map building and exploration of unknown environments. However, the most relevant to the problem at hand include topological exploration of graph-like environments by single and multiple robots. In [4], a single robot with a marker explores such an environment via a depth-first linear time algorithm. While at a node of the graph, the robot has the ability to identify the neighboring nodes, order them in a consistent way, remember the last node visited and drop a marker to designate that a given node has already been explored. Topological exploration with multiple robots is the subject of [5]. Multiple robots, each equipped with a marker, explore the map independently. They communicate with robots located at the same node. The robots start at the same node, plan partition of work and rendezvous schedule (by exchanging messages), explore a portion of the environment and return to a predetermined location where they merge their maps. The process is repeated till the maps with each of the robots is isomorphic with the the world map. Multi-robot exploration of an unknown environment while reducing the odometry error has also been studied [6]. Here, exploration proceeds via constructing partitions of the environment into triangles or quadrilaterals, depending on whether the diameter of the environment is large compared to the range of the sensor, and then moving along the dual graph of the partition.

Deployment of robotic networks

Some related works on deployment include [7], where an incremental heuristic for deployment is proposed, [8] where distributed algorithms for coverage control based on Voronoi partitions are designed, and [9], in which the relevance of random walk on graphs is discussed (the environment and its graphical representation are assumed known a priori, and general strategies are evaluated via Monte Carlo simulation). Coordinated deployment of multiple heterogeneous robots has also been studied in [10]. Deployment locations are user-specified after an initial map of the unknown environment has been built.

Illumination problems and geometric optimization

Illumination and art gallery problems are classic topics, e.g., see [11, 12, 13]. Coverage algorithms (for systems with binary, limited-range sensors) are surveyed in [14]. Next-best-view problems are discussed in [15]. Geometric optimization is a vast and exciting avenue of current research, see for example [16, 17]. Here, by geometric optimization, we mean an optimization problem induced by a collection of geometric objects. For example, in facility location problems service sites are spatially allocated to fulfill a specified request [18, 19]. These approaches mainly rely on centralized computation for a known static environment and are not applicable in a distributed, asynchronous, adaptive setting.

Distributed algorithms

The study of distributed algorithms is concerned with providing mathematical models, devising precise specifications for their behavior, and formally proving their correctness and complexity. Via an automata-theoretic approach, the reference [20] treats distributed consensus, resource allocation, communication, and data consistency problems. Numerical distributed asynchronous algorithms as networking algorithms, rate and flow control, and gradient descent flows are discussed in [21]. All these references do not typically address algorithms over ad-hoc dynamically changing networks. The recent work [22] proposes a model of distributed robotic network.

In addition, the proposed work is related to visibility-based pursuit-evasion problems, see [23, 24], although these works focus on single agents and not on distributed policies for groups of agents.

The sensing and communication abilities of each agent is attuned to the coordination problems at hand. The study of vision as a sensor in coordination problems is in its infancy; beside our work described below, only few preliminary references are available [24, 25]. Vision and, more generally, sensor-based coordination is instead a key interaction modality for animal networks.

12.3 Preliminaries and notation

We begin by introducing some basic notation. If p is a point in the polygon Q, we let $V(p)$ denote the set of visible points from p. A set S is *star shaped* if there exists $p \in S$ such that $S \subset V(p)$; if S is star shaped, we let $\ker(S)$ be its *kernel*, i.e., the set of points $k \in S$ such that $S \subset V(k)$. Finally, a *diagonal* of a polygon Q is a segment inside Q connecting two vertices of Q (and therefore splitting Q into two polygons). A vertex of a polygon Q is *nonconvex* when the internal angle is strictly greater than π.

We consider a group of robotic agents modeled as point masses, moving in a simple nonconvex polygonal environment, Q. Each agent has a unique identifier UID, say i. Let p_i refer to the position of agent i. Each agent is equipped with an omnidirectional line-of-sight range sensor. Thus, the agent can sense its star-shaped visibility set $V(p_i)$. It can communicate with any other agent within line-of-sight and less than a certain distance r. The quantity r can be adjusted by the agent but is upper bounded, say by $R > 0$.

Each agent has access to some memory $\mathcal{M}_i$. By memory, we refer to all the necessary information that is not accessible via local sensing and communication. An agent i can broadcast its UID together with its memory contents to all agents inside its communication region. Such a broadcast is denoted by BROADCAST$(i, \mathcal{M}_i)$. It can also receive broadcasts from other agents. We also assume that there is a bounded time delay, $\delta > 0$, between a broadcast and the corresponding reception.

Every agent i repeatedly performs the following sequence of actions beginning at a time instant, say T_l^i:

(i) send repeated BROADCAST$(i, \mathcal{M}_i)$ after δ time intervals, until it starts moving;
(ii) LISTEN for a time interval equal to at least 2δ before processing the information;
(iii) PROCESS the necessary information. Also continue to LISTEN during this interval;
(iv) MOVE to a desired point.

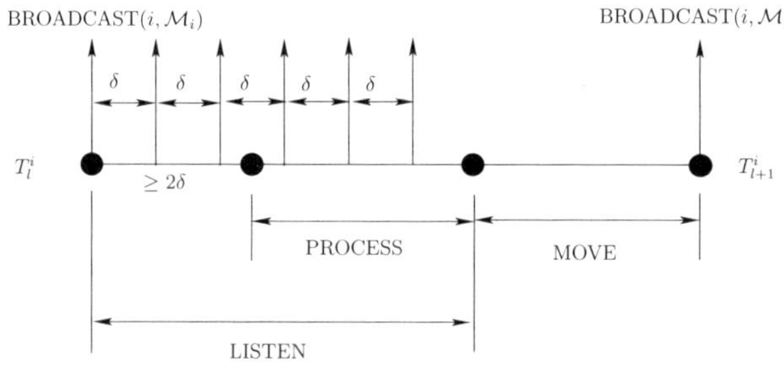

Fig. 12.2. Sequence of actions for agent i beginning at time T_l^i. Instantaneous BROADCAST$(i, \mathcal{M}_i)$ events are represented by vertical pulses. The MOVE interval might be empty if the agent does not move.The subsequent instant T_{l+1}^i is the time when the agent stops performing the MOVE action and it is not predetermined.

Agent i, in the MOVE state, is capable of moving at any time t according to the following discrete-time control system:

$$p_i(t + \Delta t) = p_i(t) + u_i,$$

where the control is bounded in magnitude by 1. The control action depends on time, on the memory $\mathcal{M}_i(t)$, and on the information obtained from communication and sensing. The subsequent wake-up instant T^i_{l+1} is the time when the agent stops performing the MOVE action and it is not predetermined. This model of visually-guided agents is similar in spirit to the **partially asynchronous network model** described in [21].

Given this model, the goal is to design a provably correct discrete-time algorithm which ensures that the agents converge to locations such that each point of the environment is visible to at least one agent. This is the **distributed art-gallery deployment problem** for visually-guided agents.

12.4 Distributed Art Gallery Deployment Problem

In this section we detail the **incremental partition and deployment algorithms** described in the introduction. We begin by describing a partition of a given simply connected nonconvex environment into star-shaped polygons and the graph that such a partition induces.

12.4.1 The vertex-induced partition and tree

Given a nonconvex polygon Q without holes and a vertex s of it, we compute a list $\{P_1, \dots, P_m\}$ of star-shaped polygons composing a partition of Q and a list $\{k_1, \dots, k_m\}$ of kernel points for each star-shaped polygon $\{P_1, \dots, P_m\}$. The computation of these quantities is discussed in the following algorithm and is illustrated in Figure 12.3.

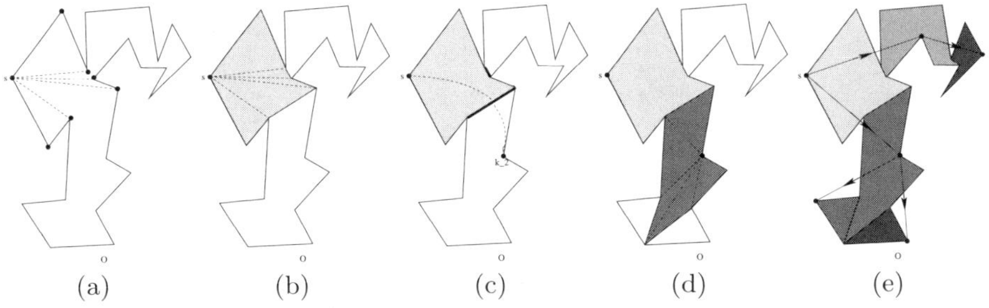

Fig. 12.3. Computation of the vertex-induced partition and tree in 5 steps.

`Vertex-Induced Partition and Tree Algorithm`

1: set $k_1 = s$, and collect all vertices of Q visible from k_1 (see Fig. 12.3(a))

2: let P_1 be the polygon determined by these vertices (by definition $k_1 \in \ker(P_1)$) (see Fig. 12.3(b))
3: identify the edges of P_1 that are diagonals of Q; call them **gaps**. For all gaps, place a new point, say k_2, across the gap at a new vertex of Q such that k_2 sees the gap (see Fig. 12.3(c))
4: repeat last three steps for new point k_2, until all gaps have been crossed (see Fig. 12.3(d))
5: define edges starting from s going to all kernel points and crossing all edges (see Fig. 12.3(e))

We refer to the list $\{P_1, \dots, P_m\}$ computed in the algorithm as the **vertex-induced partition**. The algorithm computes not only the partition and a list of kernel points, but also a collection of edges connecting the kernel points. In other words, we also computed a directed graph, the **vertex-induced tree**, denoted by $\mathcal{G}_Q(s)$: the nodes of this directed graph are $\{k_1, \dots, k_m\}$ and an edge exists between any two vertices k_i, k_j if and only if $P_i \cap P_j$ is a diagonal of Q. Note that $k_1 = s$; we refer to this node as the root of $\mathcal{G}_Q(s)$. We now state some important properties of the vertex-induced tree.

Proposition 1. *Given a polygon Q without holes and a vertex s, the following statements hold:*

(i) the directed graph $\mathcal{G}_Q(s)$ is a rooted tree;
(ii) the maximum number of nodes in the vertex-induced tree is less than or equal to $\lfloor \frac{n}{2} \rfloor$, where n is the number of vertices in Q.

Proof. The fact that $\mathcal{G}_Q(s)$ is a tree is a consequence of the fact that Q has no holes. Since s is designated as the root, $\mathcal{G}_Q(s)$ is a rooted tree. This proves statement (i). To prove statement (ii), notice the set of nodes of $\mathcal{G}_Q(s)$ belong to the vertices of Q. Also, by construction no two adjacent vertices of Q can both belong to the node set $\{k_1, \cdots, k_m\}$. Since the number of vertices of Q is n, it follows that number of nodes of $\mathcal{G}_Q(s)$ is less than or equal to $\lfloor \frac{n}{2} \rfloor$.

It is clear from the construction of the vertex-induced tree that, if we design a distributed algorithm to place agents on each node of the tree, then we will have solved the distributed art-gallery deployment problem.

Remark 1. If we can deploy the agents over the kernel points, then we will have solved the art-gallery deployment problem requiring $\lfloor n/2 \rfloor$ agents in the worst case, which is in general more than the $\lfloor n/3 \rfloor$ number required if the entire environment were known a priori. This is not surprising considering the weaker assumption of no global knowledge that we make while posing the problem.

Local node-to-node navigation algorithms

Note that by virtue of the constructions in the previous section, we have converted the original problem into a graph "navigation and deployment" problem. We now describe algorithms to plan paths between neighboring nodes of the vertex-induced tree. In a rooted tree, every neighbor of a node is either a child or the parent. Therefore, we present two simple informal descriptions.

```
Move-to-Child Algorithm
```
1: compute the mid-point of the gap between the node and the child
2: go to the mid-point
3: compute the nearest vertex from which the entire gap is visible and which is across the gap
4: go to that vertex

```
Move-to-Parent Algorithm
```
1: compute the mid-point of the gap between the node and the parent
2: go to the mid-point of the gap
3: from the mid-point, go to the vertex representing the parent node

Figure 12.4 shows paths between parents and children as computed by the previous two algorithms. It is easy to see that navigation is very simple if sufficient information is available to the agents. We address this aspect in the next subsection.

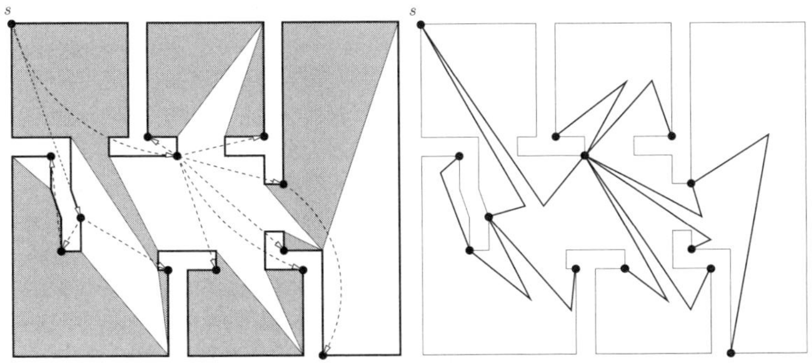

Fig. 12.4. Left figure: a vertex-induced tree and partition in a prototypical floorplan. Right figure: the planned paths between neighboring nodes.

12.4.2 Distributed information processing

¿From the previous discussion we know that the following information must be available to an agent to properly navigate from node to node. If the node is executing the `Move-to-Child Algorithm`, then it needs to know what gap to visit, i.e., what child to visit. If the node is executing the `Move-to-Parent Algorithm`, then it needs to know where the parent node is located.

This geographic information is gathered and managed by the agents via the following state transition laws and communication protocols. At this time, we make full use of the computation, communication and sensing abilities of visually-guided agents mentioned in Section 12.3.

(i) The memory content $\mathcal{M}$ of each agent is a quadruple of points in Q labeled $(p_{\text{parent}}, p_{\text{last}}, g_1, g_2)$. All four values are initialized to the initial location of the agent. During any broadcast, these values are sent over together with the agent's UID.

During run time, $\mathcal{M}$ is updated to acquire and maintain the following meaning: p_{parent} is the parent kernel point to the current agent's position, p_{last} is the last *way point*[4] visited by the agent, and (g_1, g_2) is the diagonal shared between the current cell and the parent cell, i.e., the gap toward the parent node. This is accomplished as follows:

(ii) After an agent moves from a kernel point k_i to a child kernel point k_j through a gap described by two vertices v', v'', its memory $\mathcal{M}$ is updated as follows: $p_{\text{parent}} := k_i$, $p_{\text{last}} = k_j$ and $(g_1, g_2) := (v', v'')$.
(iii) After an agent moves from a kernel point k_j to the parent kernel point k_i, its memory $\mathcal{M}$ is updated as follows: first, $p_{\text{last}} := w$, where w is the way point on the path between k_j and k_i, and second, the agent acquires updated values of $\{p_{\text{parent}}, g_1, g_2\}$ by listening to the incoming message with the highest UID.

Remark 2. At any time, at any occupied node, p_{parent} corresponding to the agent with the highest UID refers to the location of the parent of the current node. Also, (g_1, g_2) refers to the gap between the current node and the parent node. To see this, we argue as follows: Given any node k_i of $\mathcal{G}_Q(s)$ that is occupied by one or more agents, let l be the highest UID among all agents. Then, we claim that the last node visited by l is the parent of k_i. We prove this by contradiction. Let the last node visited by l be a child of k_i. To visit that child, it must have first visited k_i. Then, by the `Depth-First Navigation Algorithm`, it must have moved from k_i because of the presence of an agent with a UID greater than l. Therefore, the maximum UID at k_i must be greater than l which is a contradiction. Hence, the last node visited by l is the parent of k_i. Now according to (ii) above, the quantity p_{parent} for l refers to the parent of k_i. Also, (g_1, g_2) refers to the gap between k_i and its parent.

No common reference frame

In the description of the memory update laws, we have used a global reference frame to refer to the contents of $\mathcal{M}$. However, this assumption can be easily

[4] A way point is a mid-point of the gap between two nodes (Figure 12.4 right)

relaxed by storing the variables in $\mathcal{M}$ in a different way. For example, instead of storing the location of the parent node as a point p_{parent} relative to some global frame, the location of the parent can be stored as an integer d_{parent}, as shown in Figure 12.5. Note that such a representation does not depend on the orientation of the reference frame of an agent. The location of the

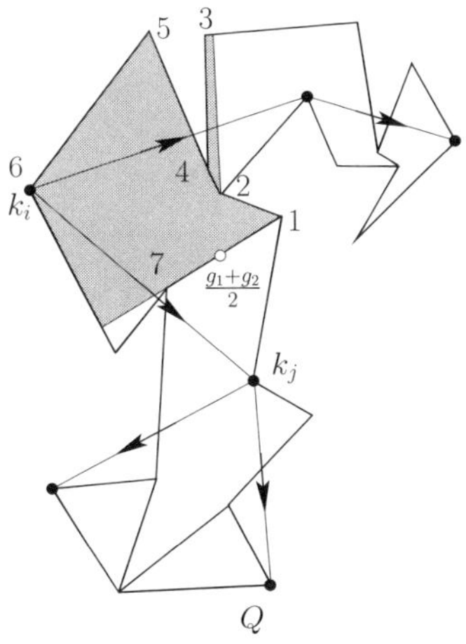

Fig. 12.5. Illustration of how the relative location of the parent of a node can be stored without the use of a common reference frame. The polygon is the environment Q. The graph with the directed edges is the vertex-induced tree in Figure 12.3. The node k_i is the parent of k_j and the point $\frac{g_1+g_2}{2}$ denoted by the white disc refers to the mid-point of the gap between k_j and k_i. The shaded region is the set of all points visible from $\frac{g_1+g_2}{2}$ on the side of the diagonal (g_1, g_2) not containing k_j. The vertices of Q in this visibility set are enumerated $(1, \cdots, 7)$ in counter-clockwise order, the vertex 1 being one of the vertices $\{g_1, g_2\}$, say g_1, and with g_2 being the last vertex in the ordering. The location of the parent can now be stored as $d_{\text{parent}} = 6$.

gap (g_1, g_2) can be stored in a similar fashion. The point p_{last} can be stored with respect to the local reference frame. We do not store p_{parent} and (g_1, g_2) in terms of local coordinates since these variables may be used as updates by other collocated robots. This would necessitate that the robots be aware of the relative orientations of their local coordinate frames or, equivalently, be equipped with compasses. By storing p_{parent} and (g_1, g_2) according to the scheme in Figure 12.5, the use of compasses is eliminated.

Remark 3. If the number of vertices of the environment visible from any point of the environment is bounded, then the amount of memory required to store p_{parent} and (g_1, g_2) is also bounded. Also if the diameter of the environment is bounded, then the memory required to store p_{last} is bounded. Thus, under the aforesaid assumptions, the memory $\mathcal{M}$ is constant irrespective of the complexity of the environment.

`Depth-First Navigation Algorithm`
All agents are initially located at root s
During each PROCESS action, each agent executes:

1: Find maximum UID received during the LISTEN action
2: **If** maximum received UID is less than its own UID
3:    **then** stay at current kernel point
4: **else**
5: **If** there are no children of the present kernel point
6:    **then** `Move-to-Parent Algorithm` towards p_{parent} via $\{g_1, g_2\}$
7: **else**
8:    Order the children in a suitable way
9:    **If** p_{last} in memory is the parent of the present node,
      then `Move-to-Child Algorithm` towards the first child in the ordering
10:    **If** the last node visited is a child that is not the last in the ordering,
      then `Move-to-Child Algorithm` towards next child in the ordering
11:    **If** (the last node visited is a child that is the last in the ordering) AND (current node is not the root),
      then `Move-to-Parent Algorithm` towards p_{parent} via $\{g_1, g_2\}$
12:    **If** (the last node visited is a child that is the last in the ordering) AND (current node is the root),
      then `Move-to-Child Algorithm` towards the first child in the ordering

Table 12.1. `Depth-First Navigation Algorithm`.

12.4.3 Global exploration and deployment algorithms

At this time, we have all the elements necessary to present a global navigation algorithm that leads the agents to deploy themselves over the nodes of the vertex-induced tree. We term this algorithm `Depth-First Navigation Algorithm`, see Table 12.1.

Note that the instruction `5:` through `11:` in `Depth-First Navigation Algorithm` essentially amount to a depth-first graph search. Alternatively, it is fairly easy to design randomized graph search algorithms, where the nodes select their motion among equally likely children/parent decisions.

The following Figures 12.6 and 12.7 show the results of the simulations of the depth-first search and randomized search algorithms respectively. The nodes of the vertex-induced tree of the environment in the simulations are precisely the locations where the agents in Figure 12.6 are located at the end of the simulation. In Figure 12.7, there are more agents than the number of nodes in the vertex-induced tree. Hence, the extra agents keep exploring the graph without coming to rest.

12.4.4 Convergence and run time analysis

In this section, we provide the results on convergence of the algorithm and we also characterize the time taken for the task to be completed. Given a

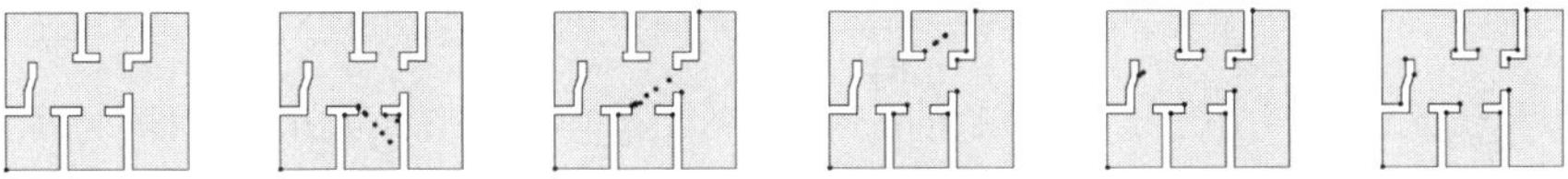

Fig. 12.6. From left to right, evolution of a network implementing depth-first search. The number of vertices of the environment is $n = 46$ and the number of agents is $N = 13 < \lfloor \frac{46}{3} \rfloor$. Each point of the environment is visible at the end of the simulation.

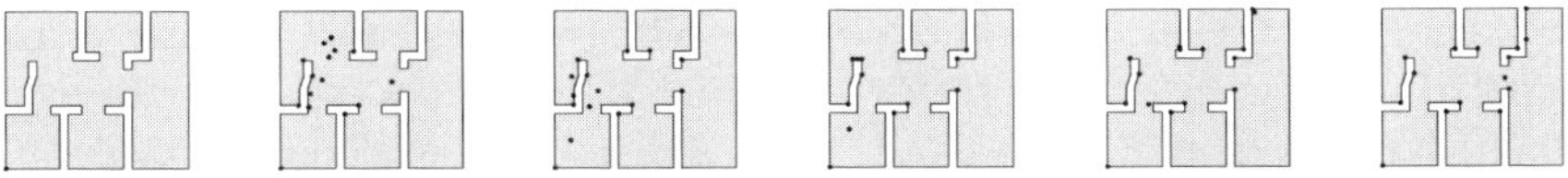

Fig. 12.7. From left to right, evolution of a network implementing randomized search. While the polygon is the same as above and therefore the vertex-induced tree still has only 13 nodes, the number of agents is 15; after each node of the tree is populated, the 2 extra agents continue to explore the vertex-induced tree.

polygon Q without holes and a vertex s, we define the following length: For each edge (k_i, k_j) of $\mathcal{G}_Q(s)$, let $d_{\text{edge}}(k_i, k_j)$ be the path length between k_i and k_j. The length of the vertex-induced tree $\mathcal{G}_Q(s)$ is defined by

$$\mathcal{L}_{\text{vit}}(\mathcal{G}_Q(s)) = \sum_{e \in \text{ edges of } \mathcal{G}_Q(s)} d_{\text{edge}}(e).$$

With these notions we can state the next result.

Theorem 1 (Convergence and Run Time Analysis). *Given a polygon without holes Q, assume that N visually-guided agents begin their motion from a vertex s of Q. Assume Q has n vertices and the vertex-induced tree $\mathcal{G}_Q(s)$ has m nodes. Assume also that there exists a bound $\lambda_{\max}$ on the* LISTEN *interval for any agent i. Then the following statements hold:*

(i) In finite time t^* there is at least one agent on $\min\{m, N\}$ nodes of $\mathcal{G}_Q(s)$.
(ii) If $N \geq \lfloor n/2 \rfloor$, then the art-gallery deployment problem is solved in finite time by the `Depth-First Navigation Algorithm`.
(iii) assuming unit speed for any agent, the time taken for task completion, t^*, obeys the following:

$$t^* = \mathcal{T}_{\text{motion}} + \mathcal{T}_{\text{nodes}},$$

where $\mathcal{T}_{\text{motion}} \leq 2\mathcal{L}_{\text{vit}}(\mathcal{G}_Q(s)) - \min\{d_{\text{edge}}(e) | \; e \in \text{ edges of } \mathcal{G}_Q(s)\}$ and $\mathcal{T}_{\text{nodes}} \leq 2(m-1)\lambda_{\max}$.

Proof. We first prove statement (i). Let us first see that at any time t, any agent is either at a node of $\mathcal{G}_Q(s)$ or on the path between two nodes. According to the `Depth-First Navigation Algorithm`, an agent always moves according to either the `Move-to-Child Algorithm` or the `Move-to-Parent Algorithm`. By the memory update laws in Section 12.4.2, during any PROCESS interval, an agent at a node always has in its memory the location of the parent node and the gap between the current node and its parent. Therefore, an agent at a node always has enough information to compute the locations of the parent and the children and, thus, always is either at a node of $\mathcal{G}_Q(s)$ or on the path between two nodes.

Now, from step 2 of `Depth-First Navigation Algorithm`, an agent stays at a node unless there is an agent with a higher UID collocated at the same node. It also follows that once a node is occupied by an agent, it continues to be occupied by at least one agent for all future times. Therefore, the number of occupied nodes is non-decreasing. Since the number of nodes are finite, there exists a finite time τ_1 such that for all time $t \geq \tau_1$, exactly w nodes are occupied. Also, the highest UID at any occupied node is non-decreasing. Since the number of agents are finite, there exists a finite time τ_2 such that for all time $t \geq \tau_2$, the highest UID at all w occupied nodes is constant. Now, let $\tau \geq \max\{\tau_1, \tau_2\}$. Now, if $w \geq N$, then we are done. If $w < N$, then at any time $t \geq \tau$, there are $N - w$ agents that either belong to w occupied nodes or belong to the paths between two nodes of $\mathcal{G}_Q(s)$. Since the UID at any occupied node is constant, this implies that the $N - w$ agents are the ones with the lowest UIDs. From the `Depth-First Navigation Algorithm`, each of the $N - w$ agents perform a depth-first search on $\mathcal{G}_Q(s)$ spending at most $\lambda_{\max}$ time at any node. If $w \geq \min\{m, N\}$, then we are done. If $w < N \leq m$, there is at least one node that is unoccupied. Therefore, each of the $N - w$ agents will reach an unoccupied node of $\mathcal{G}_Q(s)$ in finite time. Thus, the number of occupied nodes increases which is a contradiction. Therefore, $w \geq N$. If on the other hand, $w < m \leq N$, then again there is at least one node that is unoccupied. By a similar argument as before, it follows that the number of unoccupied nodes increases.

Statement (ii) follows from statement (i) and from Proposition 1 (ii) which states that $m \leq \lfloor \frac{n}{2} \rfloor$.

To prove statement (iii), let us assume that k_l be the last node to be occupied at time t^*. Clearly, k_l has to be a leaf. Let the agent first occupying k_l be j. To travel from the root to any leaf via a depth-first search, an agent traverses each edge at most twice except for the edge incident to the leaf, which has to be traversed only once. Thus, agent j travels at most $\left(\sum_{e \in \text{ edges of } \mathcal{G}_Q(s)} 2d_{\text{edge}}(e)\right) - \min\{d_{\text{edge}}(e) |\ e \in \text{ edges of } \mathcal{G}_Q(s)\}$ distance. Since the agent is assumed to move with unit speed, the time taken to travel this distance, $\mathcal{T}_{\text{motion}}$, is $2\mathcal{L}_{\text{vit}}(\mathcal{G}_Q(s)) - \min\{d_{\text{edge}}(e) |\ e \in \text{ edges of } \mathcal{G}_Q(s)\}$. Also, while travelling from the root to k_l, agent j stops at each of the remaining $m-1$ nodes at most twice. At each node, agent j spends at most $\lambda_{\max}$ time. Thus, the time spent at the nodes, $\mathcal{T}_{\text{node}}$, is $2(m-1)\lambda_{\max}$. The total time, $t*$ is equal to $\mathcal{T}_{\text{motion}} + \mathcal{T}_{\text{node}}$ and the result follows.

12.5 Conclusions

In this chapter, we pose a distributed version of the classic Art Gallery Problem for mobile robotic agents. Under assumptions of line-of-sight communication and sensing on the agents, we design a provably correct `Depth-First Navigation Algorithm` that solves the problem given that the agents are initially collocated at a vertex of the environment. The algorithm is robust to arbitrary but bounded communication delays. Under the assumptions of bounded environment diameter and bounded number of vertices visible from any point in the environment, the memory required by the agents is constant irrespective of the environment complexity. An early version of this algorithm appeared in [26].

Acknowledgment

This material is based upon work supported in part by AFOSR through Award F49620-02-1-0325, by NSF through Award CMS-0626457, and by NSF through CAREER Award ECS-0546871. The authors thank Prof. Seth Hutchinson for his kind support.

References

[1] V. Chvátal, "A combinatorial theorem in plane geometry," *Journal of Combinatorial Theory. Series B*, vol. 18, pp. 39–41, 1975.

[2] R. Honsberger, *Mathematical Gems II: The Dolciani Mathematical Expositions.* Mathematical Association of America, 1976.

[3] S. Fisk, "A short proof of Chvátal's watchman theorem," *Journal of Combinatorial Theory. Series B*, vol. 24, p. 374, 1978.

[4] I. Rekleitis and V. Dujmović, "Efficient topological exploration," in *IEEE Int. Conf. on Robotics and Automation*, (Detroit, MI), pp. 676–681, May 1999.

[5] G. Dudek, M. Jenkin, E. Milios, and D. Wilkes, "Topological exploration with multiple robots," in *International Symposium on Robotics and Applications*, (Anchorage, Alaska), May 1998.

[6] I. M. Rekleitis, G. Dudek, and E. E. Milios, "Multi-robot exploration of an unknown environment, efficiently reducing the odometry error," in *International Joint Conference in Artifical Intelligence*, vol. 2, (Nagoya, Japan), pp. 1340–1346, Aug. 1997.

[7] A. Howard, M. J. Matarić, and G. S. Sukhatme, "An incremental self-deployment algorithm for mobile sensor networks," *Autonomous Robots*, vol. 13, no. 2, pp. 113–126, 2002.

[8] J. Cortés, S. Martínez, T. Karatas, and F. Bullo, "Coverage control for mobile sensing networks," *IEEE Transactions on Robotics and Automation*, vol. 20, no. 2, pp. 243–255, 2004.

[9] J. Grace and J. Baillieul, “Stochastic algorithms for autonomous robotic surveillance,” in *IEEE Conf. on Decision and Control and European Control Conference*, (Seville, Spain), pp. 2200–2205, Dec. 2005.

[10] R. Simmons, D. Apfelbaum, D. Fox, R. Goldman, K. Haigh, D. Musliner, M. Pelican, and S. Thrun, “Coordinated deployment of multiple heterogenous robots,” in *IEEE/RSJ Int. Conf. on Intelligent Robots & Systems*, (Takamatsu, Japan), pp. 2254–2260, 2000.

[11] T. C. Shermer, “Recent results in art galleries,” *IEEE Proceedings*, vol. 80, no. 9, pp. 1384–1399, 1992.

[12] J. Urrutia, “Art gallery and illumination problems,” in *Handbook of Computational Geometry* (J. R. Sack and J. Urrutia, eds.), pp. 973–1027, Amsterdam, the Netherlands: North-Holland, 2000.

[13] J. E. Goodman and J. O’Rourke, eds., *Handbook of Discrete and Computational Geometry.* Boca Raton, FL: CRC Press, 1997.

[14] H. Choset, “Coverage for robotics - a survey of recent results,” *Annals of Mathematics and Artificial Intelligence*, vol. 31, pp. 113–126, 2001.

[15] H. H. González-Baños, A. Efrat, J.-C. Latombe, E. Mao, and T. M. Murali, “Planning robot motion strategies for efficient model construction,” in *International Symposium on Robotics Research*, (Snowbird, UT), Oct. 1999.

[16] J. S. B. Mitchell, “Shortest paths and networks,” in *Handbook of Discrete and Computational Geometry* (J. E. Goodman and J. O’Rourke, eds.), ch. 24, pp. 445–466, Boca Raton, FL: CRC Press, 1997.

[17] P. K. Agarwal and M. Sharir, “Efficient algorithms for geometric optimization,” *ACM Computing Surveys*, vol. 30, no. 4, pp. 412–458, 1998.

[18] A. Okabe, B. Boots, K. Sugihara, and S. N. Chiu, *Spatial Tessellations: Concepts and Applications of Voronoi Diagrams.* Wiley Series in Probability and Statistics, New York: John Wiley, 2 ed., 2000.

[19] Z. Drezner, ed., *Facility Location: A Survey of Applications and Methods.* Springer Series in Operations Research, New York: Springer Verlag, 1995.

[20] N. A. Lynch, *Distributed Algorithms.* San Mateo, CA: Morgan Kaufmann Publishers, 1997.

[21] D. P. Bertsekas and J. N. Tsitsiklis, *Parallel and Distributed Computation: Numerical Methods.* Belmont, MA: Athena Scientific, 1997.

[22] S. Martínez, F. Bullo, J. Cortés, and E. Frazzoli, “On synchronous robotic networks – Part I: Models, tasks, and complexity . Part II: Time complexity of rendezvous and deployment algorithms,” *IEEE Transactions on Automatic Control*, Apr. 2005. Submitted.

[23] L. Guilamo, B. Tovar, and S. M. LaValle, “Pursuit-evasion in an unkown environment using gap navigation trees,” in *IEEE/RSJ Int. Conf. on Intelligent Robots & Systems*, (Sendai, Japan), pp. 3456–3462, Sept. 2004.

[24] V. Isler, S. Kannan, and S. Khanna, “Randomized pursuit-evasion in a polygonal environment,” *IEEE Transactions on Robotics*, vol. 5, no. 21, pp. 864–875, 2005.

[25] N. Moshtagh, A. Jadbabaie, and K. Daniilidis, "Vison-based distributed coordination of multiagent systems," in *Robotics: Science and Systems* (S. Thrun, G. Sukhatme, S. Schaal, and O. Brock, eds.), pp. 41–48, Cambridge, MA: MIT Press, 2005.

[26] A. Ganguli, J. Cortés, and F. Bullo, "Distributed deployment of asynchronous guards in art galleries," in *American Control Conference*, (Minneapolis, MN), pp. 1416–1421, June 2006.

Index

Printed in the United States of America.